The Science-Music Borderlands

The Science-Music Borderlands

Reckoning with the Past and Imagining the Future

Edited by Elizabeth Hellmuth Margulis, Psyche Loui, and Deirdre Loughridge

The MIT Press
Cambridge, Massachusetts
London, England

The MIT Press would like to thank the anonymous peer reviewers who provided comments on drafts of this book. The generous work of academic experts is essential for establishing the authority and quality of our publications. We acknowledge with gratitude the contributions of these otherwise uncredited readers.

This book was set in Stone Serif and Stone Sans by Westchester Publishing Services. Printed and bound in the United States of America.

Library of Congress Cataloging-in-Publication Data

Names: Margulis, Elizabeth Hellmuth, editor. | Loui, Psyche, editor. | Loughridge, Deirdre, editor.
Title: The science-music borderlands : reckoning with the past and imagining the future / edited by Elizabeth H. Margulis, Psyche Loui, and Deirdre Loughridge.
Description: Cambridge, Massachusetts : The MIT Press, 2023. | Includes bibliographical references and index.
Identifiers: LCCN 2022014716 (print) | LCCN 2022014717 (ebook) | ISBN 9780262047647 (paperback) | ISBN 9780262373036 (epub) | ISBN 9780262373043 (pdf)
Subjects: LCSH: Music—Psychological aspects. | Musical ability. | Cognition. | Neuropsychology.
Classification: LCC ML3830 .S293 2023 (print) | LCC ML3830 (ebook) | DDC 781.1/1—dc23/eng/20220328
LC record available at https://lccn.loc.gov/2022014716
LC ebook record available at https://lccn.loc.gov/2022014717

10 9 8 7 6 5 4 3 2 1

Contents

Acknowledgments ix

Introduction 1

I Beyond Nature vs. Nurture
Volume editors

1 **Human Musicality and Gene-Culture Coevolution: Ten Concepts to Guide Productive Exploration** 15
Aniruddh D. Patel

2 **Musical Meaning in Transspecies Perspective: A Semiotic Model** 39
Gary Tomlinson

3 **Cross-Species Research in Biomusicality: Methods, Pitfalls, and Prospects** 57
Diandra Duengen, Marianne Sarfati, and Andrea Ravignani

4 **Humane Treatment, Sound Experiments** 97
Rachel Mundy

Interlude

5 **Of Sound Minds and Tuning Forks: Neuroscience's Vibratory Histories** 115
Carmel Raz

II Beyond Music as a Window into the Mind
Volume editors

6 **Music, Mind, Body, and World** 135
Jonathan De Souza

7 **Rhythmic Entrainment and Embodied Cognition** 161
Maria A. G. Witek

8 **The Musical Mind: Perspectives from Developmental Science** 183
Haley E. Kragness, Erin E. Hannon, and Laura K. Cirelli

9 **The Science of Music Is about Relations** 203
Jim Sykes

Interlude

10 **Toward Neurotechnology for Musical Creativity** 221
Eduardo Reck Miranda

III Beyond Reductionism
Volume editors

11 **Combating Reductionism in Music Neuroscience with Ecologically Valid Paradigms: What Can (and Cannot) Be Gained?** 239
Jamal Williams and Matthew Sachs

12 **Hidden Repertoires in the Brain Accessed by Music in Aging and Neurodegeneration** 263
Sarah Faber and Randy McIntosh

13 **Composing at the Border of Experimental Music and Music Experiment** 277
Grace Leslie

Interludes

14 **Music Theory and Experimental Science** 291
Diana Deutsch

15 **Conversation with Pamela Z** 303
Pamela Z, Psyche Loui, and Deirdre Loughridge

IV Beyond Musicians and Nonmusicians
Volume editors

16 **"The Musical Mind Is the Normal Mind": Remaking Musicianship for Eugenics** 315
Alexander W. Cowan

17 The Musician-Nonmusician Conundrum and Developmental Music Research 329
Beatriz Ilari and Assal Habibi

18 Building Sustainable Global Collaborative Networks: Recommendations from Music Studies and the Social Sciences 347
Patrick E. Savage, Nori Jacoby, Elizabeth H. Margulis, Hideo Daikoku, Manuel Anglada-Tort, Salwa El-Sawan Castelo-Branco, Florence Ewomazino Nweke, Shinya Fujii, Shantala Hegde, Hu Chuan-Peng, Jason Jabbour, Casey Lew-Williams, Diana Mangalagiu, Rita McNamara, Daniel Müllensiefen, Patricia Opondo, Aniruddh D. Patel, and Huib Schippers

19 Conversations with Steven Feld 367
Steven Feld, Nori Jacoby, Deirdre Loughridge, Psyche Loui, and Elizabeth H. Margulis

List of Contributors 385
Index 389

Acknowledgments

This volume sprang from a shared belief that scientific and humanistic approaches to music have much to offer each other and a desire to create spaces and conversations to realize the opportunities for greater mutual understanding and collaboration. We would like to thank the many people and institutions that helped make this endeavor possible. For financial support for the daylong workshop that brought the contributing authors together, thank you to the Princeton University Department of Music, Princeton University Council on Science and Technology, Northeastern University Department of Music, and National Science Foundation (NSF 1734025 to Elizabeth Hellmuth Margulis). Natalie Miller, Maddy Kushan, Cara Turnbull, and Dr. Parker Tichko managed the workshop's Zoom logistics and documented the discussions, and Vijay Iyer contributed challenges to the category of music and encouraged us to think about musicalities as embodied relations among beings and objects. Thank you for lending your time and expertise to the workshop.

For transcribing the interviews included in this volume, thank you to Johanna Linna and Yaen Chen. A Research, Scholarship, and Creative Activity Dissemination Grant from Northeastern University College of Arts, Media, and Design funded our writing retreat at the cottage in Chester, Connecticut, known locally as "The Recycle House." Additional summer support came from the National Science Foundation (NSF-CAREER 1945436 to Psyche Loui) and the National Institutes of Health for Sound Health: An NIH–Kennedy Center Partnership (NIH R21AG075232 to Psyche Loui). Thanks to George Lewis for posing the question "what is an experiment?" at the Musical Thought and Scientific Imagination Study Day at Harvard University in 2018 and to Emily MacGregor and Emily Dolan for organizing that inspiring event. Thank you to Philip Laughlin and the editorial team at MIT for shepherding this project through publication and to Larisa Martin and the team at Westchester Publishing Services for managing the production process. For indexing, thank you Erika Millen.

Finally, we would like to thank our families, with special gratitude to each of our husbands for moral and practical support, including bringing us lunch (because "real people need a real lunch").

Introduction

Do you hear *Laurel* or *Yanny*? The realization that one person might be hearing something very different from another person in the same sequence of sounds spurred urgent headlines about the audio clip that "divided America" (Salam & Victor, 2018). A train of questions followed: Which is the "right" way to hear it? What should we make of individuals hearing different things with equal conviction? Why is the disagreement so shocking in the first place?

Morphing the original sound clip revealed that one person's perception could switch between *Laurel* and *Yanny* and that one's current percept was dependent on one's immediate prior percept (Pressnitzer et al., 2018). This persistence of the past, known as perceptual hysteresis, tantalizingly gestures at a system's dependence on its own history. Perception, a topic typically thought to occupy the realm of psychological science, seems intertwined with history, a topic typically thought to occupy the realm of the humanities.

The Laurel-Yanny phenomenon is more than linguistic, bleeding into the musical. In fact, it is around music that one finds richly developed theories and practices—robust traditions of knowledge—relevant to explaining the Laurel-Yanny phenomenon, among them neural, cognitive, cultural, historical, and technological processes. As both a physical and a perceptual phenomenon, sound entwines the natural and cultural, and with them the sciences and the humanities. In music, the idea that perception in the moment depends on what has come before is hardly surprising, this basic feature of experience being the subject of theories of learning and memory, expectation, groove, and repetition. On a longer time scale, the idea that present perception is predicated on the past is also built into historical and anthropological approaches that assume a cultural production of the senses.

Rifts

Music studies is a rich and multifaceted domain populated by researchers with expertise in history, ethnography, music analysis, technology, psychology, linguistics, physics, neuroscience, philosophy, performance, computer modeling, and more. One might think that a sustained inquiry from multiple perspectives would help build a precise, multidimensional account of music's identity and functions or well-developed practices for sharing knowledge across specializations. Yet despite the unique opportunities for confluence afforded by the more than century-long existence of humanistic and scientific inquiry into music, and despite the potential offered by the decades-long existence of a society that strives to foster interdisciplinary collaboration (the Society for Music Perception and Cognition [SMPC], whose inaugural president Diana Deutsch reflects on the field's history in this volume), rifts between the approaches persist. Instead of finding papers about music and the brain on the syllabus of musicology classes or papers about musical culture on the syllabus of psychology classes, one is more likely to find scholars from one area eviscerating the work of the other on Twitter or—perhaps worse—entirely ignoring it.

What's the problem with this state of affairs? Can't musicology and psychology maintain the status quo of independent work, leaving interaction to the minority of scholars engaged with the SMPC? A commitment to this future misses the powerful opportunity presented by a model of interdisciplinarity that "springs from a self-conscious dialogue with, criticism of or opposition to the aesthetic, ethical or political limits of established disciplines" (Barry & Born, 2008, p. 29). This practice needn't be overly conflict based; it simply "stems from a commitment or desire to contest or transcend the given epistemological and ontological foundations of historical disciplines" (Born, 2010, p. 211). Given that the "standard reductionist agenda" underpinning conventional work in the psychology of music may never scale up to connect with the questions that animate humanistic work (Hartley & Poeppel, 2020, p. 597), failing to attempt a radical mutual reckoning amounts to a capitulation to the limits of knowledge before those limits have even been explored or tested.

But that is the palest and most distant of possible consequences. Graver and more immediate is the danger of scholarship that is wrongheaded, with impoverished designs and weak, erroneous, or actively harmful conclusions. Consider, for example, a (not atypical) goal that might be articulated for a scientific project on music: identifying the features that enable a piece of music to relax someone, make that person happy, or facilitate some sort of healing, and then using this knowledge to concoct an off-the-shelf musical prescription available to anyone. But if a person's experiences, body, and culture all contribute to that individual's sensibilities regarding music, there is no one universal set of musical features that can be expected to reliably act on the brain in the way this study's design assumes. Instead, relaxing, happiness-inducing, or healing effects must be reconceptualized in terms of a constellation of relevant factors.

As argued by Hasson, Nastase, and Goldstein (2020) and by Siegel et al. (2018), scientific and computational tools exist to facilitate these complex models, but for them to be usefully applied, scientific questions must be asked in new and humanistically informed ways. Understanding cutting-edge accounts of culture and musical interaction in the humanities is foundational to good science.

Concerning, too, are tendencies to forswear vast fields of inquiry because of particular problems or challenges. It is always possible to ask whether there are counterexamples to generalized claims, whether a concept or measure is valid across cultures, how histories of exploitation and dehumanization influence the present, who is left out and whose interests are being served. These questions are important, and historians and anthropologists of music specialize in the analyses that can answer them. In practice, however, such questions are often lobbed against scientific approaches to invalidate them a priori—with assumed answers about their limitations and complicities—rather than as a genuine inquiry into a study's aims, methodologies, implications, and possibilities. Such tendencies are especially apparent in cross-cultural comparisons, due to the understandings developed over many decades of ethnomusicological self-critique that comparative methods emerged from and supported a colonial world order wherein hierarchy was presupposed, and to compare was to measure against Western art music. The perpetuation of a colonial world order is a serious risk, but comparative work can take other forms, and there are questions about music as well as modes of collaboration that make cross-cultural comparison something to consider anew.

Humanistic and scientific approaches to music *must* interact for new and potentially transformative insights to occur within their own domains, let alone insights that transcend them. Such interaction moves beyond imagining that humanities expertise exists either to serve science or to critique it. Chapters in this volume by Witek; Tomlinson; De Souza; Kragness, Hannon, and Cirelli; and Sykes, for instance, articulate state-of-the-art accounts of the co-constitutive roles of environment and interactions in shaping musical behavior—accounts that entail the social worlds of music making to which music historians and anthropologists have long devoted attention. In contrast, much research in the psychology and neuroscience of music assumes a unidirectional causal flow from the brain to musical behavior and posits that the structure and organization of the mind and brain can be inferred from musical behavior itself. In such a landscape for music research, humanistic and scientific accounts cannot continue to proceed in parallel as if they have nothing to do with each other; in fact, they are actively contradictory. If humanistic accounts are correct, then many experiments in the psychology of music are built on faulty assumptions. If scientific accounts are correct, then the humanistic assumptions are wrong.

This volume aims to turn attention and energy away from rifts and toward the borderlands—contact zones where the generative potential of interaction can be realized. Although rifts have been the sites of outrage cycles and acrimonious exchanges in recent years, researchers have also been steadily working toward new paradigms informed by developments across disciplinary boundaries and the global conditions of the twenty-first century. The contributors to this volume include both scholars at the forefront of such developments and emerging researchers. Together, they point the way to a future where sustained interaction among disciplines can lead to richer understandings of musical life.

What Is Musicality?

An inescapable question bedevils efforts at humanistic and scientific interaction: what is musicality? There is no shortage of definitions of music or musicality in the literature; in fact, the shift to understanding musicality as a suite of capacities for music making, rather than some bounded set of actions describable as music, was important in enabling confluence among scholars who think about behavior and scholars who think about sound. By considering how a fundamental question like "what is musicality" is refracted through multiple voices from diverse disciplines, this volume charts a path forward for work at the intersection of these approaches. To Patel (chapter 1), musicality refers to the "widespread and spontaneously developing mental and physical abilities that underlie the human capacity for music"—that is, developing without explicit instruction. Feld (chapter 19) takes a somewhat different perspective, referring to musicality as an enculturated system of knowledge that is developed through repeated interactions with the culture. Kragness, Hannon, and Cirelli (chapter 8) posit that "the musical mind does not develop in isolation from other domains"; this is echoed by Ilari and Habibi (chapter 17), who see musical development resulting from "musical participation and affordances, contexts, and culture over time." These two chapters view musicality as an emergent property of the developing mind and body. Thus, the chapters in this volume represent diverse perspectives along a spectrum ranging from nativist (Patel) to emergentist (Feld) views of musicality, with others falling somewhere in between (Kragness, Hannon, and Cirelli; Ilari and Habibi). Together, they cover a variety of views on the origins of musicality. The goal is not to arrive at one definition of musicality but rather to help readers understand the multiplicity of views, providing a sense of where they are truly incompatible, where they reflect mutually consistent but different approaches, and how they might be integrated to drive research forward within individual fields.

Another aspect of musicality that gives rise to a spectrum of views concerns where musicality is situated. At issue is the relation between body and mind. How large a part does the body play in music processing? Is it a vessel, an instrument, or an active contributor? Does music originate in the mind of its conceptualizer, as suggested by Miranda (chapter 10)? He describes the paramusical ensemble, wherein a brain-computer interface enables musical ideation in the absence of bodily movement, employing a distributed set of agents to enact musical ideas. Or, as De Souza (chapter 6) discusses, does music emerge from "an interaction between sound and mind," where the mind includes mental representations of grammatical structures but is also "embodied and situated in a world, alongside objects and others"? Witek (chapter 7) reviews different claims related to embodiment, entrainment, and theories of the relationship between brain and body in musicality. Perhaps the most all-encompassing view comes from Tomlinson (chapter 2), who offers an account of the evolution of music making via the plastic responsiveness of the genome in moment-to-moment interaction with its environment, a model he terms radical niche construction.

What Is an Experiment?

Another overarching theme that cuts across the chapters in this volume relates to experimentation as a means of generating knowledge. What constitutes an experiment, and how can experiments, broadly construed, contribute to knowledge in different disciplines? Most would agree that the design and implementation of experiments are classic elements of training in the sciences. Yet experimentation is not exclusive to scientists; many researchers in the humanities and the arts have employed empirical observations or manipulated different variables to observe their effects on certain outcomes as a way to learn about musicality. Deutsch (chapter 14) provides some early examples of musical artists as scientists, and Leslie (chapter 13) describes scientist-composers whose perceptual experimentation guided their music making, and vice versa. Our interviews with Pamela Z (chapter 15) and Steven Feld (chapter 19) also reveal how the spirit of experimentation goes hand in hand with music making: Pamela Z is inspired by found objects and their usefulness for sonic art, whereas Feld describes recording sessions with Ghanaian artist Nii Otoo Annan as a way of learning about musical knowledge and how it is constantly negotiated through interactions ("acoustemology as relational ontology").

The feedback loop between experimentation and sound making precedes disciplinary boundaries between sciences and the humanities. Raz (chapter 5) chronicles how empirical experimentation led to the rise of new physiological theories of music's effects on the mind and body in the eighteenth and nineteenth centuries. By the 1920s, music psychologist Carl Seashore was pushing for empirical experimentation as a way to learn about musicality. This entailed a shift in thinking about musicality—from an exceptional state (possessed by few) to a combination of continuous variables (possessed by many). This shift, though useful, also sowed the seeds of some deep problems in the science of music, motivated by eugenic selection, as discussed by Cowan (chapter 16).

Though one core of experimentation in contemporary music psychology centers around statistical learning, mental representations, and the grammatical structures of music in humans, the field has recently shifted toward experimentation that takes into account the role of the body in enactive perceptual experiences (Witek, chapter 7) or embodied metaphors (De Souza, chapter 6). Observing embodied behaviors is a central component of experiments with nonhuman animals in the study of musicality across different species (Duengen, Sarfati, and Ravignani, chapter 3) and of experiments involving infants to study the development of the musical mind (Kragness, Hannon, and Cirelli, chapter 8).

Seeking to push the boundaries toward more flexible, inclusive, and forward-thinking experiments, Faber and McIntosh (chapter 12) use music as a metaphor to describe theoretical tools that characterize the mind and brain as an intricate system of networks where hidden states may reveal important information about injury or illness. In an effort to understand musicality through neuroscience using more flexible, experiential approaches, Williams and Sachs (chapter 11) describe naturalistic music listening studies that span the continuum between ecological validity and generalizability. And to attain a richer cross-cultural understanding of musical behaviors, Savage et al. (chapter 18) offer some best-practice suggestions for building sustainable collaborative networks that move beyond the traditional overreliance on WEIRD (Western, educated, industrialized, rich, and democratic) music or musicians.

To escape an impoverished level of interdisciplinary interaction that remains mired in the abstract, this volume moves readers into the specifics and the stakes, balancing the value of deep domain expertise with the value of accessibility and translatability so that each chapter speaks to specialists and nonspecialists alike. Each chapter reflects conversations among the authors and editors that took place before, during, and after a workshop where drafts were circulated and their contents and cross-cutting themes were earnestly and openly discussed. Thus, the chapters strive to reach a common understanding rather than merely juxtapose incompatible viewpoints.

Past, Present, and Future

In some ways, it is harder than ever to keep up to date with the cutting edge of knowledge and approaches outside any one specialty, especially across the humanities-sciences divide. Yet, as we hope to show in this volume, state-of-the-art work on music from diverse fields has opened up new possibilities for dialogue and collaboration. Where there was previously a focus on clearly differentiating biological and cultural components of music, there is now more concern with how the biological and the cultural work together. As Creanza, Kolodny, and Feldman write in a discussion of gene-culture coevolution, "attempting to answer the question of what are the extensions of human biology through culture leads to a striking conclusion: There are few aspects of human biology that have not been shaped by our culture. Human culture has also affected the biology, even the survival, of nonhuman species" (2017, p. 7785). For humanities scholars, interests have likewise extended into the natural world and nonhuman species. As Ochoa Gautier argues in a call for "acoustic multinaturalism," the climate crisis makes the early twenty-first century a time of "radical transformation of the conditions for posing questions regarding what historically in the West have been considered the differential fields of nature and culture" (2016, p. 108).

Reflecting a commitment to the notion that sensitive consideration of the past can help illuminate the present, this book is divided into four sections, each of which engages the history, current status, and future of a myth surrounding the scientific study of music. The first section addresses the apparent clash between music as a product of nature and music as a product of nurture, a false dichotomy that newer models, such as gene-culture coevolution, move beyond. Research based on the assumption of an inherent opposition between biology and culture, such as asking what kind of music the brain likes before experience intervenes, turns out to be poorly conceived when considered in light of contemporary work on the co-constitutive aspect of culture and biology. Patel (chapter 1) evaluates the repercussions for the design of experiments in music cognition, and Tomlinson (chapter 2) lays out a state-of-the-art model of meaning that places neural architectures and environmental niches in continual interaction. Duengen, Sarfati, and Ravignani (chapter 3) and Mundy (chapter 4) consider, from a humanistic and scientific perspective, how animal musicality illuminates the construction of and alternatives to a nature-nurture dichotomy.

The second section addresses the myth that the human experience of music can reflect how the mind works, a notion that becomes increasingly insufficient as one moves away from conceptions of the mind as something unitary and disembodied. Studying musical behaviors may not uncover the workings of some amodal cognitive processor because cognition emerges out of a brain in a body that is constantly interacting with its environment. De Souza (chapter 6) argues that both music and the mind are inherently relational, and Witek (chapter 7) delineates how the musical mind emerges out of embodied interactions. Kragness, Hannon, and Cirelli (chapter 8) trace how these feedback loops between mind, behavior, and environment develop in the earliest stages of life, irrefutably demonstrating that rather than being passive blank slates, babies are active participants in knowledge building. Sykes (chapter 9) interrogates the conceptions of music and self that have governed music-humanities and music-sciences alike and introduces concepts from anthropology and his own fieldwork—including sound as a gift exchanged between and beyond humans—that open up new framings. This section is bookended by interludes by Raz (chapter 5), who explores how historical links between nerves and vibration have shaped modern neural sciences, and Miranda (chapter 10), whose compositions involve overt interactions between performance and neural signals, thus presenting a test case that vividly illustrates the consequences of the other chapters' ideas.

The third section addresses the notion that specific components of music can be understood and manipulated separately, a common tenet in experimental approaches that reduce music to untenable or distorted musical experiences. Williams and Sachs (chapter 11) evaluate the extent to which using full-fledged, realistic musical stimuli addresses the dangers of reductionism. Faber and McIntosh (chapter 12) consider how tools drawn from complex systems research can help make less reductive research designs tractable. Leslie (chapter 13) argues that music cognition researchers and composers share a suite of experimental techniques, illuminating what science can learn from artistic practice. In a pair of interludes, Deutsch (chapter 14) and Pamela Z (chapter 15) offer insights based on their experimental practice. A pioneering researcher in music cognition, Deutsch reflects on how computer technology enables the empirical study of more complex musical phenomena. A musician and multimedia artist, Pamela Z reflects on the role of tools in her creative work and how she uses perceptual phenomena such as the manipulable boundary between speech and music.

The fourth section addresses the division between musicians and nonmusicians, a dichotomy common in the scientific literature that has the effect of building prior assumptions about the kinds of experiences and abilities that constitute proof of musicality into results. Cowan (chapter 16) traces the relationship between the eugenics movement and early psychological notions of musical ability, providing firm evidence that contemporary research practice can be understood only by closely studying its history. Ilari and Habibi (chapter 17) consider how naïve notions of what constitutes a musician versus a nonmusician can hold research back, and Savage et al. (chapter 18) provide practical steps to move the field forward to a more pervasively cross-cultural approach. In chapter 19, Feld joins the volume editors and Jacoby for a multidisciplinary conversation about the challenges of and opportunities for collaborative research that would integrate cognitive and cultural understandings of musical life.

By bringing scholars from the sciences and the humanities together around these four key issues, this volume encourages sustained attention to core disciplinary questions for music studies. The organizing myths map out the existing problems in music science, while going beyond them brings to light people, musics, and approaches that typically fall through the cracks of this parcellated space. This volume also provides the scientific study of music with its first major reckoning, integrating the field's past with the project of imagining its future. Grappling with the tension between the reductionist, universalizing impulses in scientific approaches to music and the commitment to the particular in humanistic approaches, for example, enables one to envision ways to negotiate between them. *The Science-Music Borderlands* thus charts a path forward for music studies that combines insights from the sciences and the humanities. Connecting these divergent branches requires that we address questions at the core of how knowledge is produced, providing an example for other disciplines facing similar issues. We hope that by immersing readers in a diverse field, this volume lays the groundwork for further conversations and collaborations.

References

Barry, A., & Born, G. (2008). Logics of interdisciplinarity. *Economy and Society*, *37*(1), 20–49.

Born, G. (2010). For a relational musicology: Music and interdisciplinarity. Beyond the practice turn. *Journal of the Royal Musical Association*, *135*(2), 205–243.

Creanza, N., Kolodny, O., & Feldman, M. W. (2017). Cultural evolutionary theory: How culture evolves and why it matters. *Proceedings of the National Academy of Sciences*, *114*(30), 7782–7789.

Hartley, C., & Poeppel, D. (2020). Beyond the stimulus: A neurohumanities approach to language, music, and emotion. *Neuron*, *108*, 597–599.

Hasson, U., Nastase, S., & Goldstein, A. (2020). Robust-fit to nature: An evolutionary perspective on biological (and artificial) neural networks. *Neuron*, *105*(3), 416–434.

Ochoa Gautier, A. M. (2016). Acoustic multinaturalism, the value of nature, and the nature of music in ecomusicology. *Boundary*, *2*, *43*(1), 107–141.

Pressnitzer, D., Graves, J., Chambers, C., de Gardelle, V., & Egré, P. (2018). Auditory perception: *Laurel* and *Yanny* together at last. *Current Biology*, *28*(13), R739–R741.

Salam, M., & Victor, D. (2018, May 15). Yanny or laurel? How a sound clip divided America. *New York Times*.

Siegel, E. H., Sands, M. K., Van den Noortgate, W., Condon, P., Chang, Y., Dy, J., Quigley, K. S., & Barrett, L. F. (2018). Emotion fingerprints or emotion populations? A meta-analytic investigation of autonomic features of emotion categories. *Psychological Bulletin*, *144*(4), 343–393.

I Beyond Nature vs. Nurture

Volume Editors

It would seem like we're moving past the nature vs. nurture debate. Gene-culture coevolution captures the idea that human cultural practices and lasting changes in human biology interact over evolutionary time (Richerson, Boyd, & Henrich, 2010, as cited in Patel, 2018). Biocultural hypotheses posit that the either-or dichotomy between culture and biology is limiting and inaccurate (van der Schyff & Schiavio, 2017). Before we declare the end of investigations into whether the musical mind is inherited or acquired, however, it behooves us to pause and reflect on whether anything has really changed. As Evelyn Fox Keller observed, "One of the most striking features of the nature-nurture debate is the frequency with which it leads to two apparently contradictory results: the claim that the debate has finally been resolved (i.e., we now know that the answer is neither nature nor nurture, but both), and the debate's refusal to die" (2010, p. 1). We are yet again at risk of declaring the debate resolved, only to see it persist, if we cannot imagine alternatives that speak across the humanities and sciences, to those doing research and those funding it, to specialists and the general public alike.

Why has the nature-nurture debate proved so stubborn? One way to address this question is to examine the disciplinary and institutional structures where the debate remains entrenched, despite persistent efforts to move beyond it. For music psychologists and cognitive scientists, the kinds of findings and claims that have high value are yoked to a larger hierarchy of knowledge. Although a wide range of research questions and methodologies are represented in specialist music cognition journals, the demands of publishing in general-interest scientific journals can push researchers to strive to link music to the biological, with the assumption that being biological means being scientific, objective, and universal and thus resistant to the winds of social and cultural change. Thus, within the institutional structures of science, incentives persist that encourage a stark approach to the relationship between nature and nurture.

On the flip side, it is a veritable taboo in many humanities fields to consider biology a determinant of cultural life or to take an interest in what might be "universal" to humanity. As Mundy writes (chapter 4), World War II revealed deeply flawed thinking about the relation between biological and cultural evolution, and as a result, music scholars retreated to a focus on the music itself. In chapter 19 of this volume, Feld discusses a heightened suspicion of "universalism" that took root in anthropology and cultural studies in the 1990s, in the context of increasing disciplinary specialization and competition for resources that further polarized the sciences and the humanities. More recently, humanities scholars have targeted human exceptionalism and anthropocentrism as culprits in the planetary climate crisis; this adds a new set of ethical liabilities to discussions of uniquely human capacities (Nayar, 2014; Ochoa Gautier, 2016).

One step forward would be recognizing the pressures that operate on both scientists and humanists to stay in their respective lanes vis-à-vis nature and nurture, as well as the constraining and distorting effects of those pressures. Additional steps require sustained interaction between thinkers in both fields. Keller, for instance, suggests reformulating the nature-nurture question to preserve what people care about knowing, without creating a false dichotomy: "how malleable is a given trait, at a specified developmental age?" (2010, p. 75).

For music research, such paradigm shifts have implications for what counts as a major or meaningful finding and also for the questions that experiments are designed to address. Attempting to isolate innate from learned, for instance, is less important than identifying a broad range of factors and how they interact. As Patel reminds us in chapter 1 of this volume, the value of music doesn't depend on its evolutionary status. Likewise, the value of research doesn't depend on falling on one side or the other of the nature-nurture dichotomy.

Three major areas of research that have been shaped by the nature-nurture debate are evolutionary origins, animal musicality, and cross-cultural comparisons. These areas most clearly demonstrate not only the problems of trying to separate nature and nurture but also the new possibilities opened up by alternative framings. The chapters in this section focus on evolutionary origins (Patel, Tomlinson) and animal musicality (Mundy; Duengen, Sarfati, and Ravignani). The topic of cross-cultural comparison is taken up elsewhere in this volume (in particular, see chapters 9, 18, and 19).

Patel and Tomlinson have published extensively on music and evolution: Patel as a neuroscientist whose research centers on music and language and cross-species musicality, and Tomlinson as a humanist whose engagement with differences across history and culture led to work on the emergence of human cultural capacities through evolutionary processes. In their contributions to this volume, they offer complementary voices on the emergence of culture in interaction with biological substrates. Tomlinson proposes that the burgeoning of culture, including behaviors we might call music and language, shares the same "indexical commons" across species. Patel uses the phrase "purely cultural" to describe capacities for which no neural specializations evolved specifically, while pointing out that "purely cultural" does not mean independent from biology, given that any cognitive capacity still "draws on specific brain regions or pathways and can be influenced by changes to those substrates."

The chapters by Duengen, Sarfati, and Ravignani and by Mundy address animal studies, showing how animals can help us think through music researchers' assumptions and how behaviors exhibited throughout the animal kingdom can provide insight into notions of musicality. As Duengen, Sarfati, and Ravignani demonstrate through their survey of the current literature on animal musicality, although testing animals for human musical capacities—especially pitch perception—has been a dominant research practice, it yields limited insight compared with what could be learned by considering a broader range of cognitive capacities and sounds of ecological relevance to various species. Mundy takes up the ethical dimensions of animal research, considering the history shaping that research in order to question its enabling human-animal binary and propose future research directions founded on different premises.

The chapters in this section illustrate a back-and-forth relationship between empirical studies and theoretical models, the latter endeavoring not only to explain the former but also to enable new testable hypotheses. Music provides a domain within which broader philosophical and scientific issues around the relationship between nature and nurture can be theorized and investigated in tractable ways.

References

Keller, E. F. (2010). *The mirage of a space between nature and nurture.* Duke University Press.

Nayar, P. K. (2014). *Posthumanism.* Polity Press.

Ochoa Gautier, A. M. (2016). Acoustic multinaturalism, the value of nature, and the nature of music in ecomusicology. *Boundary 2, 43*(1), 107–141.

Patel, A. (2018). Music as a transformative technology of the mind: An update. In H. Honing (Ed.), *The origins of musicality* (pp. 113–126). MIT Press.

Richerson, P. J., Boyd, R., & Henrich, J. (2010). Gene-culture coevolution in the age of genomics. *Proceedings of the National Academy of Sciences, 107*, 8985–8992.

van der Schyff, D., & Schiavio, A. (2017). Evolutionary musicology meets embodied cognition: Biocultural coevolution and the enactive origins of human musicality. *Frontiers in Neuroscience, 11*(519), 1–18.

1 Human Musicality and Gene-Culture Coevolution: Ten Concepts to Guide Productive Exploration

Aniruddh D. Patel

Introduction

Debates over the evolution of human musicality have a long history and are far from resolved. Impressed by music's psychological power and cultural ubiquity, Charles Darwin (1871) theorized an adaptive origin for music in *The Descent of Man*. He proposed that wordless songs and rhythms arose in our prehuman ancestors as a display to attract mates, laying the foundation for our strong emotional response to music and scaffolding the later evolution of articulate language. In contrast, William James, who generally accepted Darwin's view that our minds are replete with evolved instincts, held a nonadaptationist view of music's origins. James touched lightly on music in *The Principles of Psychology* and saw human musicality as a fortuitous by-product of how our minds work, a "mere incidental peculiarity of the nervous system . . . of no teleological significance" (1890, vol. 2, p. 419). In short, Darwin claimed that we are an inherently musical species, while James claimed we are not.

Darwin and James never debated their positions. Darwin died eight years before James's *Principles* was published, and no correspondence between the two scholars has ever been found. Yet their conceptual alternatives have persisted in evolutionary debates over music. In the modern era, support for Darwin's theory that human music originated in sexual selection has waned, but several other adaptive theories of music's origins (e.g., rooted in social bonding, parent-infant communication, or signaling of coalition strength) have been proposed by biologists and psychologists and continue to attract attention (e.g., Dunbar, 2012; Mehr et al., 2021). Many scholars remain unconvinced, however, and James's by-product view has morphed into a number of detailed theories positing that musical behavior is a purely cultural invention building on brain functions that evolved for other reasons (e.g., Sperber, 1996; Pinker, 1997; Marcus, 2012).

It is certainly true that cultural invention can give rise to psychologically and socially powerful cognitive capacities that are not specifically predisposed by evolution.

Literacy proves this point. Psychologists, neuroscientists, and evolutionary biologists agree that human beings did not evolve to read and write (e.g., Wolf, 2007; Dehaene, 2010; Henrich, 2020). We are no more an "inherently literate" species than we are an "inherently bicycling" species. Literacy is a relatively recent human invention whose historical roots are known and date to around 3000 BCE. Humans invented literacy by drawing on evolved capacities for language, fine motor control, detailed visual pattern perception, imitation, and so forth to forge a purely cultural invention that has transformed human life. Note that "purely cultural" in this context does not mean immune from biological influences. Literacy, like any cognitive capacity, draws on specific brain regions and brain pathways and can be influenced by changes in those substrates. For example, certain genetic variants influence reading ability and the risk of dyslexia (Centanni, 2020; Doust et al., in press). This is not because humans have specific genes for reading but because those variants influence the development of brain areas or pathways that happen to be important for reading. Literacy is a purely cultural invention in the sense that no brain areas, pathways, or mechanisms have evolved specifically to facilitate the acquisition of this ability. This is unlike speech, for which there is strong, converging evidence from multiple disciplines (including cognitive science, neuroscience, and cross-species research) that the human brain has been modified over time to support a child's ability to acquire spoken language (Patel, 2008, ch. 7; Hagoort, 2017). Of course, speaking also interfaces with neural circuitry that long predates the evolution of speech and is unlikely to be specialized for language, such as brain circuits for motor habit formation. (Thus, individuals with nonfluent aphasia, who cannot generate novel sentences due to cognitive problems after a stroke, can often recite previously memorized prayers or count fluently.) The key point is that there is broad consensus among researchers that *some* human cognitive and neural mechanisms have been specialized over evolutionary time for the processing of spoken language.

Relating back to music, the modern version of the Darwin-James debate offers two conceptual alternatives: either human brains evolved neural specializations for music processing, or music relies entirely on brain circuits that evolved to serve other functions. In the latter view, such circuits are neurally "recycled" during brain development to serve music processing functions (Dehaene & Cohen, 2007). A few years ago I conducted an informal survey using an international mailing list of thousands of researchers who specialize in human auditory cognitive neuroscience. I asked the researchers to indicate which of the two positions they favored. I received about 200 responses, and they were almost perfectly split between the two alternatives. This was striking, given that there has been an explosion of research on music and the brain since 2000. Darwin and James cannot both be right, and it seems that modern neuroscientific findings have not moved the needle strongly toward one position or the other.

One reason that neuroscience has had little impact on this debate (so far) is that during ontogeny the brain can acquire neural specializations for activities we did not specifically evolve to do, via experience-dependent neural plasticity. For example, when a person learns to read and write, a region in the ventral occipitotemporal cortex begins to respond more strongly to orthography in that person's language than to other visual stimuli (Hervais-Adelman et al., 2019). This visual word form area (VWFA) is involved in interpreting visual patterns as written characters and connects to language regions that process these characters phonologically, semantically, and grammatically (Dehaene, 2010). The existence of the VWFA implies that identifying brain regions that respond selectively to music rather than to other sounds (i.e., regions neurally selective for music) would not help resolve the Darwin-James debate. Indeed, there is now robust evidence for the existence of such regions. This is interesting because, to date, the only sounds for which category-selective neural regions have been found are speech and music (Norman-Haignere et al., 2015; Boebinger et al., 2021). Yet, like the VWFA, music-selective brain regions could result from experience-dependent neural plasticity because music is a complex, meaningful sound pattern heard throughout life. Indeed, humans begin hearing and learning about music before birth (Hepper, 1991; Partanen et al., 2013) and are regularly exposed to music throughout infancy and early childhood, when the brain is developing rapidly (Lewis, 2016; Trehub & Cirelli, 2018; de Almeida et al., 2020; Mendoza & Fausey, 2021). Thus, the existence of music-selective brain regions in adults is not conclusive evidence that the brain has evolved neural specializations for music processing.

About a decade ago, based on evidence from cognitive neuroscience and other disciplines, I argued for a purely cultural view of music's origins (Patel, 2008, 2010). Although I emphasized music's ability to have a psychologically transformative effect on individual human lives and to shape the brain through mechanisms of neural plasticity, I believed there was insufficient evidence to argue that humans had evolved neural specializations for music processing. In the ensuing years, however, new empirical findings about music cognition and evolutionary biologists' growing interest in theories of gene-culture coevolution have led me to reconsider my position. In terms of empirical work, surprising discoveries indicate that seemingly basic aspects of human musical rhythmic and melodic cognition (i.e., aspects that develop without explicit training and are culturally widespread) are not widely shared by other species and require surprisingly complex neural processing (for a review, see Patel, 2019; cf. chapter 3 in this volume). This hints that evolved neural specializations may underlie these abilities. Furthermore, recent large-scale genomic studies of musicality (Niarchou et al., 2022) are ushering in a new era of neurogenetic research that can help test evolutionary hypotheses about the origins of musicality in our species.

On the theoretical front, a number of evolutionary biologists are arguing that interactions between cultural invention and biological evolution played a key role in shaping the human mind (e.g., Henrich, 2016; Laland, 2017). This work is rooted in growing evidence that cultural inventions such as agriculture, fire control, and cooking have influenced the course of human biological evolution (Itan et al., 2009; Wrangham, 2009, 2017). Although this theory of gene-culture coevolution has been present in biology for some time (Lumsden & Wilson, 1981; Feldman & Laland, 1996), recent years have seen an uptick in its application to the evolution of human cognition (e.g., Fisher & Ridley, 2013; Dennett, 2017). The idea that gene-culture coevolution played a role in the emergence of human musicality was articulated 20 years ago (Cross, 2003), yet only recently has this idea been explored in detail (e.g., Tomlinson, 2015; Podlipniak, 2017; Patel, 2018, 2021; Savage et al., 2021; Shilton, 2022). Notably, this theory provides an alternative to the adaptation versus cultural invention polarities that have dominated debates over music and evolution for more than a century. For instance, perhaps musical behavior started as a purely cultural invention in our human ancestors and then later triggered processes of gene-culture coevolution due to its impact on survival. The question, of course, is how to test such a view (for suggestions, see Savage et al., 2021; Patel, 2021).

In formulating and testing gene-culture coevolutionary theories of musicality, it is essential to consider the diverse ways music is made, used, and perceived. In other words, research from ethnomusicology, anthropology, and musicology is crucial to this enterprise. However, crossing back and forth between the natural sciences, social sciences, and humanities is far from trivial. These territories are very different, with distinct vocabularies, customs, and concerns (Shelemay, 2011; Feld, 2012; Margulis, 2014; Mundy, 2018, chapter 4 in this volume; Albouy et al., 2020). The goal of this chapter is to equip the explorer with some concepts that will make these journeys productive. These concepts are based on my own experience and training in cognitive neuroscience and evolutionary biology and on my limited knowledge of how music varies across cultures, gleaned from my readings, conversations with specialists in anthropology and ethnomusicology, and field trips with ethnomusicologists (Roberts, 2014). I hope this chapter will lead to conversations with colleagues in other fields who can refine and augment these concepts to make them useful to scholars across a range of disciplines.

Music, Musicality, and Choosing Which Musical Abilities to Study

The ten concepts listed in this chapter are premised on a fundamental distinction central to modern evolutionary work on music—namely, the distinction between *music* and *musicality* (Honing et al., 2015). *Music* is a socially constructed category; this is evident even within the boundaries of western European music, where avant-garde composers have been challenging and expanding what counts as music for more than a century. Ethnomusicologists note that although music occurs in every culture (Nettl, 2015), beyond the boundaries of western Europe or WEIRD (Western, educated, industrialized, rich, and democratic) societies more generally (Henrich, 2020), many languages have no single word for *music* that aligns with the common meaning of this word in English. For example, Trehub, Becker, and Morley (2015) note that the word *nkwa* among the Igbo of Nigeria encompasses music and dance and that the Basonge of Zaire have names for individual genres of music but have no term that encompasses all their genres. Furthermore, any human musical tradition is the product of a rich historical process and is embedded in a complex web of social and political meanings (see chapter 9 in this volume). Thus, it makes no more sense to discuss the biological evolution of music than it does to discuss the biological evolution of French or any of the other approximately 7,000 extant human languages (Eberhard et al., 2021). Just as the capacity for language is the target for scientists interested in explaining the biological evolution of language, the capacity for music is the target for evolutionary biologists interested in explaining music. The human capacity for music, or human *musicality*, can be defined as the spontaneously developing cognitive, sensorimotor, and affective capacities supporting human musical behavior. (Here, "spontaneously developing" means emerging without explicit instruction.)

This cognitive sense of musicality is quite different from the colloquial meaning of the term, which connotes a special interest in or talent for music. For example, one component of musicality in the cognitive sense is the ability to recognize a melody when it is transposed up or down in pitch (e.g., being able to recognize the song "Happy Birthday" whether it is played on a piccolo or a tuba). This is subjectively effortless for human listeners, develops early in life without any special training (Plantinga & Trainor, 2005), and appears to be a universal feature of human music cognition (Nettl, 2015). This ability may seem so simple as to be neurologically trivial, yet numerous studies have shown that songbirds (which rely on complex, learned acoustic sequences for communication and can learn to recognize human melodies) do not have this ability, indicating that it is not an automatic consequence of having a complex auditory system (for a review, see Patel, 2019).[1] Furthermore, neuroimaging reveals that this human ability relies on a network of brain regions extending far outside traditional auditory regions (Foster & Zatorre, 2010a, 2010b). Another component of musicality in the cognitive sense is the ability to perceive a beat in rhythmic music and move in synchrony with that beat in a predictable and tempo-flexible fashion (e.g., via clapping, bobbing, or dancing). This ability emerges without special training in early childhood and is culturally widespread (Savage et al., 2015). Again, although humans can do this effortlessly, current research suggests that this ability is rare in other species, including nonhuman primates, and is neurally complex, engaging a distributed network of cortical and subcortical regions (for reviews, see Patel, 2014; Cannon & Patel, 2021).

An important question about musicality concerns avant-garde composers' expansion of Western notions of music. Have they also expanded the concept of musicality? For example, Leslie (in chapter 13 of this volume) describes her compositions that integrate physiological signals into music. By passing her own brain waves "through a sonification algorithm that imprints their spectral quality onto a bank of stored flute and voice sounds," she mixes her own live performance sound. Leslie has also created an "algorithmically generated musical composition that breathes softly along with its listener, either according to a predetermined ideal breathing rate or in response to data streamed online from a respiration sensor worn by the listener." This innovative work illustrates how musical avant-gardists can create new and valuable aesthetic experiences. Yet I doubt that avant-garde music has changed human musicality in the cognitive sense, because such musicality is based on musical practices that are *widespread* within a culture. Thus, even icons of the avant-garde like Arnold Schoenberg, who have appreciative and responsive listeners (Auner, 1999; Mencke et al., 2019), have a limited public reach. "Called upon to say something about my public," Schoenberg wrote in a 1930 essay, "I have to confess: I do not believe I have one" (Stein, 1975, pp. 96–99).

In addition to the ability to recognize transposed melodies and synchronize to an auditory beat, human musicality involves many more capacities. Developing a list of such abilities is an important (and thorny) issue that would benefit from multidisciplinary discussions. While some items on this list are likely to be uncontroversial, such as the ability to remember a melodic phrase and sing it back or the capacity to be emotionally moved by music, others would probably be contested among scholars from different disciplines. However, the two abilities cited are sufficient to illustrate a key point for evolutionary research on musicality. One of these abilities (recognition of transposed melodies) can plausibly be regarded as a consequence of neural specializations for spoken language. In speech, salient rises and falls in pitch are used to distinguish linguistically important categories (e.g., word meanings in tone languages and pragmatic categories, such as statement versus question, in many languages). Listeners need to be able to recognize these "speech melodies" independent of absolute pitch because the pitch registers of men, women, and children differ substantially (Ladd, 2008). Thus, neural specializations that allow the recognition of transposed melodies may have evolved to serve speech perception, and they may be employed *by* music cognition without any additional neural specialization *for* music cognition. In the parlance of evolutionary biology, our ability to recognize transposed musical melodies might be an exaptation of the neural infrastructure for speech, without any subsequent evolutionary specialization for music (Gould & Vrba, 1982). In contrast, our ability to synchronize movements to a rhythmic beat is not so easily explained as a secondary use of speech-related brain processes. Ordinary speech does not have a periodic beat; nor do we bob, tap, or dance to ordinary speech (Patel, 2008). For biologists studying whether humans have evolved neural specializations for music processing, choosing aspects of musicality that are not easily explained as secondary uses of other, more obviously adaptive brain functions (such as speech) is an important consideration.

The broader concept is that music cognition draws on a diverse set of brain mechanisms, many of which are unlikely to be specialized over evolutionary time for music cognition (Trainor, 2015). For example, subcortical sound localization mechanisms are used when we process music, but these mechanisms evolved in animals to serve important ecological functions, such as locating predators, long before humans came on the scene. It is highly unlikely that these mechanisms were modified in humans specifically to facilitate musical processing, even though listening to music (or to any sound, for that matter) engages these mechanisms. Thus, the question at the heart of this chapter is whether music engages *any* brain mechanisms that have been specialized over evolutionary time for music cognition, and if so, whether gene-culture coevolution is responsible for this specialization. With this question in mind, I now turn to ten concepts that can guide researchers through the borderlands between the sciences and the humanities and into the territories that meet at this border.

Concept 1: Music's Value Does Not Depend on Its Evolutionary Status

Many who are drawn to music research respond strongly to music themselves. When studying evolutionary questions, this can create an implicit bias toward adaptationist theories. Implicit biases are dangerous because they can lead a researcher to unwittingly focus on data that support a favored theory and ignore contradictory data or to examine supportive data with a less critical eye than unsupportive data. To minimize such biases, it is important to remember that a particular ability's value to human life is unrelated to whether neural specializations for that ability have evolved in humans. Literacy provides a clear example. Literacy has transformed human cultures and mental experiences in positive ways, yet humans have not evolved any neural specializations for reading and writing (Wolf, 2007). Another example is exercise. Exercise is enormously valuable for physical and mental health (Ratey, 2008), yet recent research in human evolutionary biology suggests that we have no evolutionary predispositions or specializations for this activity (Lieberman, 2021). Similarly, there is growing evidence that musical activity has beneficial neurological effects in individuals with a variety of disorders (Loui et al., 2018), yet these benefits will not be diminished if it turns out that human beings are not an inherently musical species. More generally, if research on music and gene-culture coevolution ends up supporting the conclusion that humans have not evolved any neural specializations for music, this would not be a failure; it would be a perfectly acceptable outcome. This is because doing this research will lead to valuable discoveries in cognitive and brain science, a point to which I return in the final section of this chapter.

Concept 2: There Are Two Types of Neural Plasticity: Experience-Expectant and Experience-Dependent

As noted in the introduction, during ontogeny human brains can acquire neural specializations for purely cultural inventions (e.g., literacy) via experience-dependent neural plasticity. However, neural plasticity is not the opposite of innateness. Neural plasticity also occurs in circuits that have become specialized over evolutionary time to serve particular functions. This is best documented by research on critical or sensitive periods—that is, times in early development when experience is known to be essential for the correct wiring and tuning of the brain's specialized systems, including those for vision, language, and social cognition (Reh et al., 2020). Greenough, Black, and Wallace (1987) used the term "experience-expectant plasticity" to refer to the experience-driven molding of specialized circuits that occurs during critical periods and has a significant influence on subsequent abilities. As a rule, such experiences (e.g., seeing with two eyes that focus on the same point in space or receiving regular social input from a caregiver) are common to all members of a species. By allowing experience to sculpt rapidly developing specialized circuits, these circuits can perform complex tasks better than hard-wired circuits with less flexibility. Greenough's group contrasted experience-expectant plasticity with experience-dependent plasticity, defined as changes in the brain that can happen throughout life and are based on experiences that are not common to all members of a species. (Literacy is an example of experience-dependent plasticity: even though it is more than 5,000 years old, it is still far from universal.[2]) The key point is that if a human ability is based on evolved neural specializations *and* is strongly influenced by culture, it will engage both types of plasticity. Spoken language is a good example: learning to speak requires the development of phonological and lexical processing mechanisms that are shaped by experience-expectant plasticity (Kuhl, 2010; Reh et al., 2020), but mastering a particular language involves learning specific words and grammatical structures that are not universal, requiring experience-dependent plasticity. Thus, a key question for gene-culture coevolutionary research on music is whether the development of any musical ability involves experience-expectant plasticity or only experience-dependent plasticity (Penhune, 2020).

In asking this question, it is important to remember that both types of plasticity build circuits that have ongoing gene-environment interactions throughout an individual's life. As entailed in Tomlinson's idea of "radical niche construction" (see chapter 2 in this volume), biological systems have gene-environment interactions even at rapid timescales. For example, when adult zebra finches hear rhythmic versus arrhythmic versions of their conspecific songs, they exhibit greater levels of immediate early gene expression in auditory forebrain areas in response to the latter. In contrast, juvenile zebra finches show exactly the opposite pattern. This developmental reversal in rapid gene-environment interactions may reflect differences in how rhythm is used in learning songs versus evaluating song structure once learning has occurred (Lampen et al., 2019; cf. Rouse et al., 2021). The important point is that biological systems represent a constant interplay of genes and environment, even in brain circuits specialized by evolution to facilitate certain abilities.

Concept 3: Musical Behavior Might Have Originated as a Purely Cultural Invention

The concept of gene-culture coevolution entails the idea that purely cultural inventions can lead to heritable genetic changes (as discussed earlier regarding fire control in human evolution). This idea is stated succinctly by Fisher and Ridely (2013): "The smallest, most trivial new habit adopted by a hominid species could—if advantageous—have led to the selection of genomic variations that sharpened that habit." Gene-culture coevolutionary theories of musicality are thus concordant with the view that early musical behavior was not spurred by some genetic change in human ancestors but arose when nonmusical behaviors in those ancestors (e.g., coordinated group rhythmic vocalizations, as occur among chimpanzees and bonobos) were culturally repurposed in ways that began to resemble musical behavior. The crucial issue is whether this resulted in a consistent advantage in survival or reproduction that ultimately led to selection for genomic variations favoring the proclivity and capacity for these new behaviors. (For a proposal on how this could have taken place, see Patel, 2021.)

Concept 4: Capacity and Proclivity Are Conceptually and Neurally Distinct Targets for Natural Selection

In the first edition of *The Descent of Man*, Darwin wrote, "the perception, if not the enjoyment, of musical cadences and of rhythm is probably common to all animals and no doubt depends on the common physiological nature of their nervous systems" (1871, vol. 2, p. 333). Darwin's assertion that the basic human capacity to perceive music is widely shared among animal species has been challenged by modern research. However, his conceptual distinction between musical capacity and the motivation to engage in musical behavior (i.e., "enjoyment") remains crucial for research on the evolutionary foundations of musicality. It is conceivable, for example, that evolution has enhanced our motivation to engage in musical behavior without any evolved neural specialization for processing specific rhythmic or melodic aspects of music (Trehub, 2003). In this case, the relevant neural changes would occur only in pathways that support intrinsic reward in response to certain activities, including deep brain structures such as the ventral striatum and the connections between these areas and the relevant cortical structures involved in music processing (Belfi & Loui, 2020; Mas-Herrero et al., 2021). Conversely, selection may have acted on both motivational mechanisms (proclivity) and brain regions and pathways supporting the capacity for musical processing (Patel, 2021; Savage et al., 2021). In either case, a gene-culture coevolutionary view of musicality is premised on the existence of some evolutionary neural specialization for brain mechanisms involved in musical behavior.

Concept 5: Abilities Based on Evolved Neural Specializations Can Vary Widely

A common objection to the idea that humans are inherently musical is the observation that musical abilities vary widely among people. Such claims often implicitly focus on musicality in the informal sense—that is, having a special interest in or talent for music. When one focuses on components of musicality in the cognitive sense, such as implicit knowledge of the norms of one's native musical system or the ability to move in time with a musical beat, variance is likely to be lower (Rohrmeier & Rebuschat, 2012; Tranchant et al., 2016). Yet even these sorts of abilities can show substantial variation (Tranchant et al., 2021). The key point is that such variance provides no evidence *against* such abilities being supported by evolved neural specializations. Facial recognition illustrates this point. There is compelling evidence that facial recognition relies on brain circuits specialized over evolutionary time and shaped by experience-expectant plasticity in early development (Moulson et al., 2009; Todorov, 2017, chs. 12, 13; Cabral et al., 2020; Kosakowski et al., 2022). Yet the capacity for facial recognition varies widely in neurologically intact individuals, ranging from those with very poor abilities to "super-recognizers" (Sacks, 2011). Notably, even though twin studies indicate a substantial genetic contribution to severe facial recognition deficits (developmental prosopagnosia; Wilmer et al., 2010), and despite the important role of facial recognition in human social interaction, individuals with developmental prosopagnosia have not been "weeded out" by natural selection. This shows that there will always be humans who struggle with certain abilities for which the human brain is specialized.

Concept 6: Ancient, Universal Cognitive Traits Are Not Necessarily Based on Evolved Neural Specializations for Those Traits

Proponents of adaptationist theories of music's origins often point out that music is far older and more widespread than purely cultural inventions such as literacy. Music occurs in every culture, and the earliest known instruments are around 40,000 years old. This is likely much younger than the origins of musical behavior in humans, as singing leaves no trace in the fossil record (Higham et al., 2012; Morley, 2013; Nettl, 2015; Savage et al., 2015, Mehr et al., 2019). While the age and ubiquity of music are certainly consistent with theories that posit evolved neural specializations for musicality, they provide no evidence of such specializations. This is because humans likely have some ancient, universal, and culturally prominent cognitive traits that are by-products of evolved aspects of cognition, as William James suggested in *Principles of Psychology*. Belief in ghosts or other spirits is one example. Such beliefs are universal in human culture (Norenzayan et al., 2016), but not because our brains evolved to encourage supernatural beliefs due to their survival value.[3] Rather, several authors have argued that certain features of our evolved psychology make us *susceptible* to belief in supernatural agents (e.g., Dennett, 2006). For example, Boyd (2018) and Henrich (2020) argue that such features include a reliance on cultural learning (to such an extent that it overrides even direct experience or intuition), tendencies to infer what others are thinking (theory of mind), and a bias for causal explanations based on the actions of agents. Supernatural beliefs may well be as old as articulate language in the human species and thus far older than the oldest known musical instruments. Such beliefs remind us that not every ancient and culturally universal human mental trait is a result of the brain evolving neural specializations for that trait.

Concept 7: Evolutionary Specialization and Adaptive Function Are Conceptually Distinct Issues

Research on musicality and evolution has often focused on music's adaptive value. Yet as pointed out by Tinbergen (1963), adaptation (what is it for?) is only one question about the evolution of a behavior. A second, equally important question concerns evolutionary history or phylogeny: how did it evolve? With respect to musicality, the evolutionary history of its components can be studied independently of questions about adaptation. Research on the human ability to synchronize movements to a musical beat provides an example. Patel (2006) hypothesized that this ability has its origins in the neural capacity for complex vocal learning, since both abilities involve sophisticated forebrain auditory-motor processing and engage some of the same neural circuits. This hypothesis yielded testable predictions and led to numerous empirical studies. More recently, Patel (2021) updated this hypothesis to suggest that vocal learning acted as a preadaptation for spontaneous beat perception and synchronization (BPS) and that subsequent evolutionary neural specialization for BPS took place via processes of gene-culture coevolution. This hypothesis assumes that synchronized movement to a beat in social contexts had adaptive value in early human groups, but it does not commit to just one specific adaptive function. Some theorists argue that synchrony to a beat in social contexts enhanced social bonding outside of those contexts (e.g., Savage et al., 2021), while others argue that it signaled coalition strength to other groups (e.g., Mehr et al., 2021): these theories are not mutually exclusive. The relevant point is that scientists can look for evidence of evolved neural specializations for BPS without waiting for these adaptationist debates to be resolved. To take an analogy from physical anthropology, researchers have provided convincing evidence that, compared with other primates, humans' bodies are specialized for bipedal locomotion, even though debates over *why* bipedalism was advantageous to our human ancestors are far from resolved (reviewed in Lieberman, 2013). In the case of bipedalism, evidence for specialization comes from research on comparative anatomy, biomechanics, development, and other disciplines (e.g., Richard et al., 2020). Similarly, determining whether BPS or other aspects of musicality are based on evolved neural specializations will require the integration of research across numerous disciplines, including neuroscience, genetics, cross-species studies, ethnomusicology, and developmental psychology. It is possible that we may one day have strong evidence of evolved neural specializations for musicality, even while debates over the original adaptive value of musical behavior remain unresolved.

Concept 8: A Trait Can Be Genetically Influenced without Being Genetically Determined

Gene-culture coevolution theories posit that certain components of musicality are genetically influenced. Crucially, modern views of how musical (or other cognitive) abilities are genetically influenced differ in important ways from deterministic views of the relation between genes and cognition. One of the early music psychologists who held such deterministic views was Carl Seashore, who developed auditory pattern perception tests with the aim of identifying children who were worthy of music education. "The gift of music is inborn," remarked Seashore, "and inborn in specific types which can be detected early in life, before time for beginning serious musical education" (quoted in Cary, 1922). In contrast, modern research posits that genetic influences on specific components of musicality are subtle, with genetic variants acting probabilistically and dynamically to influence but not determine such capacities. This is in line with modern research on the behavioral genetics of cognitive traits, which emphasizes feedback loops created by gene-environment interactions during child development (Harden, 2021).

For example, a recent large-scale genome-wide association study of musical beat synchronization revealed that variation in this ability is associated with particular genetic variants occurring at numerous positions along the genome, with 67 loci reaching genome-wide significance (Niarchou et al., 2022). That is, the ability to synchronize movement to a musical beat is a complex or polygenic trait that can be weakly influenced by many common genetic variants, rather than a Mendelian trait strongly influenced by variation at a single gene. The results of the study were virtually unchanged by controlling for general cognition, consistent with results from the twin literature showing that the genetics of rhythm are not solely attributable to generally cognitive effects. Importantly, in this new study, genetic variance explained only about 13 to 16 percent of phenotypic variance in the ability to clap in time to a musical beat, indicating that variance in this ability is genetically influenced but far from genetically determined. This means that interactions among genes, experience, and culture are essential in understanding how this trait develops in individuals.

Concept 9: Studying Cultural Variation in Music Requires Reading Primary Sources and Talking with Specialists

Recent years have seen a growth in "big data" studies of music that draw on ethnographic databases or collections of audio recordings to study cross-cultural patterns in musical structure or behavior using quantitative analyses. Such studies have provided valuable information and are well worth conducting, and when they are published in high-profile journals (e.g., Savage et al., 2015; Mehr et al., 2019), they are frequently cited by cognitive scientists seeking to understand musical diversity. Yet such studies are far from the final word on musical diversity because they draw on a limited sample of the world's cultures. For example, Savage et al. (2015) examined 304 diverse recordings of traditional music from around the world (taken from the *Garland Encyclopedia of World Music*), while Mehr et al. (2019) examined ethnographic descriptions from

60 traditional societies chosen to represent 60 cultural clusters from among the 315 cultures in the Human Relations Area Files. Mehr's group also analyzed audio recordings from 86 traditional societies around the world, taken mainly from the Archive for World Music in Harvard's music library. Although these studies examined more cultures than previous comparative work on music and applied modern methods of analysis to generate quantitative comparative data, many cultures were not represented in these samples (recall that there are around 7,000 extant languages on earth today; Eberhard et al., 2021). Theories built on limited samples inevitably reflect the biases of those samples.

For example, building on their previous cross-cultural work, Savage et al. (2021) and Mehr et al. (2021) proposed evolutionary theories of music that emphasize collective music making. Yet there are some traditional societies, not included in their earlier cross-cultural studies, where music making is primarily a solo endeavor (Patel & von Rueden, 2021). Without developing a list of such cultures and understanding why they opt for solo rather than collective musical behavior, theories of music and evolution remain incomplete. The relevant point is that researchers interested in musical diversity should not rely solely on existing cross-cultural "big data" studies. There is no substitute for reading primary research and talking with researchers who have spent years in the field with small-scale traditional societies. For example, such readings and conversations led Patel and von Rueden (2021) to learn about the predominance of solo musical behavior in multiple small-scale cultures (including several hunter-gatherer societies), none of which appeared in Savage's and Mehr's earlier cross-cultural studies. Fortunately for those who want to study musical diversity and develop hypotheses to help explain it, anthropology and ethnomusicology are thriving disciplines with primary sources and scholars available for consultation.

Concept 10: Variance in Musicality Is an Asset for Gene-Culture Coevolution Research

As noted earlier, musicality in the cognitive sense refers to the widespread and spontaneously developing mental and physical abilities that underlie the human capacity for music. Measurement of such abilities typically reveals substantial individual differences. For example, Tranchant et al. (2021, fig. 6) provided data on nonmusicians' ability to synchronize movements to a beat and to perceive beat in the absence of movement, and Jacoby et al. (2019, fig. 2) provided cross-cultural data on musically untrained individuals' ability to accurately sing back a simple pitch interval when the model is presented either within or outside their vocal range. Both studies revealed considerable individual variation. Significant variation in musicality is not limited to perception and production skills but is also apparent in responsiveness to music.

Research has revealed that a small percentage of individuals from western European cultures do not derive pleasure from music, even though they do not suffer from depression or other neuropsychological disorders, are not tone deaf, and can derive pleasure from other arts (Mas-Herrero et al., 2012, 2018). Brain imaging suggests that these "musically anhedonic" individuals have a low degree of neural coupling between auditory processing and reward regions of the brain (Martínez-Molina et al., 2016; Loui et al., 2017).[4] Importantly, it appears that such individuals are at the low end of a continuum in terms of how rewarding people find music and how strongly connected their auditory and reward regions are (Loui et al., 2017; Martínez-Molina et al., 2019; Belfi & Loui, 2020). In other words, whether measuring perceptual, motor, or affective aspects of musicality, there is considerable variability across individuals. From the standpoint of theories of music and gene-culture coevolution, this variance is good because it can be leveraged to study mechanistic links among genes, experience, brains, culture, and musicality. This will require research using measures of musical ability and responsiveness that can be adapted to different cultures, ages, and degrees of musical training. (For some current measures of musicality, see Mas-Herrero et al., 2012; Sandstrom & Russo, 2013; Harrison & Müllensiefen, 2018; Jacoby et al., 2019; Zentner & Gingras, 2019.) Ultimately, this work will need to connect with emerging methods for studying the relations among human genetics, brain structure, and natural selection (Tilot et al., 2021).

Conclusion: Why This Research Matters

How is research on musicality relevant to the broader study of human evolution? Understanding the evolution of the human mind is a central issue in determining human origins. Among biologists studying the evolution of the mind, there is growing interest in the idea that gene-culture coevolution played a key role in this process. For example, Laland argues that compared to other species, humans experienced an unusually strong interaction between cultural and genetic processes:

> Human culture is not just a magnificent end product of the evolutionary process, an entity that, like the peacock's tail or the orchid's bloom, is a spectacular outcome of Darwinian laws. For humans, culture is a big part of the explanatory process too. The evolution of the truly extraordinary characteristic of our species—our intelligence, language, cooperation, and technology—have proven difficult to comprehend, because, unlike most other evolved characters, they are not adaptive responses to extrinsic conditions. Rather, humans are creatures of their own making. The learned and socially transmitted activities of our ancestors, far more than climate, predators, or disease, created the conditions under which our intelligence evolved. Human minds are not just built *for* culture; they are built *by* culture. (2017, pp. 29–30)

Yet despite interest in the idea that gene-culture coevolutionary dynamics shaped the human mind, evidence supporting this view remains scarce. Convincing examples of gene-culture coevolution in human biology come from studies of genetic and physiological nonneural adaptations related to diet, climate, and disease (Laland et al., 2010; Richerson et al., 2010). Musicality may prove to be a more tractable domain for developing and testing theories of cognitive gene-culture coevolution than others mentioned by Laland, such as language and cooperation. This is because there are well-developed disciplines that study cultural variation in music and individual variation in musicality and because several core cognitive components of musicality (e.g., beat processing) are likely to be easier to understand in terms of neural mechanisms than are core cognitive components of language (e.g., lexical processing) or cooperation (e.g., theory of mind) (cf. Cannon & Patel, 2021; Cannon, 2021). Also, core components of musicality are likely to be more amenable to exploration in animal models, aiding in the elucidation of links among genes, brains, and cognition. In other words, there are several reasons why musicality may be a promising model for studying cognitive gene-culture coevolution in human beings. The methods and findings emerging from such research may prove valuable for gene-culture coevolutionary research on other key human mental faculties, such as language.

However, what happens if, decades from now, after many studies motivated by gene-culture coevolutionary theories of music, we conclude that music is a purely cultural invention based on brain functions that evolved for other reasons, like literacy? Even if this should transpire, research on the evolution of musicality will have been fruitful. As shown by a growing body of surprising discoveries, cross-species research on musicality (motivated by evolutionary hypotheses) is a powerful way to illuminate distinctive features of human nonlinguistic cognition. Furthermore, I suspect that evolutionary research on musicality will lead to breakthroughs in our understanding of the mechanistic links among human genes, brains, experience, and culture, regardless of whether we prove to be an inherently musical species.

Acknowledgments

I thank Elizabeth Margulis, Deirdre Loughridge, and Psyche Loui for organizing the workshop for which this chapter was written and for their helpful comments, and I thank Reyna Gordon for her input on concept 8. I also thank Gary Tomlinson, Rachel Mundy, Mathias Guenther, Sam Norman-Haignere, and Daniel Dennett for thoughtful comments on the first draft of this chapter.

Notes

1. At this point, it is unclear whether any nonhuman primates have this ability, as there are only a few studies with contradictory results.

2. For a fascinating discussion of how literacy spread in Europe between 1500 and 1900, see Henrich (2020, pp. 3–17).

3. Belief in ghosts or spirits is distinct from organized religion, and the latter is not present in all human cultures. In today's world there are many people who do not believe in the supernatural, but this is likely a relatively recent historical phenomenon associated with the rise of modern, scientifically based education. However, even in cultures where such education is prominent, there are many who still believe in the supernatural.

4. The extent to which these neurological patterns are a cause or a consequence of the lack of engagement with music remains unclear, and it has yet to be determined whether genetics plays a role in musical anhedonia.

References

Albouy, P., Benjamin, L., Morillon, B., & Zatorre, R. J. (2020). Distinct sensitivity to spectrotemporal modulation supports brain asymmetry for speech and melody. *Science, 367*(6481), 1043–1047.

Auner, J. H. (1999). Schoenberg and his public in 1930: The Six Pieces for Male Chorus, Op. 35. In W. Frisch (Ed.), *Schoenberg and his world* (pp. 85–125). Princeton University Press.

Belfi, A. M., & Loui, P. (2020). Musical anhedonia and rewards of music listening: Current advances and a proposed model. *Annals of the New York Academy of Sciences, 1464*(1), 99–114.

Boebinger, D., Norman-Haignere, S. V., McDermott, J. H., & Kanwisher, N. (2021). Music-selective neural populations arise without musical training. *Journal of Neurophysiology, 125*(6), 2237–2263.

Boyd, R. (2018). *A different kind of animal: How culture transformed our species*. Princeton University Press.

Cabral, L., Zubiaurre, L., Wild, C., Linke, A., & Cusack, R. (2020). Category-selective visual regions have distinctive signatures of connectivity in neonates. *BioRxiv*, 675421.

Cannon, J. (2021). Expectancy-based rhythmic entrainment as continuous Bayesian inference. *PLOS Computational Biology, 17*(6), e1009025.

Cannon, J. J., & Patel, A. D. (2021). How beat perception coopts motor neurophysiology. *Trends in Cognitive Sciences, 25*, 137–150.

Cary, H. (1922). Are you a musician? Professor Seashore's specific psychological tests for specific musical abilities. *Scientific American, 127*(6), 326–327.

Centanni, T. M. (2020). Neural and genetic mechanisms of dyslexia. In G. Argyropoulos (Ed.), *Translational neuroscience of speech and language disorders* (pp. 47–68). Springer.

Cross, I. (2003). Music and biocultural evolution. In M. Clayton, T. Herbert, & R. Middleton (Eds.), *The cultural study of music: A critical introduction* (pp. 19–30). Routledge.

Darwin. C. (1871). *The descent of man, and selection in relation to sex*. John Murray.

de Almeida, J. S., Lordier, L., Zollinger, B., Kunz, N., Bastiani, M., Gui, L., Adam-Darque, A., Borradori-Tolsa, C., Lazeryas, F., & Hüppi, P. S. (2020). Music enhances structural maturation of emotional processing neural pathways in very preterm infants. *NeuroImage, 207*, 116391.

Dehaene, S. (2010). *Reading in the brain: The new science of how we read*. Penguin.

Dehaene, S., & Cohen, L. (2007). Cultural recycling of cortical maps. *Neuron, 56*(2), 384–398.

Dennett, D. C. (2006). *Breaking the spell: Religion as a natural phenomenon*. Penguin.

Dennett, D. C. (2017). *From bacteria to Bach and back: The evolution of minds*. W. W. Norton.

Doust, C., Fontanillas, P., Eising, E., Gordon, S. D., Wang, Z., Gökberk, A., Molz, B., 23andMe Research Team, Quantitative Trait Working Group of the GenLang Consortium, St. Porcain, B., Francks, C., Marioni, R. E., Zhao, J., Paracchini S., Talcott, J. B., Monaco, A. P., Stein, J. F., Gruen J. R., Olson, R. K., . . . Luciano, M. (in press). Discovery of 42 genome-wide significant loci associated with dyslexia. *Nature Genetics*.

Dunbar, R. I. M. (2012). On the evolutionary function of song and dance. In N. Bannan (Ed.), *Music, language, and human evolution* (pp. 201–214). Oxford University Press.

Eberhard, D. M., Simons, G. F., & Fennig, C. D. (Eds.). (2021). *Ethnologue: Languages of the world* (24th ed.). SIL International.

Feld, S. (2012). *Sound and sentiment: Birds, weeping, poetics, and song in Kaluli expression, with a new introduction by the author* (3rd ed.). Duke University Press.

Feldman, M. W., & Laland, K. N. (1996). Gene-culture coevolutionary theory. *Trends in Ecology & Evolution, 11*(11), 453–457.

Fisher, S. E., & Ridley, M. (2013). Culture, genes, and the human revolution. *Science, 340*(6135), 929–930.

Foster, N. E., & Zatorre, R. J. (2010a). Cortical structure predicts success in performing musical transformation judgments. *Neuroimage, 53*(1), 26–36.

Foster, N. E., & Zatorre, R. J. (2010b). A role for the intraparietal sulcus in transforming musical pitch information. *Cerebral Cortex, 20*(6), 1350–1359.

Gould, S. J., & Vrba, E. S. (1982). Exaptation—a missing term in the science of form. *Paleobiology, 8*(1), 4–15.

Greenough, W. T., Black, J. E., & Wallace, C. S. (1987). Experience and brain development. *Child Development, 58*, 539–559.

Hagoort, P. (2017). The core and beyond in the language-ready brain. *Neuroscience & Biobehavioral Reviews, 81*, 194–204.

Harden, K. P. (2021). *The genetic lottery: Why DNA matters for social equality*. Princeton University Press.

Harrison, P. M., & Müllensiefen, D. (2018). Development and validation of the computerised adaptive beat alignment test (CA-BAT). *Scientific Reports*, *8*(1), 1–19.

Henrich, J. (2016). *The secret of our success: How culture is driving human evolution, domesticating our species, and making us smarter*. Princeton University Press.

Henrich, J. (2020). *The WEIRDest people in the world*. Farrar, Straus, & Giroux.

Hepper, P. G. (1991). An examination of fetal learning before and after birth. *Irish Journal of Psychology*, *12*(2), 95–107.

Hervais-Adelman, A., Kumar, U., Mishra, R. K., Tripathi, V. N., Guleria, A., Singh, J. P., & Huettig, F. (2019). Learning to read recycles visual cortical networks without destruction. *Science Advances*, *5*(9), 1–10.

Higham, T., Basell, L., Jacobi, R., Wood, R., Ramsey, C. B., & Conard, N. J. (2012). Testing models for the beginnings of the Aurignacian and the advent of figurative art and music: The radiocarbon chronology of Geißenklösterle. *Journal of Human Evolution*, *62*(6), 664–676.

Honing, H., ten Cate, C., Peretz, I., & Trehub, S. E. (2015). Without it no music: Cognition, biology and evolution of musicality. *Philosophical Transactions of the Royal Society of London B: Biological Sciences*, *370*(1664), 1–8.

Itan, Y., Powell, A., Beaumont, M. A., Burger, J., & Thomas, M. G. (2009). The origins of lactase persistence in Europe. *PLOS Computational Biology*, *5*(8), e1000491.

Jacoby, N., Undurraga, E. A., McPherson, M. J., Valdés, J., Ossandón, T., & McDermott, J. H. (2019). Universal and non-universal features of musical pitch perception revealed by singing. *Current Biology*, *29*(19), 3229–3243.

James, W. (1890). *The principles of psychology*. Henry Holt.

Kosakowski, H. L., Cohen, M. A., Takahashi, A., Keil, B., Kanwisher, N., & Saxe, R. (2022). Selective responses to faces, scenes, and bodies in the ventral visual pathway of infants. *Current Biology*, *32*(2), 265–274.

Kuhl, P. K. (2010). Brain mechanisms in early language acquisition. *Neuron*, *67*(5), 713–727.

Ladd, D. R. (2008). *Intonational phonology* (2nd ed.). Cambridge University Press.

Laland, K. N. (2017). *Darwin's unfinished symphony: How culture made the human mind*. Princeton University Press.

Laland, K. N., Odling-Smee, J., & Myles, S. (2010). How culture shaped the human genome: Bringing genetics and the human sciences together. *Nature Reviews Genetics*, *11*(2), 137–148.

Lampen, J., McAuley, J. D., Chang, S. E., & Wade, J. (2019). Neural activity associated with rhythmicity of song in juvenile male and female zebra finches. *Behavioural Processes*, *163*, 45–52.

Lewis, J. (2016). Play, music, and taboo in the reproduction of an egalitarian society. In H. Terashima & B. S. Hewlett (Eds.), *Social learning and innovation in contemporary hunter-gatherers: Evolutionary and ethnographic perspectives* (pp. 147–158). Springer.

Lieberman, D. (2013). *The story of the human body: Evolution, health, and disease*. Pantheon.

Lieberman, D. (2021). *Exercised: Why something we never evolved to do is healthy and rewarding*. Pantheon.

Loui, P., Patel, A., Wong, L. M., Gaab, N., Hanser, S. B., & Schlaug, G. (2018). Music, sound, and health: A meeting of the minds in neurosciences and music. *Annals of the New York Academy of Sciences, 1423*(1), 7–9.

Loui, P., Patterson, S., Sachs, M. E., Leung, Y., Zeng, T., & Przysinda, E. (2017). White matter correlates of musical anhedonia: Implications for evolution of music. *Frontiers in Psychology, 8*, 1664.

Lumsden, C., & Wilson, E. O. (1981). *Genes, mind, and culture: The coevolutionary process*. Harvard University Press.

Marcus, G. F. (2012). Musicality: Instinct or acquired skill? *Topics in Cognitive Science, 4*(4), 498–512.

Margulis, E. H. (2014). *On repeat: How music plays the mind*. Oxford University Press.

Martínez-Molina, N., Mas-Herrero, E., Rodríguez-Fornells, A., Zatorre, R. J., & Marco-Pallarés, J. (2016). Neural correlates of specific musical anhedonia. *Proceedings of the National Academy of Sciences, 113*(46), E7337–E7345.

Martínez-Molina, N., Mas-Herrero, E., Rodríguez-Fornells, A., Zatorre, R. J., & Marco-Pallarés, J. (2019). White matter microstructure reflects individual differences in music reward sensitivity. *Journal of Neuroscience, 39*(25), 5018–5027.

Mas-Herrero, E., Karhulahti, M., Marco-Pallares, J., Zatorre, R. J., & Rodriguez-Fornells, A. (2018). The impact of visual art and emotional sounds in specific musical anhedonia. *Progress in Brain Research, 237*, 399–413.

Mas-Herrero, E., Maini, L., Sescousse, G., & Zatorre, R. J. (2021). Common and distinct neural correlates of music and food-induced pleasure: A coordinate-based meta-analysis of neuroimaging studies. *Neuroscience & Biobehavioral Reviews, 123*, 61–71.

Mas-Herrero, E., Marco-Pallares, J., Lorenzo-Seva, U., Zatorre, R. J., & Rodriguez-Fornells, A. (2012). Individual differences in music reward experiences. *Music Perception, 31*(2), 118–138.

Mehr, S. A., Krasnow, M. M., Bryant, G. A., & Hagen, E. H. (2021). Origins of music in credible signaling. *Behavioral and Brain Sciences, 44*(e60), 23–39.

Mehr, S. A., Singh, M., Knox, D., Ketter, D. M., Pickens-Jones, D., Atwood, S., & Glowacki, L. (2019). Universality and diversity in human song. *Science, 366*(6468), 1–17.

Mencke, I., Omigie, D., Wald-Fuhrmann, M., & Brattico, E. (2019). Atonal music: Can uncertainty lead to pleasure? *Frontiers in Neuroscience, 12*, 979.

Mendoza, J. K., & Fausey C. M. (2021). Everyday music in infancy. *Developmental Science*, e13122.

Morley, I. (2013). *The prehistory of music: Human evolution, archaeology, and the origins of musicality*. Oxford University Press.

Moulson, M. C., Westerlund, A., Fox, N. A., Zeanah, C. H., & Nelson, C. A. (2009). The effects of early experience on face recognition: An event-related potential study of institutionalized children in Romania. *Child Development*, *80*(4), 1039–1056.

Mundy, R. (2018). *Animal musicalities: Birds, beasts, and evolutionary listening*. Wesleyan University Press.

Nettl, B. (2015). *The study of ethnomusicology: Thirty-three discussions*. University of Illinois Press.

Niarchou, M., Gustavson, D. E., Sathirapongsasuti, J. F., Anglada-Tort, M., Eising, E., Bell, E., McArthur, E., Straub, P., 23andMe Research Team, McAuley, J. D., Capra, J. A., Ullén, F. U., Creanza, N., Mosing, M. A., Hinds, D. A., Davis, L. K., Jacoby, N., & Gordon, R. L. (2022). Genome-wide association study of musical beat synchronization demonstrates high polygenicity. *Nature Human Behaviour*. https://www.nature.com/articles/s41562-022-01359-x.

Norenzayan, A., Shariff, A. F., Gervais, W. M., Willard, A. K., McNamara, R. A., Slingerland, E., & Henrich, J. (2016). The cultural evolution of prosocial religions. *Behavioral and Brain Sciences*, *39*, 1–65.

Norman-Haignere, S., Kanwisher, N. G., & McDermott, J. H. (2015). Distinct cortical pathways for music and speech revealed by hypothesis-free voxel decomposition. *Neuron*, *88*(6), 1281–1296.

Partanen, E., Kujala, T., Tervaniemi, M., & Huotilainen, M. (2013). Prenatal music exposure induces long-term neural effects. *PLOS ONE*, *8*(10), e78946.

Patel, A. D. (2006). Musical rhythm, linguistic rhythm, and human evolution. *Music Perception*, *24*, 99–104.

Patel, A. D. (2008). *Music, language, and the brain*. Oxford University Press.

Patel, A. D. (2010). Music, biological evolution, and the brain. In M. Bailar (Ed.), *Emerging disciplines* (pp. 91–144). Rice University Press.

Patel, A. D. (2014). The evolutionary biology of musical rhythm: Was Darwin wrong? *PLOS Biology*, *12*(3), e1001821.

Patel, A. D. (2018). Music as a transformative technology of the mind: An update. In H. Honing (Ed.), *The origins of musicality* (pp. 113–126). MIT Press.

Patel, A. D. (2019). Evolutionary music cognition: Cross-species studies. In P. J. Rentfrow & D. Levitin (Eds.), *Foundations in music psychology: Theory and research* (pp. 459–501). MIT Press.

Patel, A. D. (2021). Vocal learning as a preadaptation for the evolution of human beat perception and synchronization. *Philosophical Transactions of the Royal Society B*, *376*(1835), 20200326.

Patel, A. D., & von Rueden, C. (2021). Where they sing solo: Accounting for cross-cultural variation in collective music-making in theories of music evolution. *Behavioral and Brain Sciences*, *44*(e85), 77–78.

Penhune, V. B. (2020). A gene-maturation-environment model for understanding sensitive period effects in musical training. *Current Opinion in Behavioral Sciences, 36*, 13–22.

Pinker, W. (1997). *How the mind works*. W. W. Norton.

Plantinga, J., & Trainor, L. J. (2005). Memory for melody: Infants use a relative pitch code. *Cognition, 98*(1), 1–11.

Podlipniak, P. (2017). The role of the Baldwin effect in the evolution of human musicality. *Frontiers in Neuroscience, 11*, 542.

Ratey, J. (2008). *Spark: The revolutionary new science of exercise and the brain*. Little Brown.

Reh, R. K., Dias, B. G., Nelson, C. A., Kaufer, D., Werker, J. F., Kolb, B., & Hensch, T. K. (2020). Critical period regulation across multiple timescales. *Proceedings of the National Academy of Sciences, 117*(38), 23242–23251.

Richard, D., Liu, Z., Cao, J., Kiapour, A. M., Willen, J., Yarlagadda, S., Jagoda, E., Kolachalama, V. B., Sieker, J. T., Chang, G. H., Muthuirulan, P., Young, M., Masson, A., Konrad, J., Hosseinzadeh, S., Maridas, D. E., Rosen, V., Krawetz, R., Roach, N., & Capellini, T. D. (2020). Evolutionary selection and constraint on human knee chondrocyte regulation impacts osteoarthritis risk. *Cell, 181*(2), 362–381.

Richerson, P. J., Boyd, R., & Henrich, J. (2010). Gene-culture coevolution in the age of genomics. *Proceedings of the National Academy of Sciences, 107*(Supplement 2), 8985–8992.

Roberts, C. (2014). *Music of the Star Mountains: A naturalist's guide to the composition of songs in central New Guinea*. Institute of Papua New Guinea Studies.

Rohrmeier, M., & Rebuschat, P. (2012). Implicit learning and acquisition of music. *Topics in Cognitive Science, 4*(4), 525–553.

Rouse, A. A., Patel, A. D., & Kao, M. H. (2021). Vocal learning and flexible rhythm pattern perception are linked: Evidence from songbirds. *Proceedings of the National Academy of Sciences, 118*(29), 1–9.

Sacks, O. (2011). Face blind. In *The mind's eye* (pp. 82–110). Vintage.

Sandstrom, G. M., & Russo, F. A. (2013). Absorption in music: Development of a scale to identify individuals with strong emotional responses to music. *Psychology of Music, 41*(2), 216–228.

Savage, P. E., Brown, S., Sakai, E., & Currie, T. E. (2015). Statistical universals reveal the structures and functions of human music. *Proceedings of the National Academy of Sciences, 112*(29), 8987–8992.

Savage, P. E., Loui, P., Tarr, B., Schachner, A., Glowacki, L., Mithen, S., & Fitch, W. T. (2021). Music as a coevolved system for social bonding. *Behavioral and Brain Sciences, 44*(e59), 1–22.

Shelemay, K. K. (2011). Musical communities: Rethinking the collective in music. *Journal of the American Musicological Society, 64*(2), 349–390.

Shilton, D. (2022). Sweet participation: The evolution of music as an interactive technology. *Music & Science, 5*, 1–15.

Sperber, D. (1996). *Explaining culture: A naturalistic approach*. Blackwell.

Stein, L. (Ed.). (1975). *Style and idea: Selected writings of Arnold Schoenberg*. University of California Press.

Tilot, A. K., Khramtsova, E. A., Liang, D., Grasby, K. L., Jahanshad, N., Painter, J., Colodro-Conde, L., Bralten, J., Hibar, D. P., Lind, P. A., Liu, S., Brotman, S. M., Thompson, P. J., Medland, S. E., Macciardi, F., Stranger, B. E., Davis, L. K., Fisher, S. E., & Stein, J. L. (2021). The evolutionary history of common genetic variants influencing human cortical surface area. *Cerebral Cortex*, *31*(4), 1873–1887.

Tinbergen, N. (1963). On aims and methods of ethology. *Zeitschrift für Tierpsychologie*, *20*(4), 410–433.

Todorov, A. (2017). *Face value: The irresistible influence of first impressions*. Princeton University Press.

Tomlinson, G. (2015). *A million years of music: The emergence of human modernity*. MIT Press.

Trainor, L. J. (2015). The origins of music in auditory scene analysis and the roles of evolution and culture in musical creation. *Philosophical Transactions of the Royal Society of London B: Biological Sciences*, *370*(1664), 1–14.

Tranchant, P., Lagrois, M. É., Bellemare, A., Schultz, B. G., & Peretz, I. (2021). Co-occurrence of deficits in beat perception and synchronization supports implication of motor system in beat perception. *Music & Science*, *4*, 1–13.

Tranchant, P., Vuvan, D. T., & Peretz, I. (2016). Keeping the beat: A large sample study of bouncing and clapping to music. *PLOS ONE*, *11*(7), e0160178.

Trehub, S. E. (2003). The developmental origins of musicality. *Nature Neuroscience*, *6*(7), 669–673.

Trehub, S. E., Becker, J., & Morley, I. (2015). Cross-cultural perspectives on music and musicality. *Philosophical Transactions of the Royal Society B: Biological Sciences*, *370*(1664), 20140096.

Trehub, S. E., & Cirelli, L. K. (2018). Precursors to the performing arts in infancy and early childhood. *Progress in Brain Research*, *237*, 225–242.

Wilmer, J. B., Germine, L., Chabris, C. F., Chatterjee, G., Williams, M., Loken, E., Nakayama, K., & Duchaine, B. (2010). Human face recognition ability is specific and highly heritable. *Proceedings of the National Academy of Sciences*, *107*(11), 5238–5241.

Wolf, M. (2007). *Proust and the squid: The story and science of the reading brain*. Harper.

Wrangham, R. (2009). *Catching fire: How cooking made us human*. Basic Books.

Wrangham, R. (2017). Control of fire in the Paleolithic: Evaluating the cooking hypothesis. *Current Anthropology*, *58*(S16), S303–S313.

Zentner, M., & Gingras, B. (2019). The assessment of musical ability and its determinants. In P. J. Rentfrow & D. Levitin (Eds.), *Foundations in music psychology: Theory and research* (pp. 641–683). MIT Press.

2 Musical Meaning in Transspecies Perspective: A Semiotic Model

Gary Tomlinson

Music resists definition, notoriously. The resistance is partly a problem of the language in which definition must be couched, which struggles to overcome the sheer difference between musical and linguistic experience; this is the problem of our abiding linguocentrism, and it is endemic to every kind of musicology. More essentially, the resistance arises from the complexity of music, which comprises a broad swath of human capacities and practices showing immense cultural diversity and reflecting (all available evidence indicates) a long evolutionary history (for several approaches to this history, see Wallin et al., 2000; Bannan, 2012; Tomlinson, 2015; Savage et al., 2021). Even without an encompassing definition, however, we can clarify human musicking by thinking of it as the deployment of certain kinds of signs arrayed to create kinds of meanings distinct from linguistic ones. Here, two additional hard-to-define terms confront the theorist: *sign* and *meaning*. In what follows, I outline a general model of meaning, resting on a particular conception of signs, that frames some fundamental resources of human music. The model aims to clarify the status of music within a constellation of other human activities and to illuminate its relations to certain communicative acts of other species; at the same time, it suggests some more general insights into evolutionary history. It rests on basic propositions that I explore briefly but do not elaborate or defend fully. These are offered almost as axioms, and the resulting view of meaning as a promise of fuller elaboration to come (see Tomlinson, in press).

Two Kinds of Information

Information is a term widely applied to two distinct phenomena in the world. On the one hand, information describes correlated changes that come about through some causal interaction between or among systems: "causal covariance," as Fodor (1990) put it. ("Reliable causal covariance" was his full phrase, but reliability involves not the nature of the information but its quantity in any given situation.) Such *causal information* defines and enables us to quantify a relation between the systems involved. It can be conceived as occurring along a channel extending from a sender and bringing about some corresponding effect in a receiver (Shannon & Weaver, 1949). Information of this sort is relational: its channels can feed back from receiver to sender, they can be networked in parallel and serial fashion, and they can be doubled and redoubled or hierarchized in any number of routines and subroutines; but the fundamental relationality of covariance remains its defining feature.

All living things, even the simplest bacterium, are complex miracles of this kind of relational, causal information. But in the vast majority of organisms this information is *not about anything*. It is without content or meaning, a sheer causality that makes things happen, from the molecular level up through organismal and ecosystemic levels. The bonding relations between DNA and RNA, or those between these molecules and the various proteins that catalyze the relations of amino acids that bond to RNA, are archetypal instances of this biotic causal information.

A contrasting kind of information arises on the foundation of causal information but adds to it another dimension. It defines a phenomenon distinct from relational correspondence that results in a new kind of connection between things. To this connection we give various names: representation, referentiality, aboutness, intentionality, content, semantics, and meaning. This kind of informational connection is sometimes called semantic information, but I call it *semiotic information*, reflecting the fact that it arises only from signs, wherein one thing represents or refers to another, and hence only from the processes that give rise to signs, gathered under the term *semiosis*. All life-forms involve causal information; a few of them also involve semiotic information.

We could say of signs, as Marx said of commodities, that while they appear to be trivial and easily understood, they are in reality "abounding in metaphysical subtleties." Semiotic information depends not on relations alone, like causal information, but on a relation to a relation—a *metarelation*. Conceptualizing this situation calls for something more complex than the model of a transmitter, channel, and receiver and something different even from the model of a hierarchy of routines. Most basically, a metarelation is required for semiosis because a sign is something different from a cause of its object. It is an *analysis* of its object: it divides up the object, picking out certain aspects of it to use in re-presenting it. It is constrained in the picking both by its own nature and, reciprocally, by the nature of the object. A road sign showing that deer might be in the area represents certain aspects of deer and not others, and the aspects selected are determined by both the nature of deer and the nature of the sign. Signs never capture all aspects of their objects; that would collapse the sign into the object, collapsing at the same time aboutness and content. Identity admits no space for re-presentation.

What is it that enters into this complex relation between aspects of one entity and aspects of another? What creates the metarelation? This is a third entity, one that is capable of forming percepts of certain sorts. This entity doesn't exercise complete, sovereign control over its percepts. Instead it is *called into* the aspectual constraints of what will become sign and object—called by their nature as *affordances* in its world or niche. Its percepts are always formed not only by its own capacities but also by the entities that come to form sign and object. In this way, semiosis is a process of three-way reciprocal constraint—or reciprocal affordance, which amounts to the same thing—among a percept and things or phenomena external to it. It is irreducibly a niche-constructive process.

Those familiar with the philosophy of Charles Sanders Peirce will recognize in this approach to semiotics two basic Peircean tenets (Peirce, 1955; Atkin, 2013). The first is the *partial, aspectual nature* of all the relations in the sign-making process (for example, the partiality of the relation of sign to deer and deer to sign). Peirce offered many typologies of signs derived from the many combinatorial possibilities of this partiality; the most famous of these is the division of icon, index, and symbol. The second tenet is the calling of a perceiver into the aspectual parsing of the world, an idea for which Peirce coined the term *interpretant*. This is a fundamental term in his philosophy but one that has been much misunderstood. It requires no interpretation, as we tend to use the word, but relies instead on the perceptual capacity to divide up the world—to parse or analyze it, registering some of its parts, clumping some together with others, and relating itself to those relations.

The basic distinction between the two kinds of information, causal and semiotic, is relevant to the question of meaning because *meaning is founded in the sign metarelation*. It emerges only from semiosis. Signs are always signs *of*: they introduce aboutness in the world by virtue of this fundamental nature, introducing at the same time what we connote by all the synonyms named above: reference, content, semantics, and so forth. Signs are *representational*—in the precise sense of re-presenting one thing in another. This is what signs do, and this is what the content-full information that arises from signs does; it is not what causal information does. Signs are not the same as *signals*, as this word is loosely but pervasively used by biologists and others. A signal can be non-semiotic and often comes down to a threshold mechanism in the operation of causal information. It doesn't necessarily re-present one thing in another.

Making Signs

How do some life-forms come to form interpretants? The distinction between causal information in general and the special case of semiotic information helps clarify the evolved capacities that enable semiosis and meaning. These should not be thought of as identifiable traits possessed by particular organisms—that is, as adaptations selected for particular functions. Such *selection for* is not how natural selection works; it is not an optimization problem, notwithstanding Dennett's (1995, 2017a, 2017b) decades-long effort to make us think it is. The capacities to create meaning are ongoing processes of negotiation with an external world, so they can't be identified as discrete properties of an organism. *Processes, not properties*: this is a motto that should hang above the door of every evolutionist (it already hangs above the doors of many). These processes are involved in the niche construction to which I have already referred and will return.

Among the processes needed to create interpretants, *memory* is crucial, but not merely the simplest forms of storage and retrieval. Instead, something like an *episodic* memory is necessary, a process in which memories are retrieved as parts of whole situations—retrieved more readily because of cues to those situations—and hence can enable a cognitive reliving of the situations (Tulving, 1972; Baddeley, 2000). The best evidence suggests that a number of animals in addition to humans form episodic memories—especially some birds and mammals—but that most animals, and all other organisms, do not (Clayton & Dickinson, 1998, Salwiczek et al., 2010, Templer & Hampton, 2013). Evidence also suggests that the episodic nature of such memory depends on processing loops back and forth between the deep-brain hippocampus and cortical layers (Allen & Fortin, 2013).

Learning is also central—but learning of an advanced sort arising from episodic memory (Nuxol, 2012). Associative learning is purely relational and occurs in many creatures without any semiosis, as a function of simple neural nets involving excitation and inhibition connected with innate reward or punishment signals. A well-studied example is the sucrose sensitivity of honeybees, a simple neural net that forms associations concerning food sources and much else (Menzel & Giurfa, 2001; Gil & De Marco, 2005; Smith et al., 2008; Peng & Chittka, 2017). In contrast, the situational learning of a songbird, such as the song sparrow's capacity to learn to gauge the varying challenges from the songs of several neighbors, is semiotic through and through (Beecher et al., 1996; Beecher & Campbell, 2005; Beecher et al., 2020). It depends on the retrieval and processing of whole episodes from a bird's life. (I present a more detailed avian example later.)

Attention, finally, is a more basic process than episodic memory or learning, at least in its simple forms. It is related to the salience of a stimulus—the way an organism can single out a particular stimulus from all the stimuli coming at it. Again, honeybees' sensitivity to sucrose provides a good example. For sign and meaning making, an especially complex attentional process is required, one that can sustain a recursive analysis of incoming stimuli, taking them apart and focusing on them at both comprehensive and partial levels. This mereological or part-to-whole complexity shows the bond between complex attention and episodic memory, also made up of parts and wholes; it also points toward the aspectual nature of signs. Studies of such complex attention (Knudsen, 2007, 2018, 2020) suggest that it is the product of mediations of information in which bottom-up salience filters near the sensory input interface with further processing from top-down systems higher up in the brain. Such neural mediations are, I think, fundamental for the emergence of semiotic information, and I will touch on them again later.

Recursive attention, episodic memory, advanced learning: these foundations for semiosis are not discrete properties of certain animals but rather developing processes in their engagements with their niches. Among animals that manifest them, they help us focus on a processualism that has recently radicalized ideas of niche construction.

Radical Niche Construction

In general, niche construction refers to the reciprocal or feedback process by which organisms build their selective environments even as those environments build them (Odling-Smee et al., 2003). Alterations in the niche brought about by populations of organisms change the selective pressures the niche exerts on the organisms, eventually changing their genomes. Niche construction tends to be thought of as a dynamic operating across long timescales, with environments and populations of organisms slowly shifting in tandem with one another—the ancestors of beavers, for example, gradually adapting to aquatic life even as their behaviors created more aquatic environments (Rybczynski, 2007).

Here, the causal circuit seems clear: a population of organisms affects its environment through its phenotypic traits, which are ultimately dependent on its genome; these effects alter the environment, in turn making certain variants in the population's phenotype more advantageous; selection finally shifts the genome of the whole population toward a genome creating the advantageous traits.

Radical niche construction goes farther: it attempts to chart across *all timescales* the pervasive plasticity of niche-constructive feedback and its impact on natural selection. Niche construction in this view is not only a long-term process but also organisms' much quicker plastic responsiveness to their changing environments. Natural selection results not in fixed capacities with discrete, advantageous functions but in open systems shifting in relation to aspects of their environments. And—most important—this plasticity reaches *from the niche all the way to genes,* which turn out to be nothing like blueprints for life-forms but instead are in constant flux with the environments around them (Linksvayer et al., 2012; Laubichler & Renn, 2015).

Once again, honeybees provide an example. Dozens of pheromones are crucial in the intricate regulation of the honeybee colony or superorganism. They are produced by individual bees according to complex bioinformational pathways involving genes, RNA transcription of various sorts, proteins that genes construct, and more (Bortolotti & Costa, 2014; Ueno et al., 2015). But *the pathways run in both directions*. The genetic instructions don't run the show; instead, they are regulated by the proteins they produce, which in turn are regulated by the pheromones *they* help produce. Brief exposure to a single chemical in the cocktail that functions as an alarm pheromone alters factors inducing RNA transcription in the olfactory centers of the bee brain, shifting gene expression from moment to moment and initiating a chemical cascade that up- or down-regulates the expression of hundreds of genes. The whole network extending from niche to genes is altered almost in the blink of an eye—or the whiff of a scent (Pankiw, 2004; Alaux et al., 2010).

The effects of such plastic genomic expression enter into all aspects of honeybee existence. Similar cascades from environmental input, for example, regulate the transformation of nurse worker bees into foragers. It is now clear, in fact, that similar cascades regulate the genomic expression of all life-forms, sending chemical signals that regulate the transcription of so-called immediate early genes, which in turn regulate the expression of "late" genes.

With such radical niche construction in mind, let's return to semiosis and ask where we might seek neural substrates for the interpretant and the metarelation. We need to seek them not simply in the redoubling of parallel processing in large brains, and not even in the serial extension of processing along a sequence of distinct modules in such brains, although both probably play a role. An additional element of neural architecture is built on top of these, bringing massive parallelism and serialism to bear in modulating the processing of other connected centers. Instead of peripheral, sensory neurons directly signaling the highest processing centers along a more or less straight line of neural transmission, large vertebrate brains have processors connected to processors, all reconnected in looped feedback systems of brain centers or nuclei.

The song system in a songbird's brain, a system of nuclei enabling audition, processing, and production of song, offers an example (Bolhuis & Gahr, 2006; Jarvis, 2008; Mooney et al., 2008). This is often divided into two systems—one controlling auditory processing and another controlling song learning, memorization, and production. Figure 2.1 shows these systems in two sagittal sections of a songbird's brain, with the auditory pathway on the left and the song pathway on the right. It's easy to see that each pathway is networked in complex ways, involving many nuclei. Moreover, the

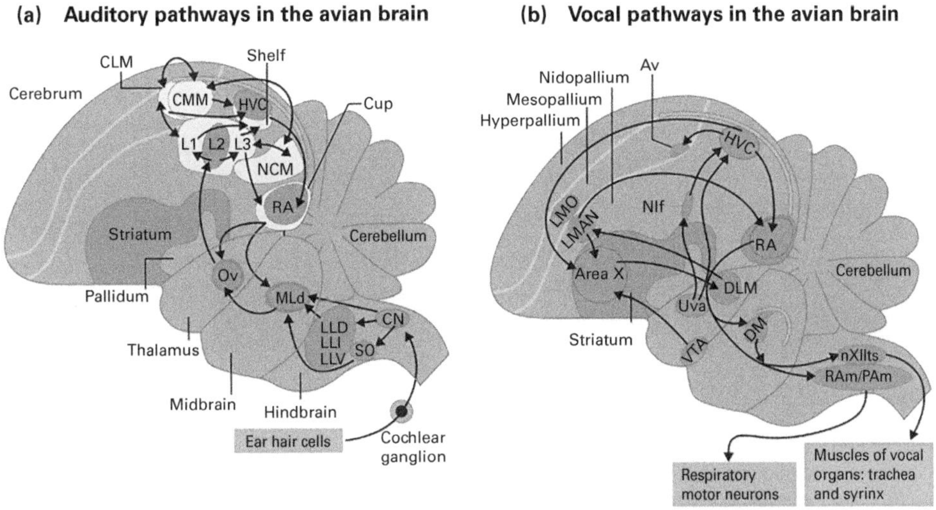

Figure 2.1
Generalized views of auditory and vocal pathways in the songbird brain. (From Berwick et al., 2012.)

two pathways are connected: the auditory nuclei are linked to the song control nuclei, linking left- and right-hand diagrams and bringing audition—the perception of niche phenomena—directly to bear on song memory and learning.

Especially important for the semiotic complexity of birdsong, I think, is the mediating role of the thalamus in these networks. It is situated between the cerebrum, with its higher processing functions, and the mid- and hindbrain. Notice that in both diagrams in the figure the thalamus doesn't merely relay processed sensory input to higher brain regions; it forms part of the feedback loops *from* higher regions involved in both song production and audition. In mammals the thalamus is a midbrain processing station mediating top-down and bottom-up networks. Recent work indicates that it is central in the complex forms of attentional focus outlined earlier, as well as being central in memory, learning, and voluntary motor behavior (Halassa & Kastner, 2017; Knudsen, 2018; Nani et al., 2019; Gu et al., 2020). The same seems to be true in songbirds.

There's nothing nearly as elaborate as these networks in the brain of a honeybee (Menzel & Giurfa, 2001; Chittka & Niven, 2009; Giurfa, 2013). As figure 2.2 shows, much of the bee's brain is taken up by the antennal lobes—the almost circular bodies near the bottom, crucial for chemosensory processing—and the optic lobes on each side, devoted to visual processing. The large areas at the top—the so-called mushroom bodies—form the integrative processing centers of the brain. The arrow starting from the bottom center and branching out into the antennal lobes and mushroom bodies (it does so bilaterally, although the diagram shows it on only one side) traces the pathway

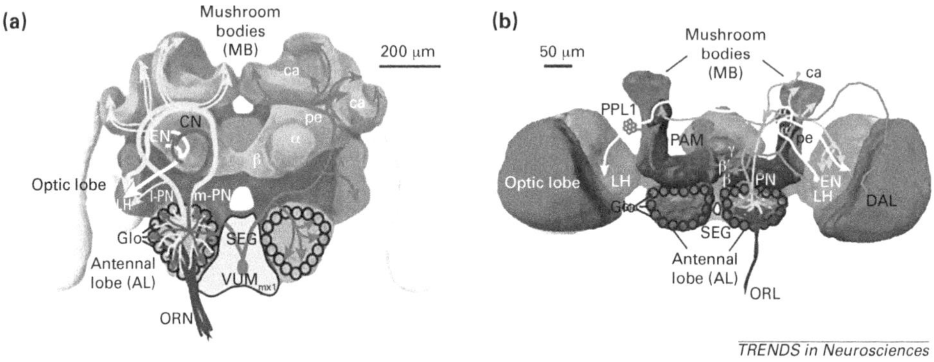

Figure 2.2
The honeybee (*Apis melifera*) brain. (From Giurfa, 2013.)

for sucrose sensitivity. It is striking in its simplicity: a single long neuron that projects from the proboscis to many regions of the brain, including motor control centers, which dictate motor responses. The complexity built into this tiny brain is amazing, but it is far outstripped by the bird brain's multiplicity of nuclei and their networks of feedback mediation through distinct levels of processing.

It is not merely the complexity of neural architecture, however, that brings about the cognitive functions of interpretant, sign, and meaning in a songbird's brain. To achieve this, the network must be set in motion as a process rather than a static architecture, a process that brings about radical niche-constructive plasticity in the face of changing affordances. The complexity of multicomponent and multilevel processing fosters an experience of input from the world proportionate to the varied complexity of the brain architecture involved. It is from this, I think, that a new fold is introduced in the experience of certain animals—and from this, they gain the parsing power that makes their percepts analytic and at the same time opens those percepts to the formative rebound from the things they analyze. From this folded, recursive percept arises the relation to a relation of things in the world, the essence of sign and meaning. Thus a radicalized niche-constructive process underlies the calling into the interpretant.

The complexities of bee behavior depend on modest attention, memory, and learning, all rendered processual and reactive to the changing environment, all transmitting causal information through the networks of radical niche construction. But the capacity of birds and some other animals to create aboutness and meaning depends on something more: complex attentional focus, meshing high- and low-level processing, and episodic memory, recalling whole situations in the past and shaping learning on the basis of them. The systems of systems exemplified by the songbird are harnessed to the plastic capacity of all organisms to remake themselves in radical, momentary negotiation with their niches.

Birdsong Meanings

Explicit here is the placement of birds and bees on different sides of the ontological divide between semiotic and causal information. This is the divide between complexity *with* and *without* meaning; on both sides there is complexity aplenty. Honeybee society comes about through the operation of immense, intricate, beautiful, finally awesome networks of relational, causal information. There are no signs; there is no metarelation, no meaning.

Even the famous waggle dances of honeybees do not operate through signs, symbols, or anything resembling language. Every aspect of their communication has yielded up its secrets more fully over the last thirty years (see Dyer, 2002; Thom et al., 2007; Grüter & Farina, 2008; Grüter et al., 2008; Hrncir et al., 2011), and each kind of information communicated turns out to involve causal networks alone. Sensory input involving cues that are mainly mechanical and chemical is connected through short, direct neural pathways to a stereotyped response. This is not meant to underplay the complexity involved. The cues are integrated across sensory modalities in the information-processing mushroom bodies; the chemosensation taps into the pheromone cascades modulating gene expression; memories, long term but not episodic, are cued and retrieved in bees that already have foraging experience; and all this enters into the balanced, niche-constructive economies extending between foraging worker bees and worker bees in the hive that off-load their nectar and tend to the brood. Nevertheless, it all works without interpretant formation, metarelation, or signs.

Birdsong is different, involving full-fledged semiosis. The pseudoduetting of Australian magpie-larks offers a quick case study of the avian interpretant at work. Mated pairs of magpie-larks sing real duets in which the two birds synchronize their own contributions of motifs, or syllables. The syllables they sing are drawn from a species-specific repertory and, in duetting, each bird contributes syllables different from those of its partner; this alternation distinguishes the duets from their solo songs, which repeat a single syllable from the repertory. The whole performance includes not only singing but also a visual aspect: wing movements alternating between the birds in coordination with the singing. Duetting pairs perform from prominent posts in their fairly open, eucalyptus-rich habitats, seemingly to ensure the clear visibility as well as audibility of their performances. The precision of the birds' synchrony increases along with their growing experience together as a pair—they learn to sing and dance together—and the heightened coordination elicits heightened responses from birds witnessing the performance. The meanings conveyed by magpie-larks' audiovisual duets, in other words, depend on the pair-bonded birds' coordination being perceived and gauged by potential rivals for mates and territory (Hall & Magrath, 2000, 2007; Rek & Magrath, 2016, 2017; Rek, 2018).

But there are also pseudoduets (Rek & Magrath, 2017). When one of the mated birds is absent, the magpie-lark remaining at home sometimes sings a solo song composed of contrasting syllables like a duet, instead of the usual repeated single syllable. The bird imitates, in other words, two birds performing a duet. Given the general meaning of true duets, the purpose of performing a pseudoduet seems to be to deceive listeners, signifying a coordinated coalition even though one bird is absent. This interpretation is strongly supported by a major difference between true and deceptive performances: pseudoduets are performed from hidden posts, rather than out in the open like true duets, and they are never accompanied by the wing movements of true performances. The whole visual component of true duets is suppressed in what seems to be a calculated attempt to make the deception more effective by concealing the absence of elements that would expose it.

The interpretant process in birds witnessing duets, whether true or pseudo, shows all the general features required for full semiosis: the recursive focus on stimuli, parsing or analyzing aspects of them; the resulting part-to-part representational relation between those aspects—aspects of the sign, aspects of the broader social situation, and aspects of the object; and the complex attention, episodic memory, and learning of situations and events within that social context. The sign that results is indexical, pointing both to the coalition in the social, niche-constructive context and to its strength and durability. The syllables in the songs also serve as indexes: magpie-larks perceive, in the context of their own learned song repertories, aspects of the duets they witness that differ from solo songs and signify the coalition of the bonded pair; additionally, their perceptions are nuanced in degree, according to the coordination of the performance.

The interpretant formation on the part of the deceiving bird is even more complex. In essence, the singing bird forms an interpretant of its own interpretant, so that its own nondeceptive percept (shared with the listener) becomes the sign vehicle for an object that now comprises both the original sign and object. The sign vehicle and object brought into relation in the listener's interpretant now function together as the object of a new sign vehicle; these are brought into relation only by the singer's interpretant. A hierarchy of interpretants, absent from the listener's semiosis, builds the singer's.

The formation of this complicated semiosis, as well as similar semiotic processes in other songbirds, is dependent on the combinatorial nature of the song: its construction from small units—variously called notes, syllables, or motifs—and the nature of their arrangement. The magpie-lark singer constructs, and the listener gauges, distinct arrangements for solo song versus duet and distinct degrees of synchronization among the syllables of the song. Such arrangements can become immensely varied and complex, even for species whose repertories contain relatively few syllables, given the syntactic rules that often govern transitions from one syllable to another. (For a finite state diagram of the possibilities for a Bengalese finch, with about ten syllables, see Okanoya, 2012.) Moreover, it is becoming clear that birdsong communication also depends on minute details of the individual gestures that make up the songs, which differentiate one gesture from another (Amy et al., 2015; Fishbein et al., 2019).

This sort of complexity is widespread in birdsong, and it has long puzzled birdsong researchers. The basic question is this: if the message of the song is either territorial ("get off my yard!") or sexual ("choose me!"), why expend considerable cognitive and physical resources on such syntactic intricacy? Territorial and mating signals in countless other animals are simpler and mostly nonsemiotic. The magpie-lark example shows what is becoming clear in other instances of birdsong: the meanings of the songs vary according to their combinatorial nuances in interaction with the situations or life episodes in which they're deployed. The meanings depend, in other words, on *the combinatorial structure as niche-constructive process*.

Hyperindexicality and Music

Meaningful syntactic arrangement in birdsong calls to mind human language, but these are not semiotically equivalent. We need to be cautious in likening birdsong to language because, although birdsong is complexly combinatorial, with nuanced differentiation of its component gestures, it seems not to be *compositional*. Linguists use this word to refer to the *two* levels of meaning in human language: meaning is found both in individual words and in words assembled in syntactic arrangement. The elements that make up birdsongs do not carry individual meanings the way human words do. Several researchers have advanced claims that they do—that there is meaning in individual syllables or motifs (Abe & Watanabe, 2011; Engesser et al., 2015; Suzuki et al., 2016). In each case, other researchers have responded with simpler, more plausible explanations of the communication observed, more in keeping with other things we know about birdsongs (Beckers et al., 2012; Bowling & Fitch, 2015; Bolhuis et al., 2018).

Birdsongs create indexical signs through syntax and combinatoriality but not compositionality. Birdsong is indexical because it forms a pointing or indicating kind of sign. It is not symbolic, because symbols derive individual meanings from their places in larger arrays of signs—like the words of language. In birdsong, meaning doesn't "stick" to the individual syllables and motifs the way it does to words; instead, it arises in the indexical pointing of the motifs. But even though the birdsong motifs don't carry meaning the way words do, their pointing creation of meaning is twofold: (1) the motifs point to one another in the song, according to their nuances and syntactic relations; and (2) by virtue of this internal pointing, the song as a whole points outward to the social situation in which it occurs. The motifs come to have meaning in their combination according to syntaxes like that of the Bengalese finch song.

Such an arrangement involves arrays of signs and some syntactical rules for combining them, both characteristic of symbolism; thus, we could call it *presymbolic* or *protosymbolic*. But this would, I think, confuse important issues and assert a false teleology: the progressive notion that birdsong is somehow on its way to symbolism. We should instead call the arrangements of indexes in birdsong *hyperindexical*.

I introduced this concept a few years ago in two books on human evolution to capture a stage of ancient hominin communication, before the last 100,000 years or so of human modernity (Tomlinson, 2015, 2018). This stage was not symbolic and was without full-fledged modern language, but it nevertheless negotiated complex social interactions by constructing arrays of indexes. In these Paleolithic societies, hyperindexicality was not only evident in the antecedents of modern language—in protolanguage, as it is often called. It was also at work in antecedents of modern music—*protomusic*—and in the structures of these societies' nascent rituals and in their increasingly elaborate, composite technologies. Hyperindexicality characterized all the most complex gestures of these societies, so we can speak of a *hyperindexical age* in the evolution of our hominin ancestors.

Hyperindexicality lives on today in all modern human societies and in all these gestures—ritual, technology, and even certain aspects of language, such as its phonology and intonational shapes. But the place where it is most highly developed, most intricate in its gestures, and most eloquently meaningful in their combinations is human musicking. Music is pervasively indexical, since its gestures point to one another, creating expectations of ongoing structures built from them. It is hyperindexical in joining arrays of gestures together under syntactic governance. To appreciate fully the indexical nature of music, we must picture music in itself, stripped of the webs of mainly linguistic symbols in which it is always embedded in today's human societies—and no doubt in human societies reaching back tens of thousands of years. In these societies the symbolic meanings of music are (and have been) deep and rich. But they are accretions built onto music's fundamental indexicality, layers of meaning added by a species that cannot help itself in this regard: *Homo loquens, Homo symbolicus*.

Think instead of an earlier hominin, our direct or indirect ancestor, presapient, without symbolic powers or modern language but adroitly deploying hyperindexical semiosis in its niche construction. Think of this ancestor vocalizing and moving and employing material prostheses ("musical instruments") to make rhythmic and sonic patterns and entraining itself and other members of its group to these patterns. This is the protomusic I describe in *A Million Years of Music*, which formed one aspect of hominin communication in the hyperindexical age, alongside combinatorial technologies, ritualized activities and spaces, and protolanguage.

To make the telos of modern language a condition of such societies is a mistake that is still made often today, and it is evident in most of the theories of early "musical languages" dating back to Darwin and beyond. This distorts both the nature of hyperindexical societies and their subsequent histories. What was at stake for these early hominins was a particular interpretant formation, proliferating a particular kind of sign—the index—in a long-lasting stage of semiotic information transmission. We can call this stage a regime of signs, borrowing a phrase from Deleuze and Guattari, but it operated on a timescale much longer than the one they imagined for such regimes.

Today, human hyperindexicality finds its most striking manifestation in musicking. Beyond humans, it finds its most striking development in birdsong, although other, less widely dispersed instances can be adduced, particularly among cetaceans (Whitehead & Rendell, 2014). Birdsong is fundamentally like human song in its hyperindexical semiotic mode; the conditions under which a syntax of indexes can arise form the broadest, deepest connection between the two. We've tended to miss this forest for the trees in comparisons of birdsong and human song, which have been dominated by questions of birds' cognitive processing of musical specifics—pitch, rhythm, and even meter. This research has discovered fascinating cognitive convergences between birds and humans and also some striking differences (Patel et al., 2009; Bregman et al., 2016; Honing, 2019; Garland & McGregor, 2020; Duengen et al., chapter 3 of this volume). It may be, for example, that birds' processing of frequencies yields nothing quite like the pitch the human brain creates; and although metric entrainment seems to occur in some birds, the question of why it is so limited in extent beyond humans remains open. (Timbral processing, meanwhile, which takes distinct if overlapping forms in human music and language and is closely related to the cognition of pitch, remains mostly a terra incognita beyond humans.)

Such questions, however, should not be allowed to obscure the *indexical semiotic commons* we share with a significant swath of the animal kingdom, including many, many birds and some mammals. Here is a place where a richly elaborated, universal human activity—musicking—is contiguous with the elaborated indexical niche constructions of some fellow creatures. These constructions are liable to fall into our conceptualization of musicking—to be called *songs*, for example, as in the case of birds and several kinds of whales—because of this general, shared semiotic nature. Beyond this indexical commons, and no less miraculous in its effects, is the realm of causal information and of life-forms operating without signs or meaning.

This highlights the large implications of the distinction between causal and semiotic information with which we began. Semiosis founds meaning, and meaning culture, even in its most expansive, nonanthropocentric definitions, which extend it well beyond *H. sapiens* in the world today (see, e.g., Wrangham et al., 1994; Laland & Galef, 2009; Whitehead & Rendell, 2014). Human language is still more self-evidently dependent on semiosis, and symbolism, the particular type of signifying that language most distinctively embodies, is nothing other than a stage of semiotic elaboration on the far side of hyper-indexicality. The burgeoning of human symbolism, language, and culture have together been identified as a "major transition" in earthly evolution—one of those branchings in biotic diversification that have loomed large in evolutionary thought since the 1990s, adding abrupt, wholesale changes to incremental ones in our thinking about natural selective process (Maynard Smith & Szathmáry, 1995). Here, too, a forest has been missed for the trees. The specifically human outgrowths of semiosis manifested in language have been taken as the major transition, when it is the far broader advent of metarelation, sign, and interpretant, today spanning thousands of species, that deserves the name.

What, finally, would a cognitive musicology be like that founded its empiricism on a semiotic model such as the one described here? It would, first of all, see that the relations between human and nonhuman musicality are not relations of musicality per se at all, but instead are relations of shared modes of indexical niche construction manifested in different behaviors. It would ask how these distinct behaviors are differently processed, understanding that nothing like the symbolic, linguistic webs in which all human musicking is entangled today entangle nonhuman indexical systems. It would see beyond the indexical commons to the far broader commons of causal information on which all life depends, a causality manifested, for all organisms, in the moment-to-moment flux of niche construction. But in seeing this, it would also discover the difference that entered the world when certain evolved capacities resulted in the possibility of interpretant, metarelation, and meaning.

References

Abe, K., & Watanabe, D. (2011). Songbirds possess the spontaneous ability to discriminate syntactic rules. *Nature Neuroscience, 14,* 1067–1074.

Alaux, C., Maisonnasse, A., & Le Conte, Y. (2010). Pheromones in a superorganism: From gene to social regulation. *Vitamins and Hormones, 83,* 401–423.

Allen, T. A., & Fortin, N. J. (2013). The evolution of episodic memory. *Proceedings of the National Academy of Sciences, 110,* 10379–10386.

Amy, M., Salvin, P., Naguib, M., & Leboucher. G. (2015). Female signalling to male song in the domestic canary, *Serinus canaria. Royal Society Open Science, 2,* 140196.

Atkin, A. (2013). Peirce's theory of signs. In E. N. Zalta (Ed.), *The Stanford encyclopedia of philosophy*. http://plato.stanford.edu/archives/sum2013/entries/peirce-semiotics.

Baddeley, A. (2000). The episodic buffer: A new component of working memory? *Trends in Cognitive Sciences, 4*, 417–423.

Bannan, N. (2012). *Music, language, and human evolution*. Oxford University Press.

Beckers, G. J. L., Bolhuis, J. J., Okanoya, K., & Berwick, R. C. (2012). Birdsong neurolinguistics: Songbird context-free grammar claim is premature. *NeuroReport, 23*, 139–145.

Beecher, M. D., Çaglar, A., & Campbell, S. E. (2020). Birdsong learning is mutually beneficial for tutee and tutor in song sparrows. *Animal Behaviour, 166*, 281–288.

Beecher, M. D., & Campbell. S. E. (2005). The role of unshared songs in singing interactions between neighboring song sparrows. *Animal Behaviour, 70*, 1297–1304.

Beecher, M. D., Stoddard, P. K., Campbell, S. E., & Horning, C. L. (1996). Repertoire matching between neighboring song sparrows. *Animal Behaviour, 51*, 917–923.

Berwick, R. C., Beckers, G. J. L., Okanoya, K., & Bolhuis, J. J. (2012). A bird's eye view of human language evolution. *Frontiers in Evolutionary Neuroscience, 13*. https://doi.org/10.3389/fnevo.2012.00005.

Bolhuis, J. J., Beckers, G. J. L., Buybregts, M. A. C., Berwick, R. C., & Everaert, M. B. H. (2018). Meaningful syntactic structure in songbird vocalizations? *PLOS Biology*. https://doi.org/10.1371/journal.pbio.2005157.

Bolhuis, J. J., & Gahr, M. (2006). Neural mechanisms of birdsong memory. *Nature Reviews Neuroscience, 7*, 347–357.

Bortolotti, L., & Costa, C. (2014). Chemical communication in the honey bee society. In C. Mucignat-Caretta (Ed.), *Neurobiology of chemical communication*. CRC Press/Taylor and Francis.

Bowling, D., & Fitch, W. T. (2015). Do animal communication systems have phonemes? *Trends in Cognitive Sciences*. htttps: doi.org/10.1016/j.tics.2015.08.011.

Bregman, M., Patel, A. D., & Gentner, T. Q. (2016). Songbirds use spectral shape, not pitch, for sound pattern recognition. *PNAS, 113*, 1666–1671.

Chittka, L., & Niven, J. (2009). Are bigger brains better? *Current Biology, 19*, R995–1008.

Clayton, N.S., & Dickinson, A. (1998). Episodic-like memory during cache recovery by scrub jays. *Nature, 395*, 272–274.

Dennett, D. C. (1995). *Darwin's dangerous idea: Evolution and the meanings of life*. Simon & Schuster.

Dennett, D. C. (2017a). A difference that makes a difference: A conversation. *Edge*. https://www.edge.org/conversation/daniel_c_dennett-a-difference-that-makes-a-difference.

Dennett, D. C. (2017b). *From bacteria to Bach and back: The evolution of minds*. Norton.

Dyer, F. C. (2002). The biology of the dance language. *Annual Review of Entomology, 47*, 917–949.

Engesser, S., Crane, J. M. S., Savage, J. L., Russell, A. F., & Townsend, S. W. (2015). Experimental evidence for phonemic contrasts in a nonhuman vocal system. *PLoS Biology, 13*, e1002171.

Fishbein, A. R., Idsardi, W. J., Ball, G. F., & Dooling R. J. (2019). Sound sequences in birdsong: How much do birds really care? *Philosophical Transactions of the Royal Society B, 375*, 20190044.

Fodor, J. (1990). *A theory of content and other essays*. MIT Press.

Garland, E. C., & McGregor, P. K. (2020). Cultural transmission, evolution, and revolution in vocal displays: Insights from bird and whale song. *Frontiers in Psychology, 11*, 3389.

Gil, M., & De Marco, R. J. (2005). Olfactory learning by means of trophallaxis in *Apis mellifera*. *Journal of Experimental Biology, 208*, 671–680.

Giurfa, M. (2013). Cognition with few neurons: Higher-order learning in insects. *Trends in Neurosciences, 36*, 285–294.

Grüter, C., Balbuena, M. S., & Farina, W. M. (2008). Informational conflicts created by the waggle dance. *Proceedings of the Royal Society B, 275*, 1321–1327.

Grüter, C.. & Farina, W. M. (2008). The honeybee waggle dance: Can we follow the steps? *Trends in Ecology and Evolution, 24*, 242–247.

Gu, Q. L., Lam, N. H., Halassa, M. M., & Murray, J. D. (2020). Circuit mechanisms of top-down attentional control in a thalamic reticular model. *bioRxiv*. https://doi.org/10.1101/2020.09.16.300749.

Halassa, M. M., & Kastner, S. (2017). Thalamic functions in distributed cognitive control. *Nature Neuroscience, 20*, 1669–1679.

Hall, M. L., & Magrath, R. D. (2000). Duetting and mate-guarding in Australian magpie-larks (*Grallina cyanoleuca*). *Behavioral Ecology and Sociobiology, 47*, 180–187.

Hall, M. L., & Magrath, R. D. (2007). Temporal coordination signals coalition quality. *Current Biology, 17*, R406–407.

Honing, H. (2019). *The evolving animal orchestra: In search of what makes us musical*. MIT Press.

Hrncir, M., Maia-Silva, C., McCabe, S. I., & Farina, W. M. (2011). The recruiter's excitement—features of thoracic vibrations during the honey bee's waggle dance related to food source profitability. *Journal of Experimental Biology, 214*, 4055–4064.

Jarvis, E. D. (2008). Brains and birdsong. In P. Marler & H. Slabbekoorn (Eds.), *Nature's music: The science of birdsong* (pp. 226–271). Elsevier.

Knudsen, E. I. (2007). Fundamental components of attention. *Annual Review of Neuroscience, 30*, 57–78.

Knudsen, E. I. (2018). Neural circuits that mediate selective attention—a comparative perspective. *Trends in Neuroscience, 41*, 789–805.

Knudsen, E. I. (2020). Evolution of neural processing for visual perception in vertebrates. *Journal of Comparative Neurology, 528*, 2888–28901.

Laland, K. N., & Galef, B. G. (2009). *The question of animal culture*. Harvard University Press.

Laubichler, M. D., & Renn, J. (2015). Extended evolution: A conceptual framework for integrating regulatory networks and niche construction. *Journal of Experimental Zoology B: Molecular and Developmental Evolution, 324*, 565–577.

Linksvayer, T. A., Fewell, J. H., Gadau, J., & Laubichler, M. D. (2012). Developmental evolution in social insects: Regulatory networks from genes to societies. *Journal of Experimental Zoology: Molecular and Developmental Evolution, 318*, 159–169.

Maynard Smith, J., & Szathmáry, E. (1995). *The major transitions in evolution*. Oxford University Press.

Menzel, R., & Giurfa, M. (2001). Cognitive architecture of a mini-brain: The honeybee. *Trends in Cognitive Sciences, 5*, 62–71.

Mooney, R., Prather, J., & Roberts, T. (2008). Neurophysiology of birdsong learning. In H. Eichenbaum (Ed.), *Learning and memory: A comprehensive reference* (vol. 3, pp. 441–474). Elsevier.

Nani, A., Manuello, J., Mancuso, L., Liloia, D., Costa, T., & Cauda, F. (2019). The neural correlates of consciousness and attention. *Frontiers in Neuroscience, 13*, 1169.

Nuxol, A. (2012). Episodic learning. In *Online encyclopedia of the sciences of learning*. https://doi.org/10.1007/978-1-4419-1428-6_1362.

Odling-Smee, F. J., Laland, K.. & Feldman, M. W. (2003). *Niche construction: The neglected process in evolution*. Princeton University Press.

Okanoya, K. (2012). Behavioural factors governing song complexity in Bengalese finches. *International Journal of Comparative Psychology, 25*, 44–59.

Pankiw, T. (2004). Cued in: Honey bee pheromones as information flow and collective decision-making. *Apidologie, 35*, 217–226.

Patel, A., Iversen, J. R., Bregman, M. R., & Schulz, I. (2009). Experimental evidence for synchronization to a musical beat in a nonhuman animal. *Current Biology, 19*, 827–830.

Peirce, C. S. (1955). *Philosophical writings of Peirce* (J. Buchler, Ed.). Dover.

Peng, F., & Chittka, L. (2017). A simple computational model of the bee mushroom body can explain seemingly complex forms of olfactory learning and memory. *Current Biology, 27*, 224–230.

Rek, P. (2018). Multimodal coordination enhances the responses to an avian duet. *Behavioral Ecology, 29*, 411–417.

Rek, P., & Magrath, R. D. (2016). Multimodal duetting in magpie-larks: How do vocal and visual components contribute to a cooperative signal's function? *Animal Behaviour, 117*, 35–42.

Rek, P., & Magrath, R. D. (2017). Deceptive vocal duets and multimodal display in a songbird. *Proceedings of the Royal Society B, 284*, 20171774.

Rybczynski, N. (2007). Castorid phylogenetics: Implications for the evolution of swimming and tree-exploitation in beavers. *Journal of Mammalian Evolution, 14*, 1–35.

Salwiczek, L. H., Watanabe, A., & Clayton, N. S. (2010). Ten years of research into avian models of episodic-like memory and its implications for developmental and comparative cognition. *Behavioural Brain Research, 215*, 221 234.

Savage, P., Loui, P., Tarr, B., Schachner, A., Glowacki, L., Mithen, S., & Fitch, W. (2021). Music as a coevolved system for social bondng. *Behavioral and Brain Sciences, 44*(e59), 1–22.

Shannon, C. E., & Weaver, W. (1949). *The mathematical theory of communication*. University of Illinois Press.

Smith, D., Wessnitzer, J., & Webb, B. (2008). A model of associative learning in the mushroom body. *Biological Cybernetics, 99*, 89–103.

Suzuki, T. N., Wheatcroft, D., & Greisser, M. (2016). Experimental evidence for compositional syntax in bird calls. *Nature Communications*. https://doi.org/10.1038/ncomms10986.

Templer, V. L., & Hampton, R. R. (2013). Episodic memory in nonhuman animals. *Current Biology, 23*, R801–806.

Thom, C., Gilley, D. C., Hooper, J., & Esch, H. E. (2007). The scent of the waggle dance. *PLOS Biology, 5*, 1862–1867.

Tomlinson, G. (2015). *A million years of music: The emergence of human modernity*. Zone.

Tomlinson, G. (2018). *Culture and the course of human evolution*. University of Chicago Press.

Tomlinson, G. (in press). *The machines of evolution and the scope of meaning*. Zone.

Tulving, E. (1972). Episodic and semantic memory. In E. Tulving & W. Donaldson (Eds.), *Organization of memory* (pp. 381–402). Academic Press.

Ueno, T., Takeuchi, H., Kawasaki, K., & Kubo, T. (2015). Changes in gene expression profiles of the hypopharyngeal gland of worker honeybees in association with worker behavior and hormonal factors. *PLOS ONE*. https://journals.plos.org/plosone/article?id=10.1371/journal.pone.0130206.

Wallin, N. L., Merker, B., & Brown, S. (2000). *The origins of music*. MIT Press.

Whitehead, H., & Rendell, L. (2014). *The cultural lives of whales and dolphins*. University of Chicago Press.

Wrangham, R. W., McGrew, W. C., de Waal, F. B. M., & Heltine, P. G. (1994). *Chimpanzee cultures*. Harvard University Press.

3 Cross-Species Research in Biomusicality: Methods, Pitfalls, and Prospects

Diandra Duengen, Marianne Sarfati, and Andrea Ravignani

> We need another and a wiser and perhaps a more mystical concept of animals. . . . We patronize them for their incompleteness, for their tragic fate of having taken form so far below ourselves. And therein we err, and greatly err. For the animal shall not be measured by man. In a world older and more complete than ours they move finished and complete, gifted with extensions of the senses we have lost or never attained living by voices we shall never hear. They are not brethren, they are not underlings; they are other nations, caught with ourselves in the net of life and time, fellow prisoners of the splendour and travail of the earth.
>
> —Henry Beston, *The Outermost House*, 1928

What Is Animal Musicality, and Why Should We Study It?

Musicality can be defined as a set of capacities that underlie music perception and production (Hoeschele et al., 2015). Fitch (2015) defines *biomusicology* as "the biological study of musicality in all its forms." Hoeschele et al. (2015) state that the origins of musicality can be investigated by searching for components of musicality in other species, thereby advocating for a comparative approach. What do we mean by *animal* in animal musicality? Many comparative papers begin by stating that *animals* refers to nonhuman animals. Here, although we adopt this meaning of *animals*, we want to emphasize the importance of considering humans as part of the animal category, not only in terms of the obvious biological classification but also in a conceptually non-anthropocentric sense (see Mundy, chapter 4 of this volume, for a complementary perspective). We consider human animals an integral part of the animal kingdom, so we use the terms *nonhuman animals* and *humans* (as an abbreviation of *human animals*).

Research on animal musicality serves a variety of purposes; one of these is to better understand the evolutionary history of musical abilities in our own species, in all its potential diversity. For a successful discipline of biomusicology, Fitch (2015) proposes to follow a pluralistic approach—including Tinbergen's (1963) method to address mechanistic (see Mundy, chapter 4), ontogenetic (see Kragness, Hannon, and Cirelli, chapter 8 of this volume), phylogenetic, and functional questions[1]—as animal communication systems provide us with information about the biology of human music, thereby underlining the importance of animal homologues or analogues. An example of a homologue (a trait shared by two species, both of which inherited it from a common ancestor) within the animal kingdom is bimanual drumming in apes, which provides us phylogenetic insight into drumming in our own species. An analogue (a trait shared by two different species, with the second species developing it independently from the first and not inheriting it from a common ancestor) can provide insight into similar evolutionary pressures; a good example is flight. Flight is a trait that evolved separately many times and in many animal classes, such as insects, birds, and mammals. To understand flight in bats, for instance, one would study not only bats but also other mammals that do not fly and other animals that have convergently evolved wings. Likewise, we argue, humans are only one data point to understand human traits, even those traits that look uniquely human at first. According to Fitch (2015), biomusicology can be divided into four core components: song, drumming, social synchronization, and dance. Dance is an almost inseparable part of biomusicality, with fascinating examples in the animal kingdom: chimpanzees (*Pan troglodytes*) rhythmically swinging to music (Hattori & Tomonaga, 2020, 2021); several bird species exhibiting impressive "dancing" skills, such as lyrebirds (*Menura novaehollandiae*) (Dalziell et al., 2013); or blue-capped cordon-bleus (*Uraeginthus cyanocephalus*), both sexes of which perform multicomponent and multimodal courtship displays, including singing, bobbing, and step dancing (Ota et al., 2015). This chapter, however, focuses on animal songs, drumming, and synchronization, leaving the less explored topic of dance for future work.

We start by discussing the anthropocentric versus biocentric approaches and critically review some examples among the plethora of past research, beginning with acoustic discrimination and categorization experiments from the twentieth century. We then discuss three aspects of biomusicality: song, instrumental music (percussion and drumming), and synchronization (entrainment, duets, and chorus). Finally, we examine the spectral and temporal parameters of music and relevant animal research and then discuss potential prospects and pitfalls in biomusicality.

Anthropocentric versus Biocentric Approach

The anthropocentric orientation places humans at the center of meaning, value, knowledge, and action (Weitzenfeld & Joy, 2014), while the biocentric perspective regards each species in its own evolutionary history (Bräuer et al., 2020). According to Shettleworth (2010), the anthropocentric approach dominates research, and Bräuer et al. (2020) report that research in animal cognition is often anthropocentric, driven by human cognition. This may seem obvious because, in (comparative) research, one species, the human animal, studies (and compares) several species, nonhuman animals. When trying to understand an animal's cognition, it is essential to consider its *Umwelt*, which is defined as a closed unit by von Uexküll (1934). The *Umwelt* consists of a *Merkwelt*, the perceptual world, and a *Wirkwelt*, the effector world. An animal's environment is determined by these two subunits, explaining how perception and production can be species specific. Naturally, immense phylogenetic contrasts exist in different species' sensory processing. Partan and Marler explained the concept nicely: "[It] is more complex than just sense organ physiology, embracing not only how animals sense and perceive their environments, both physical and social, but also what resources are proffered to the organism, how animals respond to their situation, and how those responses in turn modify both the environment and the organism's perceptions of the environment and of itself" (2002, p. 116). More recently, Bueno-Guerra (2018) suggested including another element, the *Sozialwelt*, when designing species-specific experimental setups. For example, would we expect a solitary species to succeed in a cognitive task—such as reaching a hidden food reward—that demands cooperation with conspecifics? Likely not. Although this will not provide information about the animal's cognition, it can provide information about its *Sozialwelt* as one sphere of its *Umwelt*. The same cognitive task may be successfully performed if approached in a species-tailored way, such as if it is solvable by one individual alone.

Prior to conducting cognitive studies on animals, both stimuli and responses should come naturally to the species—that is, the animal responds to the stimulus with a modality that lies within the ethogram (repertoire of capabilities) of the particular species (Bräuer et al., 2020; Bueno-Guerra, 2018; Cook, 1993). Therefore, the ecological background of the species of interest needs to be thoroughly considered, and experiments should be aligned species specifically. Bräuer et al. (2020) argue that the design of comparative studies must be ecologically valid; this is achieved by using naturalistic situations with relevant modalities and test settings that match naturally occurring contexts and by stressing the importance of both the tested skills and the experimental setup. We agree with this view: when animals are trained on unnatural skills that lie beyond the scope of their ethogram, this demonstrates cognitive flexibility but does not tell us much about their ecology. Having this in mind, we advocate inspecting musical traits in different species to elucidate the evolution of music, while pointing to the importance of homologies and analogies, and we support a biocentric approach. Cognitive capacities of species are often compared and measured according to whether and how much they match those of humans (Bräuer et al., 2020). Here, we aim to examine the evolution of music by taking a comparative perspective on biomusicality. Musicality is not a capacity unique to humans. Several statistical universals described by Savage et al. (2015) are found within the animal kingdom and provide a valuable opportunity to analyze the evolution of musicality. By tapping into homologues and analogues of musicality in other species, we may come a bit closer to understanding the nature of music.

Pioneer Studies and State of the Art

Porter and Neuringer (1984) introduced acoustic discrimination tasks when testing pigeons' ability to discriminate music by two different composers. Since then, discrimination for complex auditory stimuli has been probed in a variety of other animals. For example, Java sparrows (*Padda oryzivora*) were tested for their ability to discriminate among composers (Watanabe & Sato, 1999), koi carps (*Cyprinus carpio*) were tested for their ability to categorize blues and classical music (Chase, 2001), and rodents were tested for their ability to distinguish the Beatles' "Yesterday" from white noise, Mozart, and an altered version of "Yesterday" (Okaichi & Okaichi, 2001). Animals succeeded in most discrimination and categorization tasks. But what were the animals actually discriminating and categorizing?

Discrimination is the ability of differentiate sensory information, and categorization is the ability to put items into classes or groups. For instance, the mental representation of a chair includes any item with chair legs and a chair back; the color or material does not change the assumption that it's a chair. Removing the back, however, would turn the chair into a stool. To study perception, discrimination, or categorization abilities, researchers frequently use operant conditioning methods. In operant conditioning, animals are trained to respond to stimuli and are rewarded for correct answers (positive reinforcement) and unrewarded or punished for incorrect answers (negative reinforcement). In some studies, negative reinforcement is used to enhance learning, such as when a wrong answer is "punished" by a time-out and no new stimuli are presented (Hulse et al., 1995). In any of these tasks, if the subject has more correct answers than statistically expected by chance, it is assumed that it can reliably discriminate or categorize. Ideally, cognitive tasks should be designed so that animals can respond as naturally as possible, such as birds moving to a response perch (Watanabe et al., 2005) or monkeys touching a screen (Wright & Rivera, 2000). To study the categorization process of the animals, unrewarded generalization or transfer tests are performed: novel stimuli are presented, following the same categorical rule.

Porter and Neuringer (1984) pioneered these studies by testing the ability of pigeons (*Columba livia*) to discriminate music by two different composers. The birds were placed in a box with a response disk and a food dispenser. Excerpts of Bach served as the positive stimulus (S+, rewarded) and Hindemith as the negative stimulus (S–, unrewarded). The pieces were alternated, and the pigeons were rewarded only when they pecked during Bach. Both birds learned the task. They were then tested on excerpts of Bach and Stravinsky by means of a two-alternative forced choice (2AFC) paradigm and the addition of a second response disk. The 2AFC paradigm enables a choice between two options, in this case, excerpts by Bach or Stravinsky. Decisions can be tracked, and reaction times can be measured. Correct pecks (S+, Bach) were positively reinforced by food, and wrong pecks (S–, Stravinsky) were negatively reinforced by a time-out. After learning the task, the birds were introduced to novel excerpts and generalized consistently and independently of the instruments involved. A subsequent experiment purportedly tested whether the pigeons were discriminating musical categories. Excerpts of works from other composers were presented, that—according to human listeners—were more similar to either Bach or Stravinsky. The pigeons generalized between the different composers, except for one piece by Vivaldi that was "wrongly" attributed to Stravinsky (Porter & Neuringer, 1984). Subsequent studies had similar goals, featuring a variety of approaches and species. Chase (2001) showed that koi are capable of auditory discrimination. The overall goal of the study was to investigate whether these fish can reliably discriminate between stylistically different musical genres and whether they are capable of generalizing known musical stimuli to unfamiliar music from the same stylistic category. All koi learned to discriminate and categorize; they were also capable of discriminating melodies without timbre cues.

How do these studies vary from visual discrimination and categorization tasks, and how do they help us understand animal musicality? Crucially, in these studies, the animals likely used specific sonic features rather than the abstract concept of "music from Bach or Stravinsky" to achieve categorization. The ability to categorize such complex stimuli is remarkable, but the underlying discriminative cues remain unknown, thereby telling us little about biomusicality. However, such studies do aid in investigating the cues animals might use to discriminate or categorize complex auditory stimuli (Chase, 2001). Because the animals learned to discriminate despite an indistinguishable timbre, Chase (2001) concluded that timbre does not serve as a discriminative cue. Consequently, reducing features of complex auditory stimuli to identify discriminative cues is a fruitful approach to understanding animal cognition. Other studies approach cognition by providing valuable insight into specific components of music, such as pigeons demonstrating the ability to discriminate between chords (Brooks & Cook,

2010), or by examining the psychophysical aspects, such as the ability of bottlenose dolphins (*Tursiops truncatus*) to determine just-noticeable differences in tempo or frequency (Thompson & Herman, 1975; Yunker & Herman, 1974).

Care must be taken in interpreting these studies. For example, in the Java sparrow experiments, two of four birds preferred musical pieces by Bach over Schoenberg and by Vivaldi over Carter, while the others showed no preference; one bird preferred Bach over white noise (Watanabe & Nemoto, 1998). The authors argued that the birds seemed to prefer classical music over modern music and that their results demonstrated music by Bach had reinforcing properties on the birds. Such statements can be challenging and have misleading implications. These kinds of categorization studies cannot demonstrate musical preference because the animals are conditioned to a positive stimulus (e.g., a genre or a composer). Other musical preference studies take an importantly different approach. Mingle et al. (2014) examined whether captive chimpanzees preferred different musical styles selected from Ghana, North India, and Japan over one another or over silence. Instead of conditioning the animals to a certain stimulus, their preference behavior was investigated by observing and recording their position during playback of each musical genre. The chimpanzees' enclosure was divided into four zones, from closest to farthest from the speaker. When compared with the control condition (no music), the animals displayed a significant preference for the zone closest to the speaker when the Ghanaian or North Indian examples were played. This strongly indicates that chimpanzees preferred these sounds over silence (Mingle et al., 2014). The authors hypothesized that the chimpanzees may prefer such music because it lacks an obvious pulse, in contrast to the clear, percussive beats of the Japanese taiko drumming example used. Chimpanzees naturally demonstrate regular dominance displays by incorporating isochronous sounds, such as clapping or banging (Goodall, 1986), which suggests that they might perceive the pulsed pattern as threatening (Mingle et al., 2014). Another valuable approach is observing animals' engagement in spontaneous behavior, such as spontaneous drumming and tempo matching in some nonhuman primates (Dufour et al., 2015; Large & Gray, 2015). This provides a starting point to understand the purpose of a species' behavior. Other studies of nonhuman primate musical preference revealed that the animals preferred silence (or white noise) over Western or non-Western music, suggesting the importance of nonmusical control stimuli in music experiments (McDermott & Hauser, 2007; Ritvo & MacDonald, 2016).

A parallel strand of research uses music for animal welfare studies or, contentiously, to enrich animal production in commercial agriculture (de Jonge et al., 2008; Li et al., 2019; Piitulainen & Hirskyj-Douglas, 2020; Wallace et al., 2017). Most of these studies have shown little effect. De Jonge et al. (2008) examined the playing behavior of piglets after weaning when they were exposed to music and playtime before weaning.

Six piglets listened to music while given daily access to a playroom; a control group of six piglets had no access to the playroom but still experienced music. As expected, the playback of music facilitated play behavior post weaning, even without access to a playroom. The authors of the study correctly point to the conditioning nature of the experiment, where the animals connected a pleasant experience (playing) with a neutral one (music). Many other natural or human-produced sounds, we suggest, could have had the same effect as music. Interestingly, the control group showed an increase in play behavior when exposed to music post weaning. The authors suggest that the control group picked up on the excitement of the play group, which might have stimulated playing behavior (de Jonge et al., 2008).

Animal Song

Savage et al. (2015) describe eighteen statistical universal features of music, which include pitch, rhythm, instrumentation, and social context. When it comes to animal song, various definitions exist. In a broad sense, song is regarded as a melodic, metrical composition; it can be instrumental or vocal, and it is subjectively perceived as having an aesthetic purpose. Human songs are defined by anthropogenic constraints, which are paradoxically applied to animal songs: in human music, songs are considered songs when they have predefined characteristics; however, animal songs are often labeled songs only when they display similarities to human song, such as whale or bird song, or exhibit certain predefined characteristics (cf. Fitch, 2006; Rohrmeier et al., 2015). These animal songs are often melodic and include a sophisticated structure or hierarchy, such as the songs of humpback whales. Another good example is the musician wren (*Cyphorhinus arada*): this species' song is perceived as so extraordinarily musical that its name is "musician." By measuring the melodic intervals between successive utterances emitted by this bird, most intervals were revealed to be consonant (Doolittle & Brumm, 2012). Consonance is perceived by the human listener as pleasant or restful, but this also seems to depend on culture (Carterette & Kendall, 1999). Some authors argue that only species that are vocal learners qualify as singers, while others include nonvocal learners in their definition of song (Fitch, 2015; Geissmann, 1999; Marler & Slabbekoorn, 2004). Here, we include nonvocal learners, as our overall goal is to provide an overview of animal musicality, which, in our opinion, includes complex vocalizations such as that of gibbons (likely not vocal learners; see Geissmann, 1984, 1993). Fitch (2015) generally categorizes songs as complex, learned vocalizations, while Torti et al. define songs in indris as "complex sequence[s] of utterances emitted by group members, males and females, adults and subadults, in a co-ordinated manner" (2013, p. 596). In birds, songs are most often considered in a territorial or courtship setting

(Odom et al., 2014). Further, nonhuman primate singing has been documented among certain species of gibbons (Clarke et al., 2006; Geissmann, 2000; Marler & Mitani, 1989). To leave them out would exclude many interesting species often considered musical, according to Hoeschele et al. (2015). Far from adding another definition of song here, we make a distinction between calls and songs. While songs are complex sequences of (likely) nonreferential vocalizations, calls usually have referential properties, such as the semantically meaningful alarm calls in vervet monkeys (*Cercopithecus aethiops*; Seyfarth et al., 1980) and the elaborate signature whistles in several dolphin species (Janik et al., 2006; Sayigh et al., 2007).

Birdsong

Within the animal kingdom, birdsongs are among the most well known and studied songs. According to Fitch (2006), there is more information on the biology and evolution of birdsong than on any other animal signaling system. Birds' vocalizations can be divided into two main types of vocalizations: calls and songs, forming a broad spectrum from simple calls to complex songs (Smotherman et al., 2016). According to Catchpole and Slater (2003), calls are marked by short and discrete vocalizations, uttered irregularly or in isolation, whereas songs are characterized by longer, more complex, stereotyped sequences. Birdsongs are frequently repeated and spontaneously follow discrete daily and seasonal patterns (Catchpole & Slater, 2003), and they are most often associated with courtship or territorial battles (Odom et al., 2014). However, the long-held assumption that only male birds sing is untrue (Morton, 1996; Odom et al., 2014; Slater & Mann, 2004). Female birds seem to sing a lot more frequently in tropical regions than in temperate regions (Morton, 1996), and both females and males may sing solos or duets (Fitch, 2006). Odom et al. (2014) revealed that female song is present in 71 percent of surveyed species. Several studies investigated the function of male birdsong, identifying its role in male-male competition or mate choice (Collins, 2004), while the songs of females are thought to display territoriality, pair-bonding, mate defense, and attraction (see Langmore, 1998, 2000, and references therein). Other hypotheses exist, such as songs working as group "passwords" (Feekes, 1982) or functioning to form bonds within the family (Ritchison, 1983). Singing improves communication over long distances, thereby reducing the costs associated with territorial defense, such as injury from physical confrontations, and it enhances reproductive success (Morton, 1986; Read & Weary, 1992). Duetting (the coordination of a song in time by two participants, resulting in a joint song) is a widespread feature in birds: it appears in more than 200 different bird species (Farabaugh, 1982). Functions of duetting include territory and mate defense (Logue, 2005; Logue & Gammon, 2004; Sonnenschein & Reyer, 1983).

There are differences in interspecific song output, composition, and complexity, correlating with metabolism, foraging ecology, mating systems, and migratory behaviors (Read & Weary, 1992). In almost every songbird species studied today, the young birds need to listen to adults to learn their own species-specific songs (Soha, 2020). Thus, birdsong is a great example of animal culture, and it has been studied extensively (see Otter et al., 2020, for a continent-scale study for two decades). These cultures range from extremely stable (Pipek et al., 2018) to shifting and varying from one year to the next (Garland & McGregor, 2020).

Bat Song

Bat calls and songs span an extensive range of frequencies, up to the ultrasonic. Singing bats were found in five families (Smotherman et al., 2016). Bat songs show analogues to birdsongs, including courtship functions (Behr & von Helversen, 2004) and syntactic organization, being composed of hierarchically structured syllables and phrases (Bohn et al., 2008, 2009, 2013). Most species of bats appear to sing in defense of foraging or roosting territories and in support of courtship behaviors (Smotherman et al., 2016). Bat song remains understudied, mainly due to technical constraints; however, newly developed tools allow the study of bat songs across time, habitats, and behaviors (Behr & von Helversen, 2004; Bohn et al., 2008; Smotherman et al., 2016). Smotherman et al. (2016) propose that bats sing like birds because they fly like birds, which is energetically expensive: one of the main benefits of singing is improved long-distance communication, thereby reducing the need to fly (Morton, 1986). The authors suggest that because singing mitigates the high costs of flying, this likely explains why singing is comparatively rare among mammals.

Whale Song

Cetaceans (whales and dolphins) comprise toothed whales (Odontocetes) and baleen whales (Mysticetes), which vary significantly in their ecology and vocal behavior. Odontocetes include the family of dolphins (Delphinidae), with vocal learning representatives such as bottlenose dolphins and orcas (*Orcinus orca*; Deecke et al., 2000; Janik, 2014; Reiss & McCowan, 1993). These species are highly vocal; however, some do not consider these vocalizations songs due to the lack of complexity (Fitch, 2006; Smotherman et al., 2016). The variety of vocalizations Odontocetes emit are termed calls, pulses, whistles, and codas (Bradbury & Vehrencamp, 2011; Janik et al., 2006; Reiss & McCowan, 1993). Mysticete species, such as the fin whale (*Balaenoptera physalus*), emit species-specific calls that, as songs sung by males, seem to function in breeding displays (Croll et al., 2002). Some Mysticete species emit long and highly complex vocalizations (Payne & McVay, 1971; Rekdahl et al., 2013; Risch et al., 2014; Watkins et al., 1987), especially bowhead whales (*Balaena mysticetus*) and humpback whales (*Megaptera novaeangliae*), which produce songs that feature mature hierarchical structures (Payne & McVay, 1971). Humpback whale songs display a variety of sophisticated features, including cultural transmission, an extensive musical hierarchy, and geographic variation (Garland et al., 2011; Payne & McVay, 1971; Winn & Winn, 1978). The hierarchical structure of humpback whale songs is elaborate, consisting of units that form subphrases and phrases, which create full themes (Cholewiak et al., 2013; Payne & McVay, 1971; Winn & Winn, 1978). Apparently, similar songs are sung by all the males in a population, where song structure gradually evolves over the season (Cholewiak et al., 2013). Humpback whales show remarkable song revolutions, whereby a population-wide shared song type is replaced by a novel type introduced by a neighboring population (Garland & McGregor, 2020, and references therein). This type of cultural transmission occurs extremely fast (within one breeding season) and is far-reaching (across ocean basins), exhibiting a geographic dimension rivaled only by humans (Garland & McGregor, 2020). Bowhead whales are considerably less studied than humpback whales, but current research gives rise to the assumption that their songs might be just as complex and sophisticated, with the sharing of songs between clusters of animals (Erbs et al., 2021; Johnson et al., 2015; Stafford et al., 2018). Stafford et al. (2018) revealed that, over a three-year period, some bowhead whales displayed 184 different song types.

Seal Song

Earless seals (Phocidae) constitute one of three families within the group of pinnipeds—the others being eared seals (Otariidae) and walruses (Odobenidae). All pinniped families are vocal; vocalizations constitute a large part of their social lives, depending on the species. Although odobenids and phocids display complex vocalizations sometimes categorized as song, eared seals' vocalizations consist mainly of less complex, sometimes repetitive calls (termed, e.g., barks and screams), depending on context and sex (Fitch, 2006; Gwilliam et al., 2008; Peterson & Bartholomew, 1969). Phocids exhibit a rich vocal repertoire and use vocal displays during mating, when male vocalizations are often individually distinctive (Boness et al., 2006; Van Parijs et al., 2000, 2003). Females usually vocalize only during mother-offspring interactions, but it depends on the species whether both female and pup call or only pups emit individually distinctive calls (Insley, 1992; Renouf, 1984; Van Opzeeland et al., 2012; Van Parijs et al., 2003). Ralls, Fiorelli, and Gish (1985) showed that harbor seals (*Phoca vitulina*) are capable of vocal imitation: two adult males mimicked English words, and one of them even imitated whole phrases and engaged in formant modulation. This seal became quite popular in the media, mimicking phrases such as "hello there" and "come over here" (Ralls et al., 1985). Stansbury and Janik (2019) reported remarkable vocal learning skills in the closely related gray seals (*Halichoerus grypus*): the animals were able to match modulations in peak frequency patterns of call sequences or whole melodies—which they achieved by modifying formants of their own calls (see also Torres Borda et al., 2021).

Instrumental Music

We start this section by considering the definition of musical instruments and drawing a parallel to the classic definition of tools. We then present some examples of animals using external means for sound production, showing that both animals and humans employ physical structures to create or enhance sound. In fact, we relate the use of physical phenomena in nonhuman animal sound production to the tools classically considered musical instruments. We adopt a broad concept of instrumental music and highlight the remarkable features of sound production across species when production is facilitated by using tool-like structures or piggybacking on physical phenomena.

Instrumental music can be defined as music production facilitated by the use of musical tools. Tool use has been considered a uniquely human trait for centuries. Jane Goodall's (1964) discovery that chimpanzees use tools provoked a tremendous reexamination of what makes us unique as a species. It might be time to acknowledge that every species is unique, instead of redefining our own uniqueness each time a discovery calls it into question. At present, 284 species have been shown to use tools (Bentley-Condit & Smith, 2010). A consistent feature of the definition of tool use is manipulation, or using an object detached from a substrate (see Bentley-Condit & Smith, 2010, for a concise history of tool definition in nonhuman animals). For example, Boswall (1977) identifies a true tool as an object manipulated by its user and not part of the substrate. However, whether something, such as a leaf, is part of the substrate may be subjective. Orangutans take leaves and place them in front of their mouths while vocalizing, whereas tree crickets place themselves in a hole they dig into a leaf while vibrating their wings (Mhatre et al., 2017; van Schaik, 2003). Both species use leaves for sound production, but orangutans can pick up a leaf, whereas tree crickets cannot. Following Boswall's classification, orangutans use a true tool, while the tree crickets do not. However, in several ways, tree crickets' use of leaves is more similar to human tool use than orangutan tool use: tree crickets pick the leaves they use for sound production based on size and modify them accordingly, whereas there is no sign that orangutans behave in such a way. Nevertheless, because orangutans hold the leaves in their hands and tree crickets do not, the former are considered to use a true tool (per the definition) and the latter are not. It may be time to abandon the idea that tool use requires specific motor functions, especially when a primate with an opposable thumb becomes comparable to an insect using a leaf considerably larger than itself. Boswall (1977) defines the leaf, when used by tree crickets, as a "borderline tool"—a tool that remains part of the substrate. The distinction between "true" and "borderline" establishes a hierarchy of value, but defining and refining the vocabulary we use for animals should keep in mind its main purpose: serving questions about concepts, not establishing a hierarchy of value.

A fascinating physical phenomenon, common across species, is the resonance effect in sound production. Most musical instruments make use of this; the body of a violin, flute, and saxophone, for instance, are all resonant objects. In every resonant object, some frequencies make the object vibrate at a greater amplitude than others, resulting in these frequencies being louder relative to others. This depends on many factors, including the dimension, shape, tension, and mass of the object. In this way, the conception and construction of musical instruments consider the desired frequencies of the user. Male mole crickets (Orthoptera: family Gryllotalpidae), for example, use the resonance effects of their burrow: They build horn-like holes in which they produce sounds by rubbing their forewings against specialized anatomical structures to attract females. They gradually restructure their burrows to bring its resonance closer to their call frequency, which is constrained mainly by body size, in order to boost sound amplitude. To some extent, mole crickets tune their burrows (Forrest & Green, 1991), and they seem to do so by trial and error, making short chirps during burrow building, after which they adjust its volume and shape (Bennet-Clark, 1987). This results in a louder signal with lower frequencies at a higher intensity, closer to pure tones (Bennet-Clark, 1987). In two different species of mole crickets, louder calls attract more females (Forrest & Green, 1991; Walker & Forrest, 1989), which indicates that sound amplification and modification provide an evolutionary benefit.

Quite similarly, the short-tailed cricket (*Anurogryllus muticus*) selects its call site and body position to optimize sound amplification (Erregger & Schmidt, 2018). It uses anthropogenic calling sites (e.g., walls of buildings or concrete stairs) to amplify signals, which presumably allows its calls to be heard twice as far (Erregger & Schmidt, 2018). Male sand gobies (*Pomatoschistus minutus*) are tiny fish that produce mating calls inside cavities underneath submerged objects like stones, shells, and artificial shelters. They cover these objects by piling sand onto them; the function of this behavior has been discussed for decades. It appears that the sand pile amplifies the signal and therefore aids in attracting females (Lugli, 2013). Male Mientien tree frogs (*Kurixalus idiootocus*) modify calling sites by using anthropogenic objects with high resonance properties

(Tan et al., 2014). Spix's disc-winged bats (*Thyroptera tricolor*) also use cavities to communicate by choosing roosting structures that resemble an acoustic horn (Chaverri & Gillam, 2013). In this case, the cavity amplifies both incoming and outgoing vocalizations. While some species select or adjust their burrows to match the cavity's resonant frequency to that of their calling frequency, tree-hole frogs (*Metaphrynella sundana*) do the reverse: they adjust their calling strategy and pitch to the shape and volume of the hole from which they call (Lardner & bin Lakim, 2002). In the rain forest, cavities can fill rapidly with water, altering their resonant frequency. When this happens, frogs adapt their frequency to consistently produce a loud signal.

A second phenomenon often encountered in instrumental music is acoustic short-circuiting. Membrane vibration produces sound waves. When moving outward, one side of the membrane delivers positive pressure to the air, while the other delivers negative pressure. The converse happens when the membrane moves inward. The membrane therefore produces sound waves from each side that are out of phase. When these wavelengths meet, a destructive interference occurs and attenuates the produced sound. One way to reduce short-circuiting is delaying the confrontation of these two wavelengths by elongating the distance they travel before meeting. Many musical instruments are designed to reduce this acoustic short-circuiting by adjusting the structure, shape, or thickness of the instrument (Heller, 2013). Tree crickets (genus *Oecanthus*) deal with the same issue: They set their forewings, which are vibrating membranes, into resonant vibration while standing on leaves (Mhatre et al., 2017). Sometimes they cut a hole in the leaf, position themselves in the hole, and call. This strategy against short-circuiting is similar to the one used in loudspeakers: the leaf acts as a baffle, like a loudspeaker's membrane (Heller, 2013). The out-of-phase wavelengths must travel all over the leaf before they meet, which attenuates the destructive interference and amplifies the sound level. In decision-making experiments, crickets chose big leaves over small ones and tended to make the hole into the center of the leaf. This resulted in calls that were up to four times louder than when the cricket was standing on the edge of the leaf (Mhatre et al., 2017). Some populations of Bornean orangutans (*Pongo pygmaeus*) share a cultural trait of using vibrating membranes for sound production: they place leaves in front of their mouths while producing the so-called kiss-squeak vocalization, an agonistic signal produced in the face of a threat (Wich et al., 2008). The use of the leaves to produce sound seems analogous to that of humans using instruments such as the kazoo or mirliton (Wieczorkowska et al., 2007). Kiss-squeaks can be produced with or without leaves, but kiss-squeaks with leaves are louder and have a lower maximum call frequency than those produced using only the mouth (Hardus et al., 2009). This may be a sort of dishonest signaling: the maximum frequency of a sound provides information about its emitter, and a lower frequency indicates a larger individual (Charlton & Reby, 2016; Martin et al., 2017); therefore, lowering the frequency by using a tool may mislead the receiver about the emitter's size and potentially dissuade it from attacking.

The last element of instrumental music discussed here is the use of percussion (Savage et al., 2015). Drumming is a common behavior in a variety of nonhuman primates, such as bonobos (*Pan paniscus*) and chimpanzees (Arcadi et al., 1998; De Waal, 1988), but underlying regular beats have rarely been detected (Kugler & Savage Rumbaugh, 2002). Nonhuman primate drumming may be of particular importance when studying the origins of music due to the phylogenetic proximity to our own species (Bispham, 2006). General characteristics of drumming include structure and context (Dufour et al., 2015; Merker et al., 2009), and according to Arom (2000), these constitute intentionality, decontextualization, and formality. However, these characteristics do not appear solely in humans. Dufour et al. (2015) examined a spontaneous performance of a captive-born chimpanzee according to these characteristics: the chimpanzee's drumming seemed to be intentional (focused on drumming), was decontextualized (the animal was unaccompanied, so no social context), and displayed a formal quality (even intervals with regular beating). Chimpanzees drum with their hands and feet on tree buttresses, often accompanied by pant-hoot calls, a species-specific long-distance call (Reynolds, 2005). Babiszewska et al. (2015) suggest that chimpanzee drumming may serve to coordinate the movement and distribution of dispersed individuals within a community. Another study describes chimpanzees throwing stones against trees, mainly coupled with pant-hoot calls (Kühl et al., 2016). Male palm cockatoos (*Probosciger aterrimus*) shape tree twigs into drumsticks and use them to strike hollow tree limbs during vocal and visual displays (Heinsohn et al., 2017; Wood, 1984). This drumming is nonrandom and creates a regular pulse, and the birds have a wide range of tapping rates both within and between individuals (Heinsohn et al., 2017). Remarkably, each individual has its own consistent signature that differs significantly in intertap intervals. With the goal of identifying the significance of these differing beat rates, the authors considered other species' behavior and suggest that these drumming displays encode information about the drummer: the beat rate in palm cockatoos might be age related, as it is in humans, who produce a spontaneous regular beat rate from early childhood that decreases with age (Drake et al., 2000). Beat rate might be an identity cue, given that some bird species have recognizable songs, perhaps enabling the recognition of strangers over neighbors (Stoddard et al., 1991). Woodpeckers (Piciformes: family Picidesdrum) commonly drum by repeatedly striking their beaks against resonating surfaces. These displays seem to have a mating and territorial function (Williams, 2005) and can carry individual-level information (Budka et al., 2018). Woodpeckers also drum on anthropogenic constructions such as metallic drainpipes and gutters, likely choosing surfaces with high resonating properties (De Ernest, 2005).

These examples show parallels between the use of human musical instruments and nonhuman animal sound production. The main commonality lies in the use of external devices (tools or instruments) to manipulate sound when bodily structures (e.g., the vocal tract) fail to do so. This manipulation should constitute an improvement of the emitted sound, as it likely comes with a benefit, such as frequency-modulated kiss-squeaks to deter potential threats. Determining whether a trait constitutes an improvement is a thorny question. From a biological point of view, frequency modulation can be considered an improvement if it has a positive effect on reproductive success. In the foregoing examples, we tried to analyze as many elements as possible to assess this question of evolutionary function. When animals, such as mole crickets, adjust their sound production, probably due to auditory feedback, we are much more confident in calling the object they use a sound production tool, especially because auditory-sensorimotor integration is known to be crucial for musical performance (Bishop et al., 2014; Osmanski & Dooling, 2009). In other cases, determining whether an object used during sound production improves the signal may be less clear-cut. For example, the cohesion calls emitted by the Spix's disc-winged bat in a cavity are slightly louder with enhanced directionality, but because the signal receivers are flying conspecifics, it is debatable whether cavities constitute sound-producing tools. However, because the incoming calls are much louder, cavities might be considered hearing tools rather than sound-producing tools (Chaverri & Gillam, 2013).

The amplification of sound is crucial in the conception of many instruments, along with optimization of the resonance effect and reduction of acoustic short-circuiting (Heller, 2013). The importance of sound amplification shouldn't be underestimated in the evolution of human music: researchers found a positive correlation between the resonance properties of French caves and the number of prehistoric paintings and signs, suggesting an early and considerable interest in sound quality and resonance properties among prehistoric humans (Reznikoff, 2008). More recent examples include Greek and Roman amphitheaters, which have astonishing amplification properties evolving across time; some of the more ancient ruins date to approximatively 600 BC (Mourjopoulos, 2015).

Synchronization

Entrainment and Beat

Rhythmical entrainment is the ability to perceive a beat (i.e., the underlying musical pulse) and align one's body movements to it (Hoeschele et al., 2015; see also Witek, chapter 7 of this volume). Studies have demonstrated that entrainment is not, as previously thought, unique to humans. It may be linked to vocal production learning, although studies have proved that some animals that are unlikely to possess this ability can perceive a pulse in a regular stimulus or "keep the beat" (Cook et al., 2013; Patel et al., 2009). The vocal learning–beat perception and synchronization hypothesis states that vocal production learning, or vocal mimicry, is a species' prerequisite to perceiving a pulse in a periodic stimulus and synchronizing movements to it (Patel, 2006).

Sea lions (*Zalophus californianus*) are not considered vocal learners, but a four-year-old female sea lion named Ronan was tested on her entrainment abilities (Cook et al., 2013; Rouse et al., 2016). Cook et al. (2013) examined the sea lion's capabilities with respect to the three criteria stated in the vocal learning–beat perception and synchronization hypothesis (Patel, 2006): the ability to entrain to (1) rhythm multimodally, (2) a range of different tempos, and (3) a tempo embedded in joined rhythmic-melodic elements. These criteria were tested by six different experiments: (1) entrainment to a familiar stimulus, (2) transfer of entrainment to novel tempos, (3) assessment of beat-matching performance as a potential stimulus-response chain, (4) transfer of entrainment to complex musical stimuli, (5) entrainment to a novel complex musical stimulus, and (6) assessment of the capability for sustained entrainment. Ronan was trained to respond to a visual discrimination stimulus by bobbing her head and neck up and down. She was then trained to bob her head to auditory stimuli. She learned all six tasks, displaying the ability to entrain and synchronize motor behavior to an auditory beat. Remarkably, Ronan was able to transfer to novel tempos and stimuli, including complex settings. She kept the beat in musical pieces with both steady and unsteady tempos, meeting all three criteria of the vocal learning–beat perception and synchronization hypothesis (Patel et al., 2009). Thus, a California sea lion was the first nonhuman mammal to exhibit the ability to keep the beat (Cook et al., 2013).

Duets and Choruses

Duetting and chorusing are remarkable capacities found in insects, anurans, birds, and primates. Interestingly, chorusing seems to involve males only, while duets include male-female pairs (Yoshida & Okanoya, 2005). A prerequisite of duetting is the ability to take turns, or send an acoustic signal after the preceding signal has ended (Yoshida & Okanoya, 2005). Duetting is not restricted to song-learning species. Gibbons provide a highly interesting subject, as all species of gibbons exhibit elaborate vocalizations; in most cases, mated pairs combine their songs to create coordinated duet songs (Geissmann, 1999). Duets may have several functions: territorial advertisement, mate attraction, and maintenance of pair and family bonds (Geissmann, 1986, 1999). By investigating the acoustic variation between daughters and mother-daughter resemblance, a more recent study suggested that in agile gibbons (*Hylobates agilis agilis*), mother-daughter duets may indicate socially mediated vocal flexibility in subadults and adults, as well as suggesting that mother-daughter co-singing may enhance vocal development (Koda, Lemasson, et al., 2013).

Parameters of Sound

As a first principle of biomusicology, Fitch (2015) established the importance of identifying and studying its multiple interacting components. Music is certainly more complex than the sum of its components. On the one hand, key to understanding complex sounds is the ability to isolate individual sound features in nonhuman animal production and perception. On the other hand, one can expect the interactions of different features to produce completely different results.

Sound comprises three nonexclusive components: amplitude (sound level), temporal features, and spectral features (parameters relative to pitch, frequencies). Here we provide an overview of the latter two, presenting how they are featured in the nonhuman animal kingdom. We discuss both fieldwork and lab studies to allow comprehensive conclusions regarding musical capacities. We share the views of Shofner (Plack et al., 2005) and Fay (1994) that cross-species research should ask this question: Are the stimulus features that influence perception and production the same in human listeners and in animals? Moreover, we want to go beyond these comparisons and be open to the possibility that nonhuman animals possess categorization abilities that might be utterly different from ours.

Temporal Parameters

Rhythm Rhythm can be defined as a nonrandom temporal auditory pattern (Hoeschele et al., 2015) and can be found in a variety of modalities and species (for a review, see de Reus et al., 2021; Ravignani, 2019a). Hagmann and Cook (2010) tested pigeons' ability to discriminate between different meters (the regular recurrence of stressed and unstressed beats), rhythms, and tempos and found that pigeons can time periodic auditory events. The birds readily discriminated 8/4 and 3/4 meters, fast and slow tempos of piano sound, and novel tempos. However, they were incapable of discriminating arrhythmic and rhythmic sound patterns.

Male northern elephant seals (*Mirounga angustirostris*) exhibit an extensive and highly competitive courtship display. Males of this species fight for status in the dominance hierarchy, where alpha males mate with and control female harems and beta males only occasionally mate with females (Le Bœuf & Petrinovich, 1974). Vocalizations play an elaborate role: mature male northern elephant seals produce a rhythmic series of pulses, where the call of each individual is characterized by tempo and timbre (Casey et al., 2015). Playback experiments showed that males memorize and recognize individual rhythmic and timbral features of other males' voices to identify their competitors (Mathevon et al., 2017). In another study, the calling of a harbor seal (*Phoca vitulina*) pup was investigated with respect to antisynchronous timing and its rhythmic capacities (Ravignani, 2018, 2019b). The wild-born, seven-day-old pup showed rhythmic call characteristics. Using playback experiments, Ravignani (2019b) demonstrated that the pup adapted the timing of her calling in response to conspecifics' calls. In detail, the calls' onset was adjusted to occur at a fraction of the playback tempo, thereby displaying a relative-phase antisynchrony. Interestingly, this species displays vocal learning in male adult individuals (Ralls et al., 1985) and a very vocal (Van Parijs et al., 1999) and vocally plastic (Torres Borda et al., 2021) puppyhood.

Isochrony vs. anisochrony Isochrony describes a series of events occurring at evenly spaced time intervals. Isochrony makes subsequent events predictable, and it facilitates entrainment (Arom, 2000; Dufour et al., 2015). The perception of isochronous patterns has been tested in several nonhuman species, including European starlings (*Sturnus vulgaris*; Hulse et al., 1984), an avian species with the ability to discriminate between rhythmic and arrhythmic patterns. Hulse, Humpal, and Cynx (1984) constructed an experiment with two different sound patterns: one according to a linear rule, in which tones and intertone intervals of equal duration alternated, and one according to a hierarchical rule, in which two subpatterns alternated. The arrhythmic pattern consisted of a tone and intertone interval, both of random duration. To receive a food reward, the birds were asked to peck on one key for a rhythmic pattern and on another key for an arrhythmic pattern. The birds learned the rhythmic-arrhythmic discrimination, and their discrimination accuracy was identical in both the linear and the hierarchical rhythmic structures. Transfer tests revealed that the birds reliably discriminated even if temporal structures were transformed logarithmically or additively, interchanged, or shifted an octave in pitch. However, performance deteriorated when patterns were degraded by holding tone duration constant while intertone duration varied randomly (or vice versa) (Hulse et al., 1984). In contrast to these findings in starlings, Hagmann and Cook (2010) found no evidence that pigeons could discriminate between rhythmic and arrhythmic structures.

Spectral Parameters

Pitch Rather than being a purely physical feature, *pitch* is a perceptual attribute related to a spectral feature—the fundamental frequency and its harmonics—that enables the perception of sounds as higher or lower (Hoeschele et al., 2015). In some papers, pitch is termed a *music note*. Actually, a note can be considered a particular pitch, depending on culture. For example, a sound with a pitch of 440 Hz, according to the current English convention, is termed an A note, while a sound with a pitch of 448.35 Hz does not have a note name in English. Two main categories of pitch perception and production can be distinguished: the *absolute pitch*, without an external referent, and the *relative pitch*, resulting from an external referent. For example, when listening to a succession of notes played on a keyboard, recognizing a melody or perceiving an ascending pattern results from relative frequency perception abilities. However, if this same succession of notes is perceived only as an x-note followed by a y-note, it means that mainly absolute pitch perception abilities are mobilized. Both *pitch height* and *pitch chroma* are components of absolute pitch. When we listen to music, these attributes interact, and some of them prevail. In humans, timbre can interact with pitch: in an experiment by Warrier and Zatorre (2002), subjects perceived a larger difference in pitch between two tones if their timbre was different.

Absolute pitch

PITCH HEIGHT Pitch height amounts to the fundamental frequency of a sound; for example, 10,000 Hz may be considered high-pitched and 100 Hz low-pitched. Different species have different sensitivities and discrimination abilities for pitch height. In that context, sensitivity refers to which frequencies and which minimum amplitudes can be perceived. Audiograms are common methods of measuring sound reception, based on behavioral or neural responses. Compared with humans, some species might need louder audio stimuli, such as yellow-bellied slider turtles (*Pseudemys scripta*; Patterson, 1966), or sounds with a different frequency range, such as house mice (*Mus musculus*) and some bat species (*Eptesicus fuscus*, *Rhinolophus ferrumequinum*; (Bohn et al., 2006; Heffner et al., 2001). Different species might present different sensitivity patterns: humans hear middle-range frequencies (400–4,000 Hz) best, and their sensitivity decreases for high-range and low-range frequencies. However, various bat species have two regions of enhanced sensitivity separated by a relatively insensitive region (Bohn et al., 2006). Coqui frogs (*Eleutherodactylus coqui*) even exhibit a different sensitivity pattern between females and males (Narins & Capranica, 1976).

To identify such hearing thresholds, tones are played to animals in operant conditioning experiments, with some frequency ranges rewarded and alternated with unrewarded frequency ranges. For example, the rewarded frequency range might be 500 to 700 Hz and the unrewarded range 300 to 500 Hz. If a 550 Hz tone is played and the animal shows the correct response, it is reinforced. Some studies use three-pitch range tasks, while others use tasks up to a range of eight pitches. Like rats, humans perform

Table 3.1
Pitch height experiments

Species	Task	Results	Reference
Budgerigars (*Melopsittacus undulateus*)	Eight-pitch range	High accuracy	(Weisman et al., 2004)
White-throated sparrows (*Zonotrichia albicollis*)	Eight-pitch range	Medium accuracy	
Zebra finches (*Taeniopygia guttata*)	Eight-pitch range	High accuracy	(Lee et al., 2006)
Black-capped chickadees (*Poecile atricapillus*)	Eight-pitch range	High accuracy	
Mountain chickadees (*Poecile gambeli*)	Eight-pitch range	Medium accuracy	
Pigeons (*Columba livia*)	Eight-pitch-range	Low accuracy	(Friedrich et al., 2007)
Boreal chickadees (*Poecile hudsonicus*)	Eight-pitch range	Low accuracy	(Weisman et al., 2010)
Brown rats (*Rattus norvegicus*)	Three-pitch range	Success	(Weisman et al., 2012)
	Eight-pitch range	Fail	

poorly at these tasks, succeeding in a three-pitch range but not an eight-pitch range (Friedrich et al., 2007; Weisman et al., 2012). In a cross-species review by Weisman et al. (2012), bird species succeeded more easily than mammals. However, some birds were more accurate than others. The authors hypothesized that birds with more developed vocal learning abilities performed better.

PITCH CHROMA Pitch chroma concerns the names of notes and their overlapping partial harmonics. A 440 Hz tone has the same pitch chroma as an 880 Hz one, and they are both A notes. Depending on the experimental conditions, humans are more likely to categorize different tones as similar when they have the same pitch chroma (Hoeschele, 2017; Hoeschele, Weisman, et al., 2012). This phenomenon is called *octave generalization.*

In nonhuman animals, Blackwell and Schlosberg (1943) concluded that rats exhibit octave equivalence. They trained rats to react only when 10 kHz pure tones were played; however, the rats also reacted to previously unrewarded tones that were one octave lower than the reinforced stimulus (i.e., 5 kHz), indicating octave equivalence. This study was criticized because the stimuli might have included harmonic distortion, which could have provided octave information (Burns, 1999). Octave equivalence was later tested successfully in rhesus monkeys (*Macaca mulatta*; Wright & Rivera, 2000) and bottlenose dolphins (Richards et al., 1984). Cynx (1993) failed to show octave

Table 3.2
Pitch chroma experiments

Species	Octave equivalence	Reference
Brown rats	Successful Contested	(Blackwell & Schlosberg, 1943)
Bottlenose dolphins	Successful	(Richards et al., 1984)
European starlings	Failed Contested	(Cynx, 1993)
Rhesus monkeys	Successful	(Wright & Rivera, 2000)
Black-capped chickadees	Failed	(Hoang, 2015; Hoeschele et al., 2013)
Budgerigars	Failed	(Wagner et al., 2019)

equivalence in European starlings, but this study was criticized because it did not control for pitch height. Hoeschele, Weisman, and Sturdy (2012) made their point by replicating Cynx's experiment with humans. Participants were trained to react to a pure tone and then exposed to entirely new tones, including one having an octave interval with the probe tone. This procedure failed to prove octave equivalence in humans, despite its presence. In similar procedures, tests for octave generalization in black-capped chickadees (Hoang, 2015; Hoeschele et al., 2013) and budgerigars (Wagner et al., 2019) likewise failed. This beautifully demonstrates the importance of fine-tuned experimental design in cognitive testing.

Relative pitch Relative pitch information can be divided into the pitch contour and the frequency ratio processing (Deutsch, 2013).

PITCH CONTOUR The pitch contour refers to the directional change of frequency (e.g., is this tone higher-pitched than the previous one?). A rising intonation can change the meaning of a word in tonal languages like Mandarin or Tikuna and change a statement into a question in more than 70 percent of the world's languages (Bolinger et al., 1978; Murphy, 2013). Some species exhibit significant discrimination between rising and falling tones and can successfully generalize over new frequencies, including ferrets (*Mustela furo*; Yin et al., 2010), bottlenose dolphins (Ralston & Herman, 1995), and European starlings (Page et al., 1989). Other species, such as capuchin monkeys (*Cebus apella*; D'Amato & Colombo, 1988), cowbirds (*Molothrus ater*), and mockingbirds (*Mimus polyglottos*), fail to do so (Hulse & Cynx, 1985).

FREQUENCY RATIO Frequency ratio processing is one of the fundamentals of tonal music. In music theory, the ratio between two tones is called the *interval*. When playing a song, regardless of the tone played first, if the interval between each successive note is conserved, the song is easily recognized. The melody will be *transposed*. Some researchers trained animals to discriminate tone sequences (melodies, chords, or simple tones) with a consistent ratio, performing a generalization task with new pitch height. European starlings (Hulse et al., 1995), pigeons (Brooks & Cook, 2010), and black-capped chickadees (Hoeschele, Cook, et al., 2012) succeeded in tests of relative pitch discrimination and generalization. A complementary approach focuses on animal acoustic production and finds consistent patterns in the frequency ratios of vocalizations. Veeries (*Catharus fuscescens*), black-capped chickadees, Carolina chickadees (*Poecile carolinensis*), and white-throated sparrows exhibit consistency in their pitch ratios between vocalizations of the exact phrase, despite variability in each vocalization's absolute pitch (Hurly et al., 1991; Lohr et al., 1991; Weary et al., 1991; Weisman & Ratcliffe, 2004).

Consonance and dissonance Frequency ratios may influence human affective perception (Virtala & Tervaniemi, 2017). Consonant intervals are considered pleasant and restful, whereas dissonant intervals are considered unpleasant and tense. Nevertheless, the interval's effect depends on culture, period, and musical training (Carterette & Kendall, 1999). McDermott et al. (2016) argued that consonance could be a creation of Western culture. They asked Tsimane' participants from lowland Bolivia to rate the pleasantness of consonant and dissonant chords. Although the Tsimane' did not exhibit any variation in preferences, Bolivian and American citizens did. Bowling et al. (2017) criticized this work, highlighting both the avoidance of the most consonant interval across cultures (the octave) and the exclusion of highly dissonant tone combinations that are usually avoided but not unthinkable in music. They summarized, "These restrictions diminish the contrast between stimuli and would obscure their differentiation, especially by naive listeners" (Bowling et al., 2017, p. 119). They argued that McDermott's group did not discuss the similarity of tonal organization across musical cultures, with the most frequently used intervals (the octave, perfect fifths and fourths) corresponding to those considered the most consonant by culturally diverse listeners (Bowling & Purves, 2015; Burns, 1999).

The interplay of biology and culture in consonance remains controversial. Helmholtz (1912) argues that dissonance corresponds to a physical phenomenon: a slow periodic fluctuation in the amplitude of the sound wave leads to perceived roughness, for example, by slight frequency differences in fundamental frequencies or overtones of simultaneously played sounds. Bowling, Purves, and Gill defend the vocal similarity theory: "consonance of chords is predicted by their relative similarity to voiced speech sounds" (2018, p. 216). This controversy becomes even more complex when considering that the more tones included in a chord, the rougher it is, whereas the consonant perception does not necessarily decrease (Bowling & Purves, 2015).

Some studies look at the different processing of consonant versus dissonant intervals in nonhuman animals (for a review, see Toro & Crespo-Bojorque, 2017). Java sparrows (Watanabe et al., 2005), black-capped chickadees, European starlings, and Japanese monkeys (Izumi, 2000) successfully discriminated consonant over dissonant intervals and generalized them toward new frequencies. Pigeons and rats were capable of discrimination but not generalization. Other studies looked at preferences: the only species with a preference for consonant over dissonant melodies were domestic chicks (*Gallus gallus*) and a single infant chimpanzee (Sugimoto et al., 2010). Recently, budgerigars were tested with the same stimuli as the chicks but showed no preference (Wagner et al., 2020). Other primate species were tested with isolated chords but showed no preference, including Campbell's monkeys (*Cercopithecus campbelli*; Koda, Basile, et al., 2013) and cotton-top tamarins (*Saguinus oedipus*; McDermott & Hauser, 2004). A study of tungara frogs (*Physalaemus pustulosus*) tested the attractiveness of artificial mating calls with manipulated ratios, which did not influence female preference (Akre et al., 2014). In addition to the difficulty of experimentally proving preferences in nonhuman animals, the use of consonant intervals in such studies may be questioned. First, only Western intervals were used, and second, consonance is not always associated with pleasantness in humans, so one might wonder why it would be for nonhuman animals. Third, some studies used equal temperament as a tuning system to build their chords and melodies, which was developed during the seventeenth century for Western musicians. Obviously, human musical composition should not be unproblematically transferred to nonhuman animal experiments, given that recent instrumental adjustments (e.g., equal temperament) are unlikely to have biological significance. With this in mind, just intonation (which favors interval purity) should be the preferred tuning system for comparative research (Doolittle & Brumm, 2012; Richner, 2016). Nevertheless, consonance does result from particular frequency ratios. Some studies have concluded that nonhuman animals successfully generalized a consonance-dissonance rule over new chords (Toro & Crespo-Bojorque, 2017). However, some intervals were identical to the training intervals. Therefore, the animals might have memorized and integrated relative pitch without considering consonant or dissonant quality (Toro & Crespo-Bojorque, 2017).

Timbre and spectral shape Spectral shape is the overall pattern of spectral amplitudes across particular frequency bands. Timbre includes the spectral shape, the amplitude envelope, and how both change over time. Only a few studies provide information on timbre perception in nonhuman animals, surely because of the difficulty of working with complex sounds. However, timbre should be investigated in comparative studies due to its possible interaction with pitch perception. In humans, this effect is well documented (McLachlan, 2016). It has also been detected in other species, such as black-capped chickadees (Hoeschele et al., 2014), zebra finches, and budgerigars (Lohr & Dooling, 1998). Timbre may well be a salient cue for several species: Northern elephant seals memorize their rivals' unique timbre to adjust their behavior (Mathevon et al., 2017), starlings categorize tone sequences based on spectral shape (Bregman et al., 2016), and some bird species (zebra finches and budgerigars) are highly accurate in perceiving variations in timbre, much more so than humans (Lohr & Dooling, 1998; Amagai et al., 1999).

Conclusions, Pitfalls, and Prospects

We aimed in this chapter to provide an overview of studies on animal musical cognition and biomusicality, advocating the assessment of human musicality by adopting a comparative approach. We highlighted why comparative questions should be addressed species specifically using appropriate modalities, such as designing cognitive tasks in a way that allows animals to respond naturally (Bräuer et al., 2020). We also pointed out that presenting identical tasks to different species might lead to an underestimation of their cognitive abilities if the task is not equally relevant to each species. Psychophysical experiments may reveal cognitive abilities in animals, providing insight into music subcomponents, but it is crucial to not consider these in isolation. Holistic experiments are a valuable approach to gain insights into biomusicality (see Williams and Sachs, chapter 11 in this volume, for a parallel). Much of the reviewed research used unnatural sounds or musical instruments, which calls to mind the experiments of Bregman, Patel, and Gentner (2012) to test relative pitch perception in starlings. The birds failed to recognize artificial pitch-shifted melodies but succeeded in recognizing pitch-shifted songs of conspecifics. Marler (1982) suggests considering that animals possess species-specific hierarchies of attentional preferences for perceptual cues and hypothesizes that the relative position of stimulus features in each hierarchy may be task dependent. In the study by Hulse and Cynx (1985), starlings were able to generalize pitch contour, but not independently of absolute pitch. When transposed to an octave away from the training frequency range, the birds failed to distinguish the falling phrases from the rising ones. However, they succeeded when the transposition was just one semitone apart. Later, Page, Hulse, and Cynx (1989) showed that starlings use both absolute and relative pitch information during a pitch contour task (for a review, see Patel, 2017). Moreover, starlings use primarily spectral shape to recognize a tone sequence, rather than relative and absolute pitch (Bregman et al., 2016). This is a key difference from humans, who tend to consider pitch over spectral shape information to recognize tone sequences (Patel, 2017). Humans are likely to recognize a piece of well-known music regardless of whether it is played on a guitar or a piano. These results should inspire us to consider distinctive discriminative strategies or hierarchies in other species. Though this might not be easily achievable, the design of cognitive experiments should not be inspired solely by our own species' cognition. Rather, they should be broadened to encompass the wide variety of cognitive capacities present in the field of comparative research. We also suggest the investigation of additional features of sound; for example, starlings discriminate sound sequences with different amplitude (loudness) patterns (Bernard & Hulse, 1992), but discrimination over loudness has barely been investigated in nonhuman animals.

We recognize the benefits of combining field studies and laboratory work: observing natural, spontaneous behavior is particularly important to gain insight into a species' ecology, even though underlying mechanisms can be revealed only in controlled cognitive experiments. Both are needed for many reasons, but just as an example: a very well trained animal can trick us into believing that the trained behavior is within the animal's natural repertoire (Bräuer et al., 2020). We advocate a rethinking of animal categories: there is a plethora of definitions of *songs* or *tools* in nonhuman animals, yet giving a (human) definition to a nonhuman animal song seems somewhat paradoxical, and the same applies to tool use or instrumental music. Therefore, we try to avoid current definitions and prefer to highlight the abundance of interesting features found in nonhuman animal sound production, such as innovation, optimization, and active modification of the natural habitat, and we hope to find this broader perspective in more research. We understand that categorization is both an involuntary (instinctive in human and nonhuman animal cognition) and a necessary part of understanding, but we would like to consider more gray areas. By complementing black and white categories with shades of gray, we aim to avoid overlooking musical capacities that might provide us with valuable insights into ultimate and proximate explanations of the evolution of biomusicality.

Note

1. Fitch (2015) also adds glossogeny, the study of cultural transmission, to Tinbergen's classic four questions. While we acknowledge the importance of cultural transmission and culture-biology coevolution in biomusicology, a comprehensive overview of this literature is beyond the scope of this chapter (but see Patel, chapter 1 of this volume; Tomlinson, chapter 2 of this volume).

References

Akre, K. L., Bernal, X., Rand, A. S., & Ryan, M. J. (2014). Harmonic calls and indifferent females: No preference for human consonance in an anuran. *Proceedings of the Royal Society B: Biological Sciences, 281*(1789), 20140986.

Amagai, S., Dooling, R. J., Shamma, S., Kidd, T. L., & Lohr, B. (1999). Detection of modulation in spectral envelopes and linear-rippled noises by budgerigars (*Melopsittacus undulatus*). *Journal of the Acoustical Society of America, 105*(3), 2029–2035.

Arcadi, A. C., Robert, D., & Boesch, C. (1998). Buttress drumming by wild chimpanzees: Temporal patterning, phrase integration into loud calls, and preliminary evidence for individual distinctiveness. *Primates, 39*(4), 505–518.

Arom, S. (2000). Prolegomena to a biomusicology. In *The origins of music*. MIT Press.

Babiszewska, M., Schel, A. M., Wilke, C., & Slocombe, K. E. (2015). Social, contextual, and individual factors affecting the occurrence and acoustic structure of drumming bouts in wild chimpanzees (*Pan troglodytes*). *American Journal of Physical Anthropology, 156*(1), 125–134.

Behr, O., & von Helversen, O. (2004). Bat serenades—complex courtship songs of the sac-winged bat (*Saccopteryx bilineata*). *Behavioral Ecology and Sociobiology, 56*(2), 106–115.

Bennet-Clark, H. C. (1987). The tuned singing burrow of mole crickets. *Journal of Experimental Biology, 128*(1), 383–409.

Bentley-Condit, V., & Smith, E. O. (2010). Animal tool use: Current definitions and an updated comprehensive catalog. *Behaviour, 147*(2), 185–221.

Bernard, D. J., & Hulse, S. H. (1992). Transfer of serial stimulus relations by European starlings (*Sturnus vulgaris*): Loudness. *Journal of Experimental Psychology: Animal Behavior Processes, 18*(4), 323–334.

Bishop, L., Bailes, F., & Dean, R. T. (2014). Performing musical dynamics: How crucial are musical imagery and auditory feedback for expert and novice musicians? *Music Perception, 32*(1), 51–66.

Bispham, J. (2006). Rhythm in music: What is it? Who has it? And why? *Music Perception, 24*(2), 125–134.

Blackwell, H. R., & Schlosberg, H. (1943). Octave generalization, pitch discrimination, and loudness thresholds in the white rat. *Journal of Experimental Psychology, 33*(5), 407–419.

Bohn, K. M., Moss, C. F., & Wilkinson, G. S. (2006). Correlated evolution between hearing sensitivity and social calls in bats. *Biology Letters, 2*(4), 561–564.

Bohn, K. M., Schmidt-French, B., Ma, S. T., & Pollak, G. D. (2008). Syllable acoustics, temporal patterns, and call composition vary with behavioral context in Mexican free-tailed bats. *Journal of the Acoustical Society of America, 124*(3), 1838–1848.

Bohn, K. M., Schmidt-French, B., Schwartz, C., Smotherman, M., & Pollak, G. D. (2009). Versatility and stereotypy of free-tailed bat songs. *PLOS ONE, 4*(8), e6746.

Bohn, K. M., Smarsh, G. C., & Smotherman, M. (2013). Social context evokes rapid changes in bat song syntax. *Animal Behaviour, 85*(6), 1485–1491.

Bolinger, D., Greenberg, J. H., Ferguson, C. A., & Moravcsik, E. A. (1978). Universals of human language. *Phonology, 2*, 471–524.

Boness, D. J., Bowen, W. D., Buhleier, B. M., & Marshall, G. J. (2006). Mating tactics and mating system of an aquatic-mating pinniped: The harbor seal, *Phoca vitulina. Behavioral Ecology and Sociobiology, 61*(1), 119–130.

Boswall, J. (1977). Tool-using by birds and related behaviour. *Aviculture Magazine, 83*, 88–97, 146–159, 220–228.

Bowling, D. L., Hoeschele, M., Gill, K. Z., & Fitch, W. T. (2017). The nature and nurture of musical consonance. *Music Perception, 35*(1), 118–121.

Bowling, D. L., & Purves, D. (2015). A biological rationale for musical consonance. *Proceedings of the National Academy of Sciences, 112*(36), 11155–11160.

Bowling, D. L., Purves, D., & Gill, K. Z. (2018). Vocal similarity predicts the relative attraction of musical chords. *Proceedings of the National Academy of Sciences, 115*(1), 216–221.

Bradbury, J. W., & Vehrencamp, S. L. (2011). *Principles of animal communication*. Sinauer Associates.

Bräuer, J., Hanus, D., Pika, S., Gray, R., & Uomini, N. (2020). Old and new approaches to animal cognition: There is not "one cognition." *Journal of Intelligence, 8*(3), 28.

Bregman, M. R., Patel, A. D., & Gentner, T. Q. (2012). Stimulus-dependent flexibility in non-human auditory pitch processing. *Cognition, 122*(1), 51–60.

Bregman, M. R., Patel, A. D., & Gentner, T. Q. (2016). Songbirds use spectral shape, not pitch, for sound pattern recognition. *Proceedings of the National Academy of Sciences, 113*(6), 1666–1671.

Brooks, D. I., & Cook, R. G. (2010). Chord discrimination by pigeons. *Music Perception, 27*(3), 183–196.

Budka, M., Deoniziak, K., Tumiel, T., & Woźna, J. T. (2018). Vocal individuality in drumming in great spotted woodpecker—a biological perspective and implications for conservation. *PLOS ONE, 13*(2), e0191716.

Bueno-Guerra, N. (2018). How to apply the concept of umwelt in the evolutionary study of cognition. *Frontiers in Psychology, 9*, 2001.

Burns, E. M. (1999). Intervals, scales, and tuning. In *The psychology of music* (pp. 215–264). Elsevier.

Carterette, E. C., & Kendall, R. A. (1999). Comparative music perception and cognition. In *The psychology of music* (pp. 725–791). Elsevier.

Casey, C., Charrier, I., Mathevon, N., & Reichmuth, C. (2015). Rival assessment among northern elephant seals: Evidence of associative learning during male–male contests. *Royal Society Open Science, 2*(8), 150228.

Catchpole, C. K., & Slater, P. J. B. (2003). *Bird song: Biological themes and variations*. Cambridge University Press.

Charlton, B. D., & Reby, D. (2016). The evolution of acoustic size exaggeration in terrestrial mammals. *Nature Communications*, *7*(1), 12739.

Chase, A. R. (2001). Music discriminations by carp (*Cyprinus carpio*). *Animal Learning and Behavior*, *29*(4), 336–353.

Chaverri, G., & Gillam, E. H. (2013). Sound amplification by means of a horn-like roosting structure in Spix's disc-winged bat. *Proceedings of the Royal Society B: Biological Sciences*, *280*(1772), 20132362.

Cholewiak, D. M., Sousa-Lima, R. S., & Cerchio, S. (2013). Humpback whale song hierarchical structure: Historical context and discussion of current classification issues. *Marine Mammal Science*, *29*(3), E312–E332.

Clarke, E., Reichard, U. H., & Zuberbühler, K. (2006). The syntax and meaning of wild gibbon songs. *PLOS ONE*, *1*(1), e73.

Collins, S. (2004). Vocal fighting and flirting: The functions of birdsong. In P. R. Marler & H. Slabbekoorn (Eds.), *Nature's music: The science of birdsong* (pp. 39–79). Elsevier.

Cook, P., Rouse, A., Wilson, M., & Reichmuth, C. (2013). A California sea lion (*Zalophus californianus*) can keep the beat: Motor entrainment to rhythmic auditory stimuli in a non vocal mimic. *Journal of Comparative Psychology*, *127*(4), 412–427.

Cook, R. G. (1993). The experimental analysis of cognition in animals. *Psychological Science*, *4*(3), 174–178.

Croll, D. A., Clark, C. W., Acevedo, A., Tershy, B., Flores, S., Gedamke, J., & Urban, J. (2002). Only male fin whales sing loud songs. *Nature*, *417*(6891), 809.

Cynx, J. (1993). Auditory frequency generalization and a failure to find octave generalization in a songbird, the European starling (*Sturnus vulgaris*). *Journal of Comparative Psychology*, *107*(2), 140–146.

Dalziell, A. H., Peters, R. A., Cockburn, A., Dorland, A. D., Maisey, A. C., & Magrath, R. D. (2013). Dance choreography is coordinated with song repertoire in a complex avian display. *Current Biology*, *23*(12), 1132–1135.

D'Amato, M. R., & Colombo, M. (1988). Representation of serial order in monkeys (*Cebus apella*). *Journal of Experimental Psychology: Animal Behavior Processes*, *14*(2), 131–139.

Deecke, V. B., Ford, J. K. B., & Spong, P. (2000). Dialect change in resident killer whales: Implications for vocal learning and cultural transmission. *Animal Behaviour*, *60*(5), 629–638.

de Jonge, F. H., Boleij, H., Baars, A. M., Dudink, S., & Spruijt, B. M. (2008). Music during playtime: Using context conditioning as a tool to improve welfare in piglets. *Applied Animal Behaviour Science*, *115*(3–4), 138–148.

de Reus, K., Soma, M., Anichini, M., Gamba, M., de Heer Kloots, M., Lense, M., . . . Ravignani, A. (2021). Rhythm in dyadic interactions. *Philosophical Transactions of the Royal Society B, 376*(1835), 20200337.

Deutsch, D. (2013). *Psychology of music*. Elsevier.

de Waal, F. B. M. (1988). The communicative repertoire of captive bonobos (*Pan paniscus*), compared to that of chimpanzees. *Behaviour, 106*(3–4), 183–251.

Doolittle, E. L., & Brumm, H. (2012). O Canto do Uirapuru: Consonant intervals and patterns in the song of the musician wren. *Journal of Interdisciplinary Music Studies, 6*(1).

Drake, C., Jones, M. R., & Baruch, C. (2000). The development of rhythmic attending in auditory sequences: Attunement, referent period, focal attending. *Cognition, 77*(3), 251–288.

Dufour, V., Poulin, N., Curé, C., & Sterck, E. H. M. (2015). Chimpanzee drumming: A spontaneous performance with characteristics of human musical drumming. *Scientific Reports, 5*(1), 11320.

Erbs, F., van der Schaar, M., Weissenberger, J., Zaugg, S., & André, M. (2021). Contribution to unravel variability in bowhead whale songs and better understand its ecological significance. *Scientific Reports, 11*(1), 168.

Erregger, B., & Schmidt, A. K. D. (2018). Anthropogenic calling sites boost the sound amplitude of advertisement calls produced by a tropical cricket. *Animal Behaviour, 142*, 31–38.

Farabaugh, S. M. (1982). The ecological and social significance of duetting. *Acoustic Communication in Birds, 2*, 85–124.

Fay, R. R. (1994). Comparative auditory research. In *Comparative hearing: Mammals* (pp. 1–17). Springer.

Feekes, F. (1982). Song mimesis within colonies of Cacicus c. Cela (Icteridae, Aves): A colonial password? *Zeitschrift für Tierpsychologie, 58*(2), 119–152.

Fitch, W. T. (2006). The biology and evolution of music: A comparative perspective. *Cognition, 100*(1), 173–215.

Fitch, W. T. (2015). Four principles of bio-musicology. *Philosophical Transactions of the Royal Society B: Biological Sciences, 370*(1664), 20140091.

Forrest, T. G., & Green, D. M. (1991). Sexual selection and female choice in mole crickets (scapteriscus: Gryllotalpidae): Modelling the effects of intensity and male spacing. *Bioacoustics, 3*(2), 93–109.

Friedrich, A., Zentall, T., & Weisman, R. (2007). Absolute pitch: Frequency-range discriminations in pigeons (*Columba livia*)—comparisons with zebra finches (*Taeniopygia guttata*) and humans (*Homo sapiens*). *Journal of Comparative Psychology, 121*(1), 95–105.

Garland, E. C., Goldizen, A. W., Rekdahl, M. L., Constantine, R., Garrigue, C., Hauser, N. D., Poole, M. M., Robbins, J., & Noad, M. J. (2011). Dynamic horizontal cultural transmission of humpback whale song at the ocean basin scale. *Current Biology, 21*(8), 687–691.

Garland, E. C., & McGregor, P. K. (2020). Cultural transmission, evolution, and revolution in vocal displays: Insights from bird and whale song. *Frontiers in Psychology, 11*, 544929.

Geissmann, T. (1984). Inheritance of song parameters in the gibbon song, analysed in 2 hybrid gibbons (*Hylobates pileatus* × *H. lar*). *Folia Primatologica, 42*(3–4), 216–235.

Geissmann, T. (1986). Mate change enhances euetting activity in the siamang gibbon (*Hylobates syndactulus*). *Behaviour, 96*(1–2), 17–27.

Geissmann, T. (1993). *Evolution of communication in gibbons (Hylobatidae)*. [Unpublished doctoral dissertation]. Anthropological Institute, Philosoph. Faculty II, Zürich.

Geissmann, T. (1999). Duet songs of the siamang, *Hylobates syndactylus*: II. Testing the pair-bonding hypothesis during a partner exchange. *Behaviour, 136*(8), 1005–1039.

Geissmann, T. (2000). Gibbon songs and human music from an evolutionary perspective. In *The origins of music*. MIT Press.

Goodall, J. (1964). Tool-using and aimed throwing in a community of free-living chimpanzees. *Nature, 201*(4926), 1264–1266.

Goodall, J. (1986). *The chimpanzees of Gombe: Patterns of behavior*. Belknap Press.

Gwilliam, J., Charrier, I., & Harcourt, R. G. (2008). Vocal identity and species recognition in male Australian sea lions, *Neophoca cinerea*. *Journal of Experimental Biology, 211*(14), 2288–2295.

Hagmann, C. E., & Cook, R. G. (2010). Testing meter, rhythm, and tempo discriminations in pigeons. *Behavioural Processes, 85*(2), 99–110.

Hardus, M. E., Lameira, A. R., Van Schaik, C. P., & Wich, S. A. (2009). Tool use in wild orang-utans modifies sound production: A functionally deceptive innovation? *Proceedings of the Royal Society B: Biological Sciences, 276*(1673), 3689–3694.

Hattori, Y., & Tomonaga, M. (2020). Rhythmic swaying induced by sound in chimpanzees (*Pan troglodytes*). *Proceedings of the National Academy of Sciences, 117*(2), 936–942.

Hattori, Y., & Tomonaga, M. (2021). Reply to Bertolo et al.: Rhythmic swaying in chimpanzees has implications for understanding the biological roots of music and dance. *Proceedings of the National Academy of Sciences of the United States of America, 118*(2), e2017986118.

Heffner, R. S., Koay, G., & Heffner, H. E. (2001). Audiograms of five species of rodents: Implications for the evolution of hearing and the perception of pitch. *Hearing Research, 157*(1), 138–152.

Heinsohn, R., Zdenek, C. N., Cunningham, R. B., Endler, J. A., & Langmore, N. E. (2017). Tool-assisted rhythmic drumming in palm cockatoos shares key elements of human instrumental music. *Science Advances, 3*(6), e1602399.

Heller, E. J. (2013). *Why you hear what you hear: An experiential approach to sound, music, and psychoacoustics*. Princeton University Press.

Helmholtz, H. von. (1912). *On the sensations of tone as a physiological basis for the theory of music*. Longmans, Green.

Hoang, J. (2015, Fall). *Pitch perception is not unitary: Evidence for the perception of pitch chroma in black-capped chickadees*. [Unpublished doctoral dissertation]. University of Alberta.

Hoeschele, M. (2017). Animal pitch perception: Melodies and harmonies. *Comparative Cognition and Behavior Reviews, 12*, 5–18.

Hoeschele, M., Cook, R. G., Guillette, L. M., Brooks, D. I., & Sturdy, C. B. (2012). Black-capped chickadee (*Poecile atricapillus*) and human (*Homo sapiens*) chord discrimination. *Journal of Comparative Psychology, 126*(1), 57–67.

Hoeschele, M., Cook, R. G., Guillette, L. M., Hahn, A. H., & Sturdy, C. B. (2014). Timbre influences chord discrimination in black-capped chickadees (*Poecile atricapillus*) but not humans (*Homo sapiens*). *Journal of Comparative Psychology, 128*(4), 387–401.

Hoeschele, M., Merchant, H., Kikuchi, Y., Hattori, Y., & ten Cate, C. (2015). Searching for the origins of musicality across species. *Philosophical Transactions of the Royal Society B: Biological Sciences, 370*(1664), 20140094.

Hoeschele, M., Weisman, R. G., Guillette, L. M., Hahn, A. H., & Sturdy, C. B. (2013). Chickadees fail standardized operant tests for octave equivalence. *Animal Cognition, 16*(4), 599–609.

Hoeschele, M., Weisman, R. G., & Sturdy, C. B. (2012). Pitch chroma discrimination, generalization, and transfer tests of octave equivalence in humans. *Attention, Perception, and Psychophysics, 74*(8), 1742–1760.

Hulse, S. H., Bernard, D. J., & Braaten, R. F. (1995). Auditory discrimination of chord-based spectral structures by European starlings (*Sturnus vulgaris*). *Journal of Experimental Psychology: General, 12*, 409–423.

Hulse, S. H., & Cynx, J. (1985). Relative pitch perception is constrained by absolute pitch in songbirds (*Mimus, Molothrus*, and *Sturnus*). *Journal of Comparative Psychology, 99*, 176–196.

Hulse, S. H., Humpal, J., & Cynx, J. (1984). Discrimination and generalization of rhythmic and arrhythmic sound patterns by European starlings (*Sturnus vulgaris*). *Music Perception, 1*(4), 442–464.

Hurly, T. A., Weisman, R. G., Ratcliffe, L., & Johnsrude, I. S. (1991). Absolute and relative pitch production in the song of the white-throated sparrow (*Zonotrichia albicollis*). *Bioacoustics, 3*(2), 81–91.

Insley, S. J. (1992). Mother-offspring separation and acoustic stereotypy: A comparison of call morphology in two species of pinnipeds. *Behaviour, 120*(1–2), 103–122.

Izumi, A. (2000). Japanese monkeys perceive sensory consonance of chords. *Journal of the Acoustical Society of America, 108*(6), 3073–3078.

Janik, V. M. (2014). Cetacean vocal learning and communication. *Current Opinion in Neurobiology, 28*, 60–65.

Janik, V. M., Sayigh, L. S., & Wells, R. S. (2006). Signature whistle shape conveys identity information to bottlenose dolphins. *Proceedings of the National Academy of Sciences, 103*(21), 8293–8297.

Johnson, H. D., Stafford, K. M., George, J. C., Ambrose, W. G., & Clark, C. W. (2015). Song sharing and diversity in the Bering-Chukchi-Beaufort population of bowhead whales (*Balaena mysticetus*), spring 2011. *Marine Mammal Science, 31*(3), 902–922.

Koda, H., Basile, M., Olivier, M., Remeuf, K., Nagumo, S., Blois-Heulin, C., & Lemasson, A. (2013). Validation of an auditory sensory reinforcement paradigm: Campbell's monkeys (*Cercopithecus campbelli*) do not prefer consonant over dissonant sounds. *Journal of Comparative Psychology, 127*(3), 265–271.

Koda, H., Lemasson, A., Oyakawa, C., Rizaldi, Pamungkas, J., & Masataka, N. (2013). Possible role of mother-daughter vocal interactions on the development of species-specific song in gibbons. *PLOS ONE, 8*(8), e71432.

Kugler, K., & Savage Rumbaugh, S. (2002). Rhythmic drumming by Kanzi an adult male bonobo (*Pan paniscus*) at the language research center. *American Journal of Primatology, 57*, 62.

Kühl, H. S., Kalan, A. K., Arandjelovic, M., Aubert, F., D'Auvergne, L., Goedmakers, A., Jones, S., Kehoe, L., Regnaut, S., Tickle, A., Ton, E., van Schijndel, J., Abwe, E. E., Angedakin, S., Agbor, A., Ayimisin, E. A., Bailey, E., Bessone, M., Bonnet, M., . . . Boesch, C. (2016). Chimpanzee accumulative stone throwing. *Scientific Reports, 6*(1), 22219.

Langmore, N. E. (1998). Functions of duet and solo songs of female birds. *Trends in Ecology and Evolution, 13*(4), 136–140.

Langmore, N. E. (2000). Why female birds sing. In *Animal signals (pp. 317–327)*. Tapir Academic Press.

Lardner, B., & bin Lakim, M. (2002). Tree-hole frogs exploit resonance effects. *Nature, 420*(6915), 475.

Large, E. W., & Gray, P. M. (2015). Spontaneous tempo and rhythmic entrainment in a bonobo (*Pan paniscus*). *Journal of Comparative Psychology, 129*(4), 317–328.

Le Bœuf, B. J., & Petrinovich, L. F. (1974). Elephant seals: Interspecific comparisons of vocal and reproductive behavior. *Mammalia, 38*(1), 16–32.

Lee, T. T. Y., Charrier, I., Bloomfield, L. L., Weisman, R. G., & Sturdy, C. B. (2006). Frequency-range discriminations and absolute pitch in black-capped chickadees (*Poecile atricapillus*), mountain chickadees (*Poecile gambeli*), and zebra finches (*Taeniopygia guttata*). *Journal of Comparative Psychology, 120*(3), 217–228.

Li, X., Zhao, J. N., Zhao, P., Zhang, X., Bi, Y. J., Li, J. H., Liu, H. G., Wang, C., & Bao, J. (2019). Behavioural responses of piglets to different types of music. *Animal, 13*(10), 2319–2326.

Logue, D. M. (2005). Cooperative defence in duet singing birds. *Cognition, Brain, Behavior, 9*, 497–510.

Logue, D. M., & Gammon, D. E. (2004). Duet song and sex roles during territory defence in a tropical bird, the black-bellied wren, *Thryothorus fasciatoventris*. *Animal Behaviour, 68*(4), 721–731.

Lohr, B., & Dooling, R. J. (1998). Detection of changes in timbre and harmonicity in complex sounds by zebra finches (*Taeniopygia guttata*) and budgerigars (*Melopsittacus undulatus*). *Journal of Comparative Psychology, 112*(1), 36–47.

Lohr, B., Nowicki, S., & Weisman, R. (1991). Pitch production in Carolina chickadee songs. *Condor, 93*(1), 197–199.

Lugli, M. (2013). Sand pile above the nest amplifies the sound emitted by the male sand goby. *Environmental Biology of Fishes, 96*(8), 1003–1012.

Marler, P., & Mitani, J. C. (1989). A phonological analysis of male gibbon singing behavior. *Behaviour, 109*(1–2), 20–45.

Marler, P. R. (1982). Avian and primate communication: The problem of natural categories. *Neuroscience and Biobehavioral Reviews, 6*(1), 87–94.

Marler, P. R., & Slabbekoorn, H. (Eds.). (2004). *Nature's music: The science of birdsong*. Elsevier.

Martin, K., Tucker, M. A., & Rogers, T. L. (2017). Does size matter? Examining the drivers of mammalian vocalizations. *Evolution, 71*(2), 249–260.

Mathevon, N., Casey, C., Reichmuth, C., & Charrier, I. (2017). Northern elephant seals memorize the rhythm and timbre of their rivals' voices. *Current Biology, 27*(15), 2352–2356.e2.

McDermott, J., & Hauser, M. (2004). Are consonant intervals music to their ears? Spontaneous acoustic preferences in a nonhuman primate. *Cognition, 94*(2), B11–B21.

McDermott, J., & Hauser, M. D. (2007). Nonhuman primates prefer slow tempos but dislike music overall. *Cognition, 104*(3), 654–668.

McDermott, J. H., Schultz, A. F., Undurraga, E. A., & Godoy, R. A. (2016). Indifference to dissonance in native Amazonians reveals cultural variation in music perception. *Nature, 535*(7613), 547–550.

McLachlan, N. M. (2016). *Timbre, pitch, and music*. Oxford Handbooks Online.

Merker, B. H., Madison, G. S., & Eckerdal, P. (2009). On the role and origin of isochrony in human rhythmic entrainment. *Cortex, 45*(1), 4–17.

Mhatre, N., Malkin, R., Deb, R., Balakrishnan, R., & Robert, D. (2017). Tree crickets optimize the acoustics of baffles to exaggerate their mate-attraction signal. *eLife, 6*, e32763.

Mingle, M. E., Eppley, T. M., Campbell, M. W., Hall, K., Horner, V., & de Waal, F. B. M. (2014). Chimpanzees prefer African and Indian music over silence. *Journal of Experimental Psychology: Animal Learning and Cognition, 40*(4), 502–505.

Morton, E. S. (1986). Predictions from the ranging hypothesis for the evolution of long distance signals in birds. *Behaviour, 99*(1–2), 65–86.

Morton, E. S. (1996). A comparison of vocal behavior among tropical and temperate passerine birds. In *Ecology and evolution of acoustic communication in birds* (pp. 258–268). Cornell University Press.

Mourjopoulos, J. (2015). The origins of building acoustics for theater and music performances. *Journal of the Acoustical Society of America, 137*(4), 2427.

Murphy, K. (2013). Universals and variation in question intonation: A comparative study of Hawaiian and HCE speech melodies. In *Proceedings of the 21st Annual Symposium About Language and Society–Austin* (pp. 34–41).

Narins, P., & Capranica, R. (1976). Sexual differences in the auditory system of the tree frog *Eleutherodactylus coqui*. *Science, 192*(4237), 378–380.

Odom, K. J., Hall, M. L., Riebel, K., Omland, K. E., & Langmore, N. E. (2014). Female song is widespread and ancestral in songbirds. *Nature Communications, 5*(1), 3379.

Okaichi, Y., & Okaichi, H. (2001). Music discrimination by rats. *Japanese Journal of Animal Psychology, 51*(1), 29–34.

Osmanski, M. S., & Dooling, R. J. (2009). The effect of altered auditory feedback on control of vocal production in budgerigars (*Melopsittacus undulatus*). *Journal of the Acoustical Society of America, 126*(2), 911–919.

Ota, N., Gahr, M., & Soma, M. (2015). Tap dancing birds: The multimodal mutual courtship display of males and females in a socially monogamous songbird. *Scientific Reports, 5*(1), 16614.

Otter, K. A., Mckenna, A., LaZerte, S. E., & Ramsay, S. M. (2020). Continent-wide shifts in song dialects of white-throated sparrows. *Current Biology, 30*(16), 3231–3235.

Page, S. C., Hulse, S. H., & Cynx, J. (1989). Relative pitch perception in the European starling (*Sturnus vulgaris*): Further evidence for an elusive phenomenon. *Journal of Experimental Psychology: Animal Behavior Processes, 15*(2), 137–146.

Partan, S., & Marler, P. (2002). The *Umwelt* and its relevance to animal communication: Introduction to special issue. *Journal of Comparative Psychology, 116*(2), 116–119.

Patel, A. D. (2006). Musical rhythm, linguistic rhythm, and human evolution. *Music Perception, 24*(1), 99–104.

Patel, A. D. (2017). Why doesn't a songbird (the European starling) use pitch to recognize tone sequences? The informational independence hypothesis. *Comparative Cognition and Behavior Reviews, 12*, 19–32.

Patel, A., Iversen, J., Bregman, M., & Schulz, I. (2009). Studying synchronization to a musical beat in nonhuman animals. *Annals of the New York Academy of Sciences, 1169*(1), 459–469.

Patterson, W. C. (1966). Hearing in the turtle. *Journal of Auditory Research, 6*(4), 453–464.

Payne, R. S., & McVay, S. (1971). Songs of humpback whales. *Science, 173*(3997), 585–597.

Peterson, R. S., & Bartholomew, G. A. (1969). Airborne vocal communication in the California sea lion, *Zalophus californianus*. *Animal Behaviour, 17*, 17–24.

Piitulainen, R., & Hirskyj-Douglas, I. (2020). Music for monkeys: Building methods to design with white-faced sakis for animal-driven audio enrichment devices. *Animals, 10*(10), 1768.

Pipek, P., Petrusková, T., Petrusek, A., Diblíková, L., Eaton, M. A., & Pyšek, P. (2018). Dialects of an invasive songbird are preserved in its invaded but not native source range. *Ecography, 41*(2), 245–254.

Plack, C. J., Oxenham, A. J., & Fay, R. R. (2005). *Pitch: Neural coding and perception*. Springer.

Porter, D., & Neuringer, A. (1984). Music discriminations by pigeons. *Journal of Experimental Psychology: Animal Behavior Processes, 10*(2), 138–148.

Ralls, K., Fiorelli, P., & Gish, S. (1985). Vocalizations and vocal mimicry in captive harbor seals, *Phoca vitulina*. *Canadian Journal of Zoology, 63*(5), 1050–1056.

Ralston, J. V., & Herman, L. M. (1995). Perception and generalization of frequency contours by a bottlenose dolphin (*Tursiops truncatus*). *Journal of Comparative Psychology, 109*(3), 268.

Ravignani, A. (2018). Spontaneous rhythms in a harbor seal pup calls. *BMC Research Notes, 11*(1), 1–4.

Ravignani, A. (2019a). Rhythm and synchrony in animal movement and communication. *Current Zoology, 65*(1), 77–81.

Ravignani, A. (2019b). Timing of antisynchronous calling: A case study in a harbor seal pup (*Phoca vitulina*). *Journal of Comparative Psychology, 133*(2), 272–277.

Read, A. F., & Weary, D. M. (1992). The evolution of bird song: Comparative analyses. *Philosophical Transactions of the Royal Society of London. Series B: Biological Sciences, 338*(1284), 165–187.

Reiss, D., & McCowan, B. (1993). Spontaneous vocal mimicry and production by bottlenose dolphins (*Tursiops truncatus*): Evidence for vocal learning. *Journal of Comparative Psychology, 107*(3), 301.

Rekdahl, M. L., Dunlop, R. A., Noad, M. J., & Goldizen, A. W. (2013). Temporal stability and change in the social call repertoire of migrating humpback whales. *Journal of the Acoustical Society of America, 133*(3), 1785–1795.

Renouf, D. (1984). The vocalization of the harbour seal pup (*Phoca vitulina*) and its role in the maintenance of contact with the mother. *Journal of Zoology, 202*(4), 583–590.

Reynolds, V. (2005). *The chimpanzees of the Budongo forest: Ecology, behaviour and conservation*. Oxford University Press.

Reznikoff, I. (2008). Sound resonance in prehistoric times: A study of Paleolithic painted caves and rocks. *Journal of the Acoustical Society of America, 123*(5), 3603.

Richards, D. G., Wolz, J. P., & Herman, L. M. (1984). Vocal mimicry of computer-generated sounds and vocal labeling of objects by a bottlenosed dolphin, *Tursiops truncatus*. *Journal of Comparative Psychology, 98*(1), 10–28.

Richner, H. (2016). Interval singing links to phenotypic quality in a songbird. *Proceedings of the National Academy of Sciences, 113*(45), 12763–12767.

Risch, D., Gales, N. J., Gedamke, J., Kindermann, L., Nowacek, D. P., Read, A. J., Siebert, U., Van Opzeeland, I. C., Van Parijs, S. M., & Friedlaender, A. S. (2014). Mysterious bio-duck sound attributed to the Antarctic minke whale (*Balaenoptera bonaerensis*). *Biology Letters, 10*(4), 20140175.

Ritchison, G. (1983). The function of singing in female black-headed grosbeaks (*Pheucticus melanocephalus*): Family-group maintenance. *Auk, 100*(1), 105–116.

Ritvo, S. E., & MacDonald, S. E. (2016). Music as enrichment for Sumatran orangutans (*Pongo abelii*). *Journal of Zoo and Aquarium Research, 4*(3), 156–163.

Rohrmeier, M., Zuidema, W., Wiggins, G. A., & Scharff, C. (2015). Principles of structure building in music, language and animal song. *Philosophical Transactions of the Royal Society B: Biological Sciences, 370*(1664), 20140097.

Rouse, A. A., Cook, P. F., Large, E. W., & Reichmuth, C. (2016). Beat keeping in a sea lion as coupled oscillation: Implications for comparative understanding of human rhythm. *Frontiers in Neuroscience, 10*, 257.

Savage, P. E., Brown, S., Sakai, E., & Currie, T. E. (2015). Statistical universals reveal the structures and functions of human music. *Proceedings of the National Academy of Sciences, 112*(29), 8987–8992.

Sayigh, L., Esch, C., Wells, R., & Janik, V. (2007). Facts about bottlenose dolphin signature whistles. *Animal Behaviour, 74*, 1631–1642.

Seyfarth, R. M., Cheney, D. L., & Marler, P. (1980). Monkey responses to three different alarm calls: Evidence of predator classification and semantic communication. *Science, 210*(4471), 801–803.

Shettleworth, S. J. (2010). Clever animals and killjoy explanations in comparative psychology. *Trends in Cognitive Sciences, 14*(11), 477–481.

Slater, P. J. B., & Mann, N. I. (2004). Why do the females of many bird species sing in the tropics? *Journal of Avian Biology, 35*(4), 289–294.

Smotherman, M., Knörnschild, M., Smarsh, G., & Bohn, K. (2016). The origins and diversity of bat songs. *Journal of Comparative Physiology A, 202*(8), 535–554.

Soha, J. A. (2020). Nature, nurture, and noise in bird song ontogeny as determinants of phenotypic and functional variation among dialects. In *Phenotypic switching* (pp. 483–510). Academic Press.

Sonnenschein, E., & Reyer, H.-U. (1983). Mate guarding and other functions of antiphonal duets in the slate-coloured boubou (*Laniarius funebris*). *Zeitschrift für Tierpsychologie, 63*(2–3), 112–140.

Stafford, K. M., Lydersen, C., Wiig, Ø., & Kovacs, K. M. (2018). Extreme diversity in the songs of Spitsbergen's bowhead whales. *Biology Letters, 14*(4), 20180056.

Stansbury, A. L., & Janik, V. M. (2019). Formant modification through vocal production learning in gray seals. *Current Biology, 29*(13), 2244–2249.e4.

Stoddard, P. K., Beecher, M. D., Horning, C. L., & Campbell, S. E. (1991). Recognition of individual neighbors by song in the song sparrow, a species with song repertoires. *Behavioral Ecology and Sociobiology, 29*(3), 211–215.

Sugimoto, T., Kobayashi, H., Nobuyoshi, N., Kiriyama, Y., Takeshita, H., Nakamura, T., & Hashiya, K. (2010). Preference for consonant music over dissonant music by an infant chimpanzee. *Primates, 51*(1), 7–12.

Tan, W.-H., Tsai, C.-G., Lin, C., & Lin, Y. K. (2014). Urban canyon effect: Storm drains enhance call characteristics of the Mientien tree frog. *Journal of Zoology, 294*(2), 77–84.

Thompson, R. K. R., & Herman, L. M. (1975). Underwater frequency discrimination in the bottlenosed dolphin (1–140 kHz) and the human (1–8 kHz). *Journal of the Acoustical Society of America, 57*(4), 943–948.

Tinbergen, N. (1963). On aims and methods of ethology. *Zeitschrift für Tierpsychologie, 20*(4), 410–433.

Toro, J. M., & Crespo-Bojorque, P. (2017). Consonance processing in the absence of relevant experience: Evidence from nonhuman animals. *Comparative Cognition and Behavior Reviews, 12*, 33–44.

Torres Borda, L., Jadoul, Y., Rasilo, H., Salazar Casals, A., & Ravignani, A. (2021). Vocal plasticity in harbour seal pups. *Philosophical Transactions of the Royal Society B, 376*, 20200456. https://doi.org/10.1098/rstb.2020.0456.

Torti, V., Gamba, M., Rabemananjara, Z. H., & Giacoma, C. (2013). The songs of the indris (Mammalia: Primates: Indridae): contextual variation in the long-distance calls of a lemur. *Italian Journal of Zoology, 80*(4), 596–607.

Van Opzeeland, I. C., Parijs, S. M. V., Frickenhaus, S., Kreiss, C. M., & Boebel, O. (2012). Individual variation in pup vocalizations and absence of behavioral signs of maternal vocal recognition in Weddell seals (*Leptonychotes weddellii*). *Marine Mammal Science, 28*(2), E158–E172.

Van Parijs, S. M., Hastie, G. D., & Thompson, P. M. (1999). Geographical variation in temporal and spatial vocalization patterns of male harbour seals in the mating season. *Animal Behaviour, 58*(6), 1231–1239.

Van Parijs, S. M., Hastie, G. D., & Thompson, P. M. (2000). Individual and geographical variation in display behaviour of male harbour seals in Scotland. *Animal Behaviour, 59*(3), 559–568.

Van Parijs, S. M., Lydersen, C., & Kovacs, K. M. (2003). Vocalizations and movements suggest alternative mating tactics in male bearded seals. *Animal Behaviour, 65*(2), 273–283.

van Schaik, C. P. (2003). Orangutan cultures and the evolution of material culture. *Science, 299*(5603), 102–105.

Virtala, P., & Tervaniemi, M. (2017). Neurocognition of major-minor and consonance-dissonance. *Music Perception, 34*(4), 387–404.

von Uexküll, J. (1934). *Streifzüge durch die Umwelten von Tieren und Menschen*. Springer.

Wagner, B., Bowling, D. L., & Hoeschele, M. (2020). Is consonance attractive to budgerigars? No evidence from a place preference study. *Animal Cognition, 23*(5), 973–987.

Wagner, B., Mann, D. C., Afroozeh, S., Staubmann, G., & Hoeschele, M. (2019). Octave equivalence perception is not linked to vocal mimicry: Budgerigars fail standardized operant tests for octave equivalence. *Behaviour, 156*(5–8), 479–504.

Walker, T. J., & Forrest, T. G. (1989). Mole cricket phonotaxis: Effects of intensity of synthetic calling song (Orthoptera: Gryllotalpidae: Scapteriscus acletus). *Florida Entomologist, 72*(4), 655–659.

Wallace, E. K., Altschul, D., Körfer, K., Benti, B., Kaeser, A., Lambeth, S., Waller, B. M., & Slocombe, K. E. (2017). Is music enriching for group-housed captive chimpanzees (*Pan troglodytes*)? *PLOS ONE, 12*(3), e0172672.

Warrier, C. M., & Zatorre, R. J. (2002). Influence of tonal context and timbral variation on perception of pitch. *Perception and Psychophysics, 64*(2), 198–207.

Watanabe, S., & Nemoto, M. (1998). Reinforcing property of music in Java sparrows (*Padda oryzivora*). *Behavioural Processes, 43*(2), 211–218.

Watanabe, S., & Sato, K. (1999). Discriminative stimulus properties of music in Java sparrows. *Behavioural Processes, 47*(1), 53–57.

Watanabe, S., Uozumi, M., & Tanaka, N. (2005). Discrimination of consonance and dissonance in Java sparrows. *Behavioural Processes, 70*(2), 203–208.

Watkins, W. A., Tyack, P., Moore, K. E., & Bird, J. E. (1987). The 20-Hz signals of finback whales (*Balaenoptera physalus*). *Journal of the Acoustical Society of America, 82*(6), 1901–1912.

Weary, D. M., Weisman, R. G., Lemon, R. E., Chin, T., & Mongrain, J. (1991). Use of the relative frequency of notes by veeries in song recognition and production. *Auk, 108*(4), 977–981.

Weisman, R. G., Balkwill, L.-L., Hoeschele, M., Moscicki, M. K., Bloomfield, L. L., & Sturdy, C. B. (2010). Absolute pitch in boreal chickadees and humans: Exceptions that test a phylogenetic rule. *Learning and Motivation, 41*(3), 156–173.

Weisman, R. G., Mewhort, D. J. K., Hoeschele, M., & Sturdy, C. B. (2012). *New perspectives on absolute pitch in birds and mammals*. Oxford University Press.

Weisman, R. G., Njegovan, M. G., Williams, M. T., Cohen, J. S., & Sturdy, C. B. (2004). A behavior analysis of absolute pitch: Sex, experience, and species. *Behavioural Processes, 66*(3), 289–307.

Weisman, R. G., & Ratcliffe, L. (2004). Relative pitch and the song of black-capped chickadees. *American Scientist, 92*(6), 532–539.

Weitzenfeld, A., & Joy, M. (2014). An overview of anthropocentrism, humanism, and speciesism in critical animal theory. *Counterpoints, 448*, 3–27.

Wich, S. A., Utami Atmoko, S. S., Setia, T. M., & van Schaik, C. P. (2008). *Orangutans*. Oxford University Press.

Wieczorkowska, A. A., Ras, Z. W., Zhang, X., & Lewis, R. (2007). Multi-way hierarchic classification of musical instrument sounds. International Conference on Multimedia and Ubiquitous Engineering (pp. 897–902).

Williams, E. H. (2005). *The nature handbook: A guide to observing the great outdoors*. Oxford University Press.

Winn, H. E., & Winn, L. K. (1978). The song of the humpback whale *Megaptera novaeangliae* in the West Indies. *Marine Biology, 47*(2), 97–114.

Wood, G. A. (1984). Tool use by the palm cockatoo *Probosciger aterrimus* during display. *Corella, 8*(4), 94–95.

Wright, A., & Rivera, J. (2000). Music perception and octave generalization in rhesus monkeys. *Journal of Experimental Psychology: General, 129*(3), 291–307.

Yin, P., Fritz, J. B., & Shamma, S. A. (2010). Do ferrets perceive relative pitch? *Journal of the Acoustical Society of America, 127*(3), 1673–1680.

Yoshida, S., & Okanoya, K. (2005). *Evolution of turn-taking: A bio-cognitive perspective*. Japanese Cognitive Science Society.

Yunker, M. P., & Herman, L. M. (1974). Discrimination of auditory temporal differences by the bottlenose dolphin and by the human. *Journal of the Acoustical Society of America, 56*(6), 1870–1875.

4 Humane Treatment, Sound Experiments

Rachel Mundy

Introduction

In the science-music borderlands, the music of animals brings with it profound questions about the sentience of other species that cross disciplinary boundaries. In this chapter, I explore the implications of "humane" laboratory ethics for questions related to animal musicality. I frame these questions in a broad historical context and ask how thinking carefully about ethics in the science-music borderlands can help researchers engage with multispecies contexts that cannot be accounted for by emphasizing human uniqueness.

In traditional science, animals' rights are approached through an ethic known as "humane treatment." Humane treatment in scientific research is governed in the United States by the Animal Welfare Act (AWA), which first became law in 1966 and subsequently led to the standardization of oversight committees in the 1980s. Although the AWA and related legal guidelines use *humane* intervention or assessment as the central tenet of research ethics, they do not define the term. Instead, many guidelines offer specific examples or cases. For example, the US Department of Agrigulture's (2020) *Blue Book* on animal welfare, which reproduces the AWA and related legislation, uses *humane* treatment as the standard of care without defining the term, relying instead on detailed explanations of standards for housing, experimental design, and so forth. This approach is reflected in institutional applications of the legal guidelines, such as the University of California–Los Angeles's website on laboratory oversight, which instructs researchers to assess whether humane intervention is needed based on the quality of housing, the need for veterinary treatment, or the need for euthanasia.[1]

The word itself is compelling, for it suggests the very boundaries that so many of this book's contributors question. It has historical roots in notions of a *humanizing* education—what we call *the humanities* today—which was once thought to be the anchor of both rational science and aesthetics.[2] The word was adopted in the American literature on experimental oversight in the mid-twentieth century, against the backdrop of social movements and changing discourses concerned with human rights.

As veterinarian Richard Haynes (2008) notes, there are two schools of thought about animal laboratory ethics: the welfare school and the liberation school. The welfare school accepts the premise that nonhuman animals can be used as human resources but seeks to minimize their suffering. The liberation school does not accept the use of animals as human resources. The language of humane treatment is primarily (but not exclusively) used by proponents of the animal welfare approach. In the context of this chapter, I engage primarily with the traditions and practices of the welfare tradition.

By drawing attention to the tension between humane ethics and the boundary-crossing methods in studies of animal musicality, I'm also highlighting a more fundamental problem that lurks in the humane ethics we aspire to. Intellectuals have been writing about a crisis in the humanities since the early 1900s, expressing surprisingly consistent worries about the professional viability of a college education in fields such as literature, philosophy, music, and art (Jay, 2014, p. 8). More recently, writer Amitav Ghosh (2016) questioned whether there is a crisis in the sciences as well, noting that decades of compelling research on climate change have not meaningfully affected public policy. Both diagnoses suggest a crisis of faith—faith in the ability of the arts and sciences to prepare us for the present and future. Are the rudiments of modern science and humanism aligned with the realities of the twenty-first century? Sometimes they are, but the answer must be no, not yet, when we ask whether institutionalized conceptions of the humane can answer the ethical dilemmas of a world whose problems transcend biological boundaries.

Humane Treatment and Its Origins

Although today the idea of humane treatment applies primarily to animals, the notion was popularized by nineteenth-century aid societies and newsletters in the United States and Great Britain. Humane societies of the nineteenth century championed the protection of domestic animals and addressed the living conditions, medical needs, and education of poor communities of color, prisoners, and the mentally ill.

Being humane in this context meant displaying a kindness and compassion that were unique to human beings but not "natural" to humans. Instead, humane behavior came from one's religious or secular education. As nineteenth-century British theologian Nathaniel Dimrock explained in a sermon titled "On Humanizing Humanity":

> Even the very existence of such a verb as *to humanize* (in its ordinary sense) bears witness—that men regard man in the condition with least affinity to the lower animals in creation, as the most truly human. And if this be right, then this necessary conclusion follows, that man's condition is natural just in proportion as it is least human, and that man's nature is human just in proportion as it is least natural. (1885, p. 16)

Nineteenth-century dictionaries defined the words *humanize* and *humane* with reference to the civilizing or softening effects of cultural and religious education, further emphasizing the connection between culture and human identity (see, e.g., Stormonth et al., 1895; Webster, 1895). As historian Janet Davis (2016) has pointed out, the purview of the humane was primarily the work of the educated middle and upper classes, particularly white women in British and American society. To be humane was to possess a kindness that distinguished human beings from nature; but only some human beings had the resources to achieve this ethical standing.

By the twentieth century, the language of the humane was increasingly being associated with the subject of animal welfare through the emergence of public debates about animal vivisection. According to Davis, in the 1910s and 1920s, the language of humane treatment represented an essentially conservative social ethic that was first associated with the antivivisection movement and later appropriated by those in favor of medical vivisection (2016, pp. 80–83). Both opponents and advocates of animal vivisection argued that humane behavior was at stake, whether that meant protecting animals from experimentation or defining standards of treatment when they were used as scientific research subjects.

By 1966, when the Animal Welfare Act was passed, the language of humane treatment was applied far less frequently to human beings, with the exception of prisoners of war.[3] This may be due in part to another important backdrop of the period: the American civil rights movement. The year the Animal Welfare Act became law, the Black Panthers were founded; three years prior, Americans had seen Martin Luther King's March on Washington, and throughout the 1960s Congress passed a series of acts to reform Americans' access to voting rights, education, and other opportunities. Although I do not address this context in depth here, it is worth noting that although humane ethics and its implications of civilized stewardship were central to animal rights legislation of the 1960s, the language of humane behavior was rarely used in civil rights activism and legislation. Instead, civil rights literature expressed the ethical treatment of human beings through language such as *person, human being,* or *citizen*.[4] Implicit in this distance from humane treatment was the suggestion of action between equals, an approach that was significantly less common in the language of animal protection.

Musical Animals

Just as the humane treatment of animals has important precedents in nineteenth-century notions of the human, the scientific study of animal musicality has similar roots in claims to human identity. I explore the latter history in greater detail in my book *Animal Musicalities* (Mundy, 2018) but briefly summarize some of that history here. Around the same time that humane societies were being established in Britain and the United States, a surge of interest in the animal origins of musical ability was inspired by widely read debates about Darwin's theories of aesthetics. Music was believed to be a contested ability, one that might—or might not—be unique to human beings. Understanding whether animals could make music was a way to clarify what it meant to be human.

Between 1880 and 1950, hundreds of collections of folk songs, bird songs, insect songs, and national and indigenous music were published under the broad rubric of this evolutionary quest as part of an attempt to document and compare music's diversity across a spectrum of national, racial, and species differences. Like contemporary notions of humanizing education, to many intellectuals, being human in this context meant being *fully* human. While Bach, Beethoven, and Brahms were unequivocally heard as human in this tradition, other sounds, such as jazz, folk songs, and indigenous drumming, were compared to the primitive songs of birds and other animals. Some special birds, like the hermit thrush and nightingale, were even praised for having songs that were comparable to European classical music (see Doolittle, 2020; Rothenberg, 2019). Attempts to document animal music included a translation of a seventeenth-century essay on the music of bees (Hayes, 1925), a letter to the *Musical Times* arguing that the blackbird sang "old-time melodies" (Andrews, 1930), and a book-length treatise on the evolution of British bird music (Witchell, 1896).

In the aftermath of World War II, such comparisons between species and cultures were rejected as tools of racial stereotyping (see Mundy, 2014). Music scholars were particularly shocked to discover that German culture, which they had considered the most evolved musical culture in the world (epitomized by Bach and Beethoven), had fostered genocide (Mundy, 2014, p. 750). Suddenly, comparisons that had once seemed neutral were full of loaded metaphors. For example, a study of song sparrow phrases by ornithologist Aretas Saunders (1951) meticulously notated and analyzed 174 sparrows' songs to construct a typology of the species' song style. Although the project set an unusually high standard for research, it also implied comparisons with human cultural development, delineating primitive song types that raised echoes of older comparisons between the imagined sounds of human prehistory and the present-day songs of animals.

The response that emerged in the 1950s to these kinds of comparisons was twofold. Biologists rejected the old science of eugenics in favor of its cousin, genetics, arguing that genetic data would protect the sciences against racial stereotyping (see the introduction to Mayr & Provine, 1998; also see Cowan, chapter 16 of this volume, on these issues in the history of music psychology). Music scholars, in turn, tried to study music through its structures and forms—that is, "the music itself" divorced from social, historical, and racial claims (Mundy, 2014, pp. 749–758).

One unintended result of this shift was a division of labor between human and animal knowledge filtered through the humanities and the sciences. While postwar humanities scholars treated music as an exclusively human area of cultural study—a humanity—scientists focused on animal voices in the laboratory and the field as a source of bodily information about language acquisition. Studies of birdsong in particular adopted a language of objectivity, turning from music to language and emphasizing laboratory practices over field research (see Radick, 2007).[5]

A final element of this history is the unique status of visual evidence in scientific research. In contrast to sound, visual images had a long and trusted history of serving objective quantification within the sciences (Bull, 2018). Furthermore, images were printable—whether one did research in the 1920s or the 1980s, sounds couldn't be published in a research journal, but pictures could be (on visual vs. acoustic evidence in the history of music science, see Deutsch, chapter 14 of this volume).

One of the more curious side effects of this emphasis on images was an interlocking history that connected early psychoacoustic research and sound recording technology to the same debates about vivisection that had had such an impact on the language of humane treatment in animal welfare circles. As I have shown elsewhere, the technology behind early phonographs, which consisted of a rotating cylinder attached to a playback armature, derived from the cylinder-and-armature devices used to record visual images of animal vivisections in the mid-1800s. In the 1920s, several prominent music scholars and critics, some of whom were familiar with the phonograph's counterpart in vivisection, appropriated the language of vivisection as well as its technology, urging their peers to objectively study music as it "died" under the knife (see Mundy, 2018, ch. 4). Because these musical vivisections were framed by the period's evolutionary outlook, the songs "killed" in this type of research were those of non-Western, colonial, or otherwise othered subjects.

As a result of these intertwined histories, studying animal musicality after 1950 meant navigating a path between science and humanism, and there were many elements at stake. Central to that pathway was the distinction between human and animal subjects, particularly in the case of domesticated animals such as mice, rats, canaries, and other species with long-standing use in laboratory research. Highly refined methods of visual analysis, in addition to the nonhuman status of animals and the histories that made some species more likely research subjects than others, set the stage for a version of science that was deliberately distanced from the practice that we call humanism, yet one that was reliant on the outcome that American and British intellectuals had once believed would come only from studying the humanities: a humane ethic. As scientists and humanists today attempt to reengage with each other, they face this dual history, particularly in the context of animal-based research.

Animal Welfare

If today's humane ethic is connected to this complex history of racial science and its aftershocks, what does it mean to be an animal in this context—especially an animal worthy of care? The 2020 version of the Animal Welfare Act defines animal as follows:

> The term "animal" means any live or dead dog, cat, nonhuman primate, guinea pig, hamster, rabbit, or such other warm-blooded animal, as the Secretary may determine is being used, or is intended for use, for research, teaching, testing, experimentation, or exhibition purposes, or as a pet; but such term excludes (1) birds, rats of the genus Rattus, and mice of the genus Mus, bred for use in research; (2) horses not used for research purposes; and (3) other farm animals such as but not limited to livestock or poultry, used or intended for use as food or fiber, or livestock or poultry used or intended for use in improving animal nutrition, breeding, management, or production efficiency, or for improving the quality of food or fiber. With respect to a dog, the term means all dogs including those used for hunting, security, or breeding purposes.

Here, animals are defined as animals—as nonhuman beings worthy of a standard of care—according to conditions that might seem quite bizarre at first. They must be warm-blooded; they cannot be birds, mice, or rats; and they cannot be any kind of animal that is made into food or clothing. Dogs, in this odd rubric, are always animals, even though other creatures are not.

Domesticated breeds exist at the margins of the "animal" as it is defined by the AWA, occupying a space that is neither nature nor nurture. They represent what Philip Howell and Olga Petri (2020) call the "unnatural" history of animal culture, a hybrid world in which modern human technologies and colonial history intersect with animal agency. There are layers within these layers—for what makes an animal domesticated? How do we ascertain the history of a domesticated species? What does domestication signify in experimental contexts, where familiar residents of the laboratory such as the canary, zebra finch, and macaque are so often the result of forced migration by colonial powers? (Mundy, 2022).

The AWA's mention of "livestock or poultry" and "food or fiber" reveals the underlying conflict between human benefit and animal welfare that underpins these bizarre classifications and omissions. The purpose of the act is not to define animals' rights but to define exceptional cases in which animals are entitled to a particular standard of care. As the word *humane* suggests, human beings are not animals in this assessment; they are in a separate and unique category. And their rights are sometimes in conflict with notions of animal welfare.

Within these limitations, the concept of humane treatment in the United States is applied primarily through committee oversight. The process was formalized in a 1987 amendment to the AWA that established the role, duties, and constitution of Institutional Animal Care and Use Committees (IACUCs), which are still the primary method of review for the care of research animals. Each institution or university has its own IACUC, appointed by the head of the institution. Each committee must have at least three members; at least one member must be a veterinarian, and at least one member must be from outside the institution. The IACUC reviews proposals for research involving animal care and inspects research facilities every six months. Just as researchers studying human subjects have to submit proposals to an institutional review board for approval, researchers studying animals have to submit paperwork to their institution's IACUC.

Although the paperwork and the review process differ slightly from one institution to another, the guidelines are standardized by the Animal Welfare Act, and the questions are similar. Here is an example from the University of Southern Maine's IACUC:

> List and describe the animals to be studied. Indicate the anticipated number of animals to be used in each Pain Category of Research and the total number of animals involved during the three-year approval period of this protocol. Indicate whether Pain Category B: Breeding or Holding Colony Protocols. Pain Category C: No more than momentary or slight pain or distress and no use of pain-relieving drugs, or no pain or distress. For example: euthanized for tissues; just observed under normal conditions; positive reward projects; routine procedures; injections; and blood sampling. Pain Category D: Pain or distress appropriately relieved with anesthetics, analgesics, and/or tranquilizer drugs or other methods for relieving pain or distress. Pain Category E: Pain or distress or potential pain or distress that is not relieved with anesthetics, analgesics and/or tranquilizer drugs or other methods for relieving pain or distress.[6]

Readers unfamiliar with this system, myself included, might recoil from a standard that treats "euthanized for tissues" and "positive reward projects" as comparable. Yet many scientists confronted with such reactions respond, correctly, that the IACUC process provides a higher standard of care than that used for "food and fiber" contexts. Those who use animals in the laboratory are part of a complex, detailed, and ongoing conversation about what it means to pursue the ethical treatment of animals.

Ethics

At the center of the IACUC's review process is the belief in an intrinsic difference between the human and the animal. That belief is at the core of standards of humane treatment, the concept of animal welfare, and many of the questions we ask about music as well. "Is our musical predisposition unique, like our linguistic ability?" asks Henkjan Honing (2019, p. xi), echoing Darwin. How, wonders Steven Brown (2022), did human beings come to dance in a way that is unique from other animals? What can singing seals tell us about early hominids, asks a team of European scientists? (Ravignani et al., 2016). These questions are not just about the musical traits of a particular species; they are about how animals might reveal something about being human.

Animals, in this system, are not "humans and *other* animals" but rather a catchall category against which human uniqueness can be compared.[7] This catchall category is such a foundational concept in Western culture that it is difficult to summarize, trace, reduce to a few sentences, or even question. To acknowledge it as a concept cracks open a hundred passageways leading to places unknown. Once "the animal" is questioned, it raises questions about the uniqueness of "the human" and about whether the human as we know it can even exist if it does not cast a shadow that covers every other species in the cloak of the animal.

Although the human-animal binary is not a topic that has been widely explored in the sciences, it has been considered in depth in two closely related niche fields in the humanities: animal studies and posthumanism. Founded in the 1980s and 1990s, both fields emerged from a series of questions about the legal status of animals and machines and the viability of the category of the human. Foundational works drew heavily on American feminism and explored colonial, racial, and gendered biases in natural science (Haraway, 1989; Bal, 1992), the legal and philosophical status of animals (Singer, 1975; Callicott, 1980), and definitions of the human in the computer age (Haraway, 1985; Hayles, 1999). Later texts in the late 1990s and early 2000s drew inspiration from continental philosophy (Latour, 1991; Derrida, 2002; Wolfe, 2003). More recently, a backlash against this later movement has reoriented the field closer to its origins in feminism and black studies (Weheliye, 2014; Boisseron, 2018; Jackson, 2020; Fraiman, 2012; Probyn-Rapsey et al., 2019).

In the last several years, scholars in animal studies have engaged questions of the human largely through the context of interlocking cases of race and animal rights, such as divergent perceptions of pit bulls adopted by white and black owners or differing representations of animality in African and European art (Dayan, 2018; Boisseron, 2018; Jackson, 2020). This movement, which draws on methods in black studies, history, and art criticism, has argued that the human-animal binary is not just about labeling some sounds—and some singers—as more human than others. It is about a network of ideas in which categories such as species, race, gender, nation, and other forms of difference are mutually necessary to define one another within Western culture. According to Sylvia Wynter (2003) and Zakkiyah Iman Jackson (2020), categories such as race, gender, and animal are interdependent within the social and historical contexts in which both science and humanism operate. In those contexts of taken-for-granted assumptions and implied beliefs, categories of difference such as race, gender, and species tell us who, exactly, is fully human in the modern world.

Drawing on these perspectives developed in animal studies and posthumanism, let me suggest some of the elements that make up the ethic of humane treatment:

1. The animal in this context is a category premised on the exclusion of human beings.
2. The human in this context is a category informed by particular exclusionary traditions of western Europe and the United States, even if actors using this terminology are not citizens of those regions.
3. Historically, those exclusions include both the animal and humans beings who were deemed "natural," uneducated, or uncivilized in nineteenth- and early-twentieth-century scientific circles.
4. If one accepts 1, 2, and 3 even in part, it is necessary to reconsider what is at stake in studies of animal musicality, human musicality, and the humane ethics that guides them.

If this account is even partially correct, the postwar division of music into human culture, on the side of the humanities, and animal bodies, on the side of the sciences, obscures a daunting debris of unexamined beliefs about race, gender, nationality, sexuality, species, and other forms of difference. These issues have substantial implications for the way we ask questions about music and what we hope to gain from the answers, particularly in collaborative experimental contexts.

Questions

At the broadest scale, the human-animal binary at the center of humane ethics is also at the center of the humanities and the humanities-science divide. Rethinking that binary is not a simple thing. As I and other scholars have shown, the human is defined by, and defining for, notions of race, gender, nationality, sexuality, species, and other otherness; rethinking the human means rethinking all these categories and how we use difference to compare and evaluate life and selfhood. This creates a domino effect, where one change leads to another and yet another. Rethinking the human is not something that will be worked out in this chapter or even in this book. It is the work of a generation and a movement.

With that in mind, what are some of the implications for studies of musical capacity? Several contributors to this book touch on some of the stakes in play. To name just a few: Duengen, Sarfati, and Ravginini examine the limitations of using nonhuman animals to make anthropocentric claims. Patel argues that the value of music in human life is not contingent on evolutionary claims of uniqueness. Ilari and Habibi show that the musician-nonmusician paradigm carries colonial and Western baggage that implicates the broader category of the human. Such approaches point the way to new ways and reasons to measure and assess the thing we call musicality.

An issue that arises in studies of animal musicality is the goal of translation. In research contexts involving nonhuman animals, translation is the idea that discoveries and experimental results for one species will be applicable to another species. The concept is often used in medical research, where scientists use biological similitude to hypothesize whether the effects of a chemical or medical intervention on a specific animal (such as a dog or a mouse) will result in similar effects on human subjects. This, too, is part of a humane ethic codified in the 1947 Nuremberg Code, when medical experts involved in the Nuremberg trials drafted ten principles meant to set ethical boundaries for research on human subjects. Among other things, the code required voluntary consent from human subjects and a reasonable expectation that the research would not harm them; it also required that medical experiments on human beings be based on the results of animal experimentation.[8]

In studies of music, the concept of translation is used to hypothesize behavioral rather than biological similitude. For example, in classic studies of song learning in chaffinches by W. H. Thorpe, one of the most valued outcomes was the parallel between how baby finches learn to sing and how human infants learn language (Soha & Peters, 2015). These translations can lead to some very unlikely connections, such as comparisons of macaques, finches, and seals; a more informal connection has been made between human musical expressions and birds, a mismatch between one species and an entire class of them (see, e.g., Honing, 2019). Furthermore, such species translations often rely on loose notions of musicality that are themselves deeply steeped in historical associations and measurements that originate with Western culture (see chapter 17). Those historical measures and associations valorize an "evolved" human musical ability as it was defined by scientists in the colonial and racialized atmosphere of nineteenth-century Europe (see Mundy, 2014, 2018; Zon, 2014).

In this environment, researchers may feel external pressure to pursue funding opportunities framed by human achievement or human development, language grounded in an anthropocentric ethos that rewards comparisons between chaffinches and human beings but deters comparisons of song learning in the common chaffinch (*Fringilla coelebs*) and its cousin the Madeiran chaffinch (*Fringilla coelebs maderensis*). There are, as Margaret Landi, Jeffrey Everitt, and B. Berridge (2021) recently pointed out, no standard guidelines for determining which cases translate between species and which do not.

Rethinking concepts like translation outside of a human-animal binary opens up new possibilities for thinking about how scientists and scholars can collaborate and obtain financial support. Why do we study music? If we set aside the idea of studying music to find out what makes us human (who are we? what do we mean by this?), what kinds of things would we like to know? What will be useful, interesting, and powerful to hear in the twenty-first century and its contexts? If the canary's song and body are not resources for research, what are they, and why?

Today, questions about ethics occupy a startling amount of the literature on animal laboratory research, raising questions about translation, animal welfare, the emotional and physical health impact of human-animal relationships on researchers, and even philosophy and virtue (Landi et al., 2021; Nobis, 2019; Walker, 2019). This is a moment with considerable potential for ongoing discussions about ethics in the science-music borderlands. With passionate interest, active discourse, and meaningful history in play, scholars and scientists are well equipped to rethink how ethics should be structured in music-science collaborations.

By examining ethical questions as historical artifacts, it is possible to develop a rich context for these collaborations. That context puts in play profound questions about who we consider human and how we hear them. These questions date to the middle of the twentieth century, when scientists and humanists were redefining their work in response to the Holocaust. Today in the twenty-first century, we are again redefining what we do in order to account for a future that includes global research and ecological crisis. Humanists, who have become so skilled at mapping our ignorance, and scientists, who excel at mapping what we can know, each bring traditions to the table that can inspire us to develop better and more truthful musical relationships with those who are unlike ourselves.

Notes

1. See https://rsawa.research.ucla.edu/arc/humane-treatment-and-endpoints/.

2. See, for example, Vartanian (1999); Riskin (2002). For an overview of the connections among the emergence of specialized musical knowledge, colonial conquest, and the humanizing education of the eighteenth century, see Agnew (2008).

3. See, for example, the language of the 1949 Geneva Convention governing prisoners of war (Article III, "Humane Treatment of Prisoners"). Interestingly, unlike most material on animal welfare in the US, the convention defines the word *humane*: "In accordance with the ordinary meaning of the word 'humane,' what is called for is treatment that is 'compasisonate or benevolent'" (para. 1573).

4. See, for example, the language of the 1964 Civil Rights Act, Martin Luther King Jr.'s 1963 "I Have a Dream" speech, and Newton and Seale (1967, p. 3).

5. The notable exception is the discovery of humpback whale song in 1969–1970, in which Payne and McVay (1971) adopted a very different approach to musical listening. Their approach affected subsequent studies of cetacean vocalization but seemingly had no significant impact on birdsong research such as Nottebohm's (1981).

6. University of Southern Maine, IACUC Form Submissions template (May 17, 2018), https://usm.maine.edu/sites/default/files/orio/IACUC_Form_Submissions_template.pdf.

7. As I describe the challenges presented by this approach to human uniqueness, I want to note a separate but related practice that plays a role in the ethics of IACUC oversight: the idea that human needs take priority over the needs of other animals. In practice, this often functions more as a claim about kinship than a fundamental claim about human primacy, or what Frances Bartkowski (2008) calls the relationship of "kissing cousins." Consider, for example, how debates about animal trials in early research on COVID-19 vaccines were framed in the popular culture in 2020: although the animal experiments were quickly approved and conducted, rumors that such animal trials had been skipped prompted some to reject the use of vaccines for human beings (see, e.g., Dupuy, 2020; Reuters, 2021). There were no debates about whether it was *ethical* to test vaccines on animals. The principle here seems not to be about the uniqueness of human beings (indeed, animal trials are worthless if they cannot translate to human bodies) but about what we owe one another. Although in this case only humans were judged "kin," the same principle is easy to extend across species: an individual might conduct an experiment on a dog in the laboratory that he or she would never perform on the pet dog at home, not because humans are unique but because one dog is family and the other isn't. For the sake of simplicity, I don't address this position in the chapter, but it is an important alternative perspective.

8. The Nuremberg Code is reproduced in *British Journal of Medicine, 1448*(December 1996), 313, https://doi.org/10.1136/bmj.313.7070.1448.

References

Agnew, V. (2008). *Enlightenment Orpheus: The power of music in other worlds*. Oxford University Press.

Andrews, E. R. G. (1930). Bird songs. *Musical Times, 71*(1047), 446.

Bal, M. (1992). "Telling, showing, showing off." *Critical Inquiry, 18*(3), 556–594.

Bartkowski, F. (2008). *Kissing cousins: A new kinship bestiary*. Columbia University Press.

Boisseron, B. (2018). *Afro-dog: Blackness and the animal question*. Columbia University Press.

Brown, S. (2022). Group dancing as the evolutionary origin of rhythmic entrainment in humans. *New Ideas in Psychology, 64*, 1–12.

Bull, M. (2018). *The Routledge companion to sound studies*. Routledge.

Callicott, J. B. (1980). Animal liberation: A triangular Affair. *Environmental Ethics, 2*(4), 311–338.

Davis, J. M. (2016). *The gospel of kindness: Animal welfare and the making of America*. Oxford University Press.

Dayan, C. (2018). *With dogs at the edge of life*. Columbia University Press.

Derrida, J. (2002). The animal that therefore I am (more to follow) (D. Wills, Trans.). *Critical Inquiry, 29*(2), 369–418.

Dimrock, N. (1885). *Some discourses bearing on the nature of man*. Vivish.

Doolittle, E. (2020). Hearken to the hermit thrush: A case study in interdisciplinary listening. *Frontiers in Psychology, 9*. https://doi.org/10.3389/fpsyg.2020.613510.

Dupuy, B. (2020, November 25). Pfizer and Moderna did not skip animal trials. Associated Press. https://apnews.com/article/fact-checking-afs:Content:9792931264.

Fraiman, S. (2012). Pussy panic versus liking animals: Tracking gender in animal studies. *Critical Inquiry, 39*(1), 89–115.

Ghosh, A. (2016). *The great derangement: Climate change and the unthinkable*. University of Chicago Press.

Haraway, D. (1985). A cyborg manifesto. *Socialist Review, 80*, 65–108.

Haraway, D. (1989). *Primate visions: Gender, race, and nature in the world of modern science*. Routledge.

Hayes, G. R. (1925). Charles Butler and the music of bees. *Musical Times, 66*(988), 512–515.

Hayles, K. (1999). *How we became posthuman: Virtual bodies in cybernetics, literature, and informatics*. University of Chicago Press.

Haynes, R. P. (2008). *Animal welfare: Competing conceptions and their ethical implications*. Springer.

Honing, H. (2019). *The evolving animal orchestra: In search of what makes us musical* (S. MacDonald, Trans). MIT Press.

Howell, P., & Petri, O. (2020). From the dawn chorus to the canary choir: Notes on the unnatural history of birdsong. *Humanalia, 11*(2), 1–21.

Jackson, Z. I. (2020). *Becoming human: Matter and meaning in an antiblack world*. New York University Press.

Jay, P. (2014). *The humanities "crisis" and the future of literary studies*. Macmillan, 2014.

Landi, M., Everitt, J., & Berridge, B. (2021). Bioethical, reproducibility, and translational challenges of animal models. *Institute for Laboratory Animal Research*. https://doi.org/10.1093/ilar/ilaa027.

Latour, B. (1991). *We have never been modern* (C. Porter, Trans.). Harvard University Press.

Mayr, E., & Provine, W. B. (Eds.). (1998). *The evolutionary synthesis: Perspectives on the unification of biology*. Harvard University Press.

Mundy, R. (2014). Evolutionary categories and musical style from Adler to America. *Journal of the American Musicological Society*, *67*(3), 735–768.

Mundy, R. (2018). *Animal musicalities: Birds, beasts, and evolutionary listening*. Wesleyan University Press.

Mundy, R. (2022). The cecotrope: A reply to Amitav Ghosh [Unpublished manuscript].

Newton, H. P., & Seale, B. (1967). What we want now! What we believe. *Black Panther*, *3*.

Nobis, N. (2019). Why IACUCs need ethicists. *Institute for Laboratory Animal Research*, *60*(3), 324–333.

Nottebohm, F. (1981). A brain for all seasons: Cyclical anatomical changes in song control nuclei of the canary brain. *Science*, *214*(4527), 1368–1370.

Payne, R. S., & McVay, S. (1971). Songs of humpback whales. *Science*, *173*(3997), 585–597.

Probyn-Rapsey, F., O'Sullivan, S., & Watt, Y. (2019). Pussy panic and glass elevators: How gender is shaping the field of animal studies. *Australian Feminist Studies*, *34*(100), 198–215.

Radick, G. (2007). *The simian tongue: The long debate about animal language*. University of Chicago Press.

Ravignani, A., Fitch, W. T., Hanke, F. D., Heinrich, T., Hurgitsch, B., Kotz, S. A., Scharff, C., Stoeger, A. S., & de Boer, B. (2016). What pinnipeds have to say about human speech, music, and the evolution of rhythm. *Frontiers in Neuroscience*, *10*, 1–9.

Reuters. (2021, June 1). Fact check: COVID-19 vaccines did not skip animal trials because of animal deaths. https://www.reuters.com/article/factcheck-covid-vaccine-animal/fact-check-covid-19-vaccines-did-not-skip-animal-trials-because-of-animal-deaths-idUSL2N2NJ1IK.

Riskin, J. (2002). *Science in the age of sensibility: The sentimental empiricists of the French Enlightenment*. University of Chicago Press.

Rothenberg, D. (2019). *Nightingales in Berlin: Searching for the perfect sound*. University of Chicago Press.

Saunders, A. (1951). The song of the song sparrow. *Wilson Bulletin*, *63*(3), 99–109.

Singer, P. (1975). *Animal liberation: A new ethics for our treatment of animals*. New York Review.

Soha, J. A., & Peters, S. (2015). Vocal learning in songbirds and humans: A retrospective in honor of Peter Marler. *Ethology*, *121*(10), 933–945.

Stormonth, J., Phelp, P. H., & Bayne, W. (1895). *A dictionary of the English language*. Blackwood and Sons.

United States Department of Agriculture (2020). *Blue book: Animal Welfare Act and animal welfare regulations*. USDA.

Vartanian, A. (1999). *Science and humanism in the French Enlightenment*. Rookwood.

Walker, R. L. (2019). Virtue ethics and laboratory animal research. *Institute for Laboratory Animal Research, 60*(3), 415–423.

Webster, N. (1895). *Webster's academic dictionary*. American Book Company.

Weheliye, A. (2014). *Habeus viscus: Racializing assemblages, biopolitics, and black feminist theories of the human*. Duke University Press.

Witchell, C. A. (1896). *The evolution of bird-song, with observations on the influence of heredity and imitation*. Adam and Charles Black.

Wolfe, C. (2003). *Animal rites: American culture, the discourse of species, and posthumanist theory*. University of Chicago Press.

Wynter, S. (2003). Unsettling the coloniality of being/power/truth/freedom: Towards the human, after man, its overrepresentation—an argument. *CR: The New Centennial Review, 3*, 257–337.

Zon, B. (2014). *Evolution and Victorian musical culture*. Cambridge University Press.

Interlude

5 Of Sound Minds and Tuning Forks: Neuroscience's Vibratory Histories

Carmel Raz

One intriguing image stands out in the *Iconographie photographique de la Salpêtrière* (1876–1880), a collection of photographs of the psychiatric wards of Paris's leading public hospital, the Salpêtrière, taken by physicians Désiré-Magloire Bourneville and Paul Régnard. A woman sits next to an immense tuning fork, her head tilted to one side and an expression of deep concentration on her face (figure 5.1). The size of the instrument is astonishing, as is the claim that the photograph depicts catalepsy provoked by the sound of a tuning fork. But I find the image remarkable for another reason: I see in it a literal snapshot of two intersecting ideas—the notion that music directly influences our minds, bodies, and emotions; and the historical theory that vibration plays an essential role in nervous transmission.

Bourneville and Régnard documented a number of Jean-Martin Charcot's neurological experiments with sound. Charcot, generally regarded as the founder of modern neurology, held the first chair in nervous diseases at the Salpêtrière. He is remembered today for his work on multiple sclerosis, epilepsy, and Parkinson's disease, as well as for his more controversial studies of hysteria (Kumar et al., 2011; Hustvedt, 2011). His students, a veritable who's who of pioneering neurologists, included Sigmund Freud, Gilles de la Tourette, and Joseph Babinski.

Charcot provides the following account of tuning fork trials, which were conducted on patients whose minds and bodies were particularly susceptible to suggestion:

> The patients are seated over the sounding box of a strong tuning fork, made of bell metal, vibrating sixty-four times in a second. It is set in vibration by means of a wooden rod. After a few moments the patients become cataleptic, their eyes remain open, they appear absorbed, are no longer conscious of what passes around them, and their limbs preserve the different attitudes which have been given them. (1890, pp. 262–263)

Charcot's understanding of catalepsy differs from modern-day applications of the term. According to the 1876 edition of the *Encyclopaedia Britannica*, catalepsy was "a nervous affection characterized by the sudden suspension of sensation and volition,

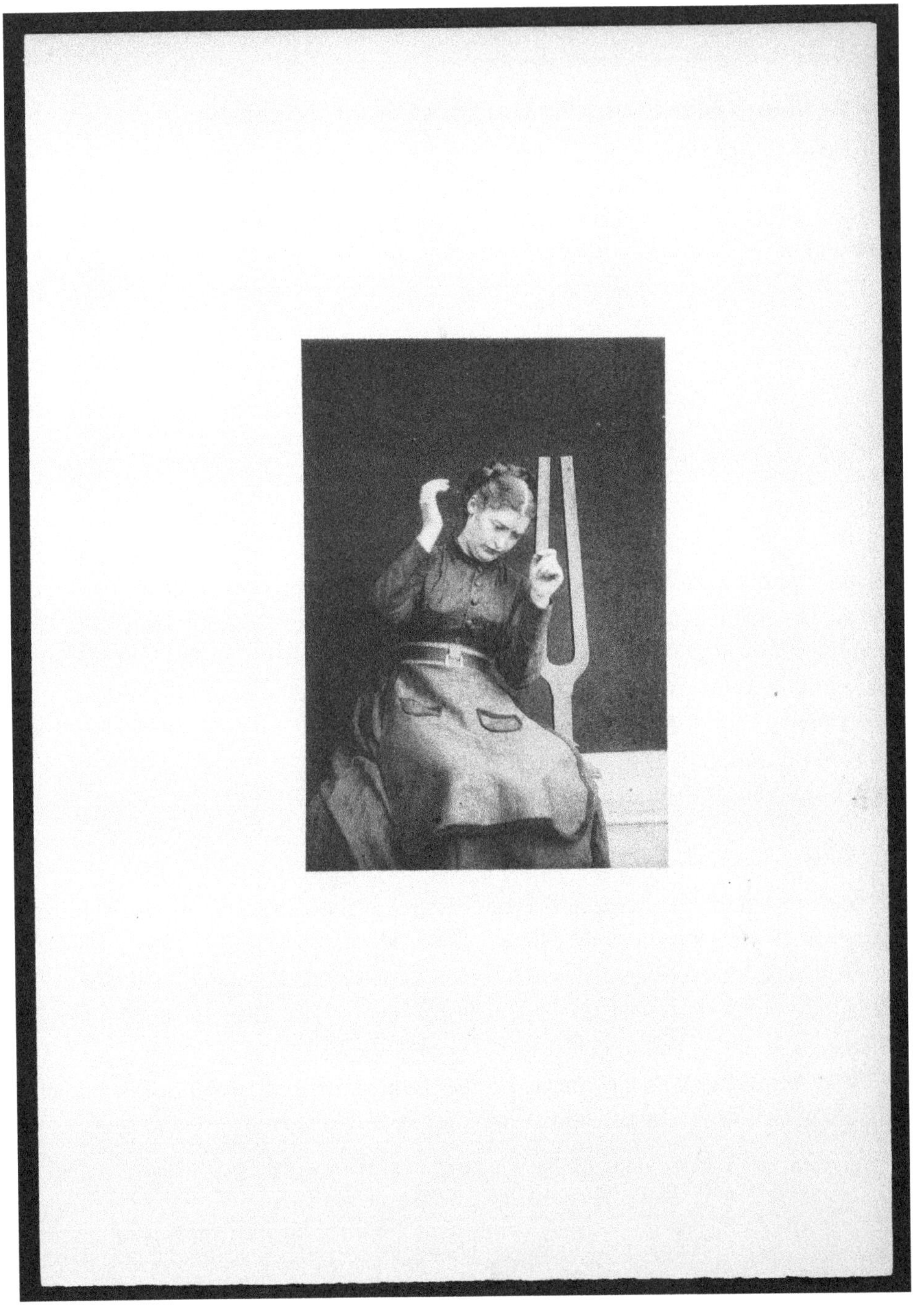

Figure 5.1
"Catalepsy provoked by the sound of a tuning fork." (From Bourneville & Régnard, 1876–1880, plate 20.)

accompanied with a peculiar rigidity of the whole or of certain muscles of the body" (Chisolm, 1876, p. 216).

As Charcot himself noted, his investigation of mechanisms for inducing catalepsy by means of sounds (or other stimuli) takes up where James Braid, the "discoverer" of hypnosis, and other neurologists left off (Charcot, 1890, p. 253; Charcot, 1879, p. 276). Yet the French physician was not convinced that suggestion alone was responsible, declaring "that every phenomenon of the natural order, whatever be its appearance of complication or of mystery, is none the less a subject of methodical observation" (Charcot, 1890, pp. 253–254; Charcot, 1879, p. 276). In what follows, I demonstrate how Charcot's experiments reflect vestiges of vibrating nerve theory, an underlying assumption of nineteenth-century neural science that originated with Isaac Newton. Showing how the reception of this theory depended on, and was intertwined with, specific features of sound, I argue that music played an important role in the emergence of what eventually became the modern neural sciences.

Up until the late seventeenth century, the dominant model of nervous transmission in Europe involved some version of the ancient Galenic theory of animal spirits (Kassler, 1995, pp. 33–36). Accounts of the manner in which the mind communicated with the heart, or any other organ, tended to refer to the lungs as bellows and the nerves as hollow pipes, as in the writings of René Descartes (1998, p. 140) and Thomas Willis (1965, p. 108), among others. Yet changing conceptions of physiology, and of electricity and action at a distance, contributed to the rise of a different model of nervous transmission in the early eighteenth century: vibrating nerve theory (Kassler, 1995; Wardhaugh, 2008; Cannon & Dostrovsky, 1981).

The notion that nervous transmission might relate to vibration was first articulated by Isaac Newton in his *Opticks*, where he hypothesized the existence of an "Aethereal Medium" pervading the universe. This same almost infinitely subtle substance, or "spirit," was, he suggested, responsible for sensation and muscle movement by acting as the medium for vibrations traveling within and between the nerves and the brain. Analogizing between the function of the optical and auditory nerves, Newton broke with the doctrine of hollow, animal-spirit-bearing nerve pipes to suggest that solid nerves, acting as the routes of travel of this ethereal medium, could enable vibratory transmission:

> Is not Animal motion performed by the Vibrations of this Medium, excited in the Brain by the Power of the Will, and propagated through the solid, pellucid and uniform Capillamenta of the optick Nerves into the Muscles, for contracting and dilating them? (Newton, 1718, p. 328)

Newton here linked muscular motion to psychic motivation via the physical vibrations propagating through the ethereal medium in the nerves. He further developed this scheme in the *General Scholium* appended to the second edition of his *Principia* in 1713. In the famous concluding passage, he suggested that the ethereal medium pervading the nerves and linking the mental faculties and their corporeal instrumentalities—will and perception; muscles and sense organs—must be in some way vibratory, elastic, and electric:

> All sensation is excited, and the members of animal bodies move at the command of the will, namely, by the vibrations of this spirit, mutually propagated along the solid filaments of the nerves, from the outward organs of sense to the brain, and from the brain into the muscles. But these are things that cannot be explained in few words, nor are we furnished with that sufficiency of experiments which is required to an accurate determination and demonstration of the laws by which this electric and elastic spirit operates. (Newton, 1819, p. 314)

Newton's statements catapulted vibrating nerve theory to the center of scientific and philosophical discourse. Vibration and sympathetic resonance—that is, the acoustical fact that a resonating body vibrates in response to external sound vibrations to which it stands in a harmonic (i.e., simple proportional) relationship—were readily evident to anyone with access to a string instrument. Thus, rather than wind chests and hollow organ pipes, strings became the dominant musical analogy for this theory of nervous transmission (Kassler, 1984). This idea caught on very quickly, as evidenced by the care contemporaneous writers took to refer to both animal spirits and Newtonian ethereal vibrations or to some amalgamation of the two. Thus, John Locke clarifies that either model can be applied to his model of perception in *Remarks upon Some of Mr. Norris' Books* (1706):

> And whether or no God has appointed that a certain modified motion of the fibres, or spirits in the optic nerve, should excite, or produce, or cause them in us, call it what you please, it is all one as if it did; since where there is no such motion there is no such perception or idea. (1812, p. 254)

Similar equivocation can be found in English physician William Cheselden's *Anatomy of the Human Body*, which proposes that "perhaps sensations may be conveyed either, or both ways. However, it being usually taken for granted, that it must be one of these ways at least" (1740, p. 248). By the middle of the eighteenth century, however, vibration had clearly won out as the primary mechanism of neural transmission, giving rise to a number of competing models in Britain and on the Continent (Krüger, 1748; Bonnet, 1782; Sulzer, 1773).

Perhaps the most influential of the so-called vibrationalist theories was devised by the English physician and philosopher David Hartley in *Observations on Man, His Frame, His*

Duty, and His Expectations, first published in 1749 (Hartley published an earlier version of this theory in 1730 as *Conjecturae quaedam de sensu, motu, & idearum generatione*). Explicitly referring to Newton's hypothesis that an electric, ethereal substance enables nervous transmission, he proposed a "doctrine of vibrations," according to which

> external objects impressed upon the senses occasion, first in the nerves on which they are impressed, and then in the brain, vibrations of the. . . . infinitesimal, medullary particles . . . [these] sensory vibrations, by being often repeated, beget, in the medullary substance of the brain, a disposition to diminutive vibrations, which may also be called vibratiuncles . . . corresponding to themselves respectively. (Hartley, 1843, pp. 8, 37)

In Hartley's scheme, mental operations such as perception, thought, and memory were caused by infinitesimally tiny neural vibrations either experienced immediately by the brain as sensation or, in the case of thought and memory, called up as faint reverberations of associative neural pathways, enabling the brain to replicate or compound traces of earlier experience. He regarded the vibrations that enabled nervous transmission as manifestly electric in some way, observing:

> The effluvia of electric bodies seem to have vibrating motions. . . . Their motions along hempen strings resemble the motions along the nerves in sensation and muscular contraction, and their attractive powers, at the end of such strings, resemble the powers of the sensations over the muscles for contracting them. So that electricity is also connected in various ways with the doctrine of vibrations. (Hartley, 1843, p. 18)

For Hartley, these electric vibrations were obviously distinct from the material, mechanical vibrations generated by musical strings. As he emphasizes: "For that the nerves themselves should vibrate like musical strings, is highly absurd; nor was it ever asserted by *Sir Isaac Newton*, or any of those who have embraced his notion of the performance of sensation and motion, by means of *vibrations*" (Hartley, 1843, p. 8; emphasis in original).

Yet in spite of Hartley's explicit warning, this mechanical version of vibrating nerve theory rapidly proliferated both within and beyond the medical domain, where it was often conflated with generic notions about sympathetic resonance. In his article on the effects of music in the *Encyclopédie*, for example, French physician Ménuret de Chambaud (1765) asserted:

> If one considers the human body simply as an assemblage of fibers under varying degrees of tension, and fluids of various kinds, disregarding their sensitivity, life, and movement, it will be quite clear that music must have the same effect on the fibers which it has on the strings of nearby instruments; that all the fibers of the human body will be set in motion; that those which are more tense, fine, and slender will be more moved by it, & that those which are in unison will preserve [that motion] longer.

Similar ideas were expressed by James Beattie in his *Essays: On Poetry and Music, as They Affect the Mind*. He likewise ascribed music's effects to the mechanical resonance of the "finer fibres"—that is, the nerves—of the body:

> If a person who sneezes, or speaks loud, in the neighbourhood of a harpsichord, often hears the strings of the instrument murmur in the same tone, we need not wonder, that some of the finer fibres of the human frame should be put in a tremulous motion, when they happen to be in unison with any notes proceeding from external objects. (Beattie, 1776, p. 4)

The currency of such ideas also attained popularity in more distant domains such as philosophy and literature, where the notion of "resonating" nerves was metaphorically applied to explain mental or psychic events. In a striking passage, for example, the German philosopher and critic Johann Gottfried Herder borrowed the notion of sympathetic resonance to explain the affective power of certain genres of poetry on the reader:

> Since the comparison of the nerve structure of feeling to a harp is very accurate, let me emphasize that just as a string only resonates with another with which it is harmoniously attuned, so the cries of the elegy demand the reader's soul be pitched to the same tone. (1985, p. 488; my translation)

Moving even further from the sphere of the nerves, the French philosopher Denis Diderot applied the same acoustic features to model the association of ideas within the mind itself. "Vibrating strings," he wrote, "have yet another property, that of making other strings vibrate; and that is how the first idea recalls a second, the two of them a third, these three a fourth and so on, so that there is no limit to the ideas awakened and interconnected in the mind of the philosopher" (Diderot, 1979, p. 56).

The English poet Samuel Taylor Coleridge (who was so taken with Hartley's ideas that he named his firstborn Hartley Coleridge) explicitly likened mental function to the sympathetic resonance between strings in his influential poem "The Eolian Harp" (1796). Trower (2012, pp. 13–36) has analyzed the following lines from this poem in view of Hartley's theories:

> Full many a thought uncalled and undetained,
> And many idle flitting phantasies,
> Traverse my indolent and passive brain,
> As wild and various as the random gales
> That swell and flutter on this subject Lute!
>
> And what if all of animated nature
> Be but organic Harps diversely framed,
> That tremble into thought, as o'er them sweeps
> Plastic and vast, one intellectual breeze,
> At once the Soul of each, and God of all? (Coleridge, 2001, p. 118)

These examples are just a tiny sample of the widespread metaphoric transference of ideas about vibration and resonance from acoustic to neural to mental domains of feeling and thought in the late eighteenth and early nineteenth centuries. Indeed, scholars have convincingly argued that the "cult of sensibility" arose directly from the early Romantics' fascination with Enlightenment neural science (Richardson, 2001). To this day, our language reflects the absorption of these ideas by philosophy and aesthetics—consider the personal descriptors "highly strung" or "low key." For countless thinkers throughout Europe, music provided a rich storehouse of metaphors by which to conceptualize subjective and neurophysiological experience alike, at least in part because it was deeply imbricated in generating altered emotional and physical states in the first place.

One arena in which this interaction played out in practice was incipient music therapy, which became closely associated with the employment of novel instruments and timbres (Raz & Finger, 2018). In the late eighteenth century, reports of Mesmer's reliance on the ethereal tones of the newly invented glass harmonica in his seances, ostensibly because these particular vibrations enhanced the nerves' receptivity to his "magnetic fluid," soon inspired various music-therapeutic theories and case studies documenting the effects of the timbres of other novel musical instruments that were believed to strongly affect the nerves.[1] This was particularly true of females, whose nervous systems, during most of the eighteenth and nineteenth centuries, were regarded as more delicate and hence more susceptible than those of men, and many more women than men were the subjects of case reports of conditions such as catalepsy (Raz, 2014; Hustvedt, 2011).

The conflation of electrical and acoustical nerve stimuli noted earlier is also evident in the domain of music criticism. The decades between 1830 and 1850 saw a rise in metaphorical comparisons of the nerves' susceptibility to music and the transmission of electricity. Reporting on the great piano virtuoso Franz Liszt, for example, French music critic Paul Scudo rhapsodized over "his iron fingers, which diffuse nervous energy as the voltaic pile diffuses electric force. . . . Liszt stimulates the nerves" (1841, p. 32). The composer Hector Berlioz similarly described the experience of conducting an orchestra using electrical metaphors, noting that the conductor's "inner flame will warm [the musicians], his electricity will charge them, his drive will propel them. He will radiate the vital spark of music" (2003, p. 490; MacDonald, 2002, p. 337). Many comparable examples could be given here, as musicologist Francesca Brittan (2020) has recently and compellingly demonstrated. But whereas Scudo saw Liszt's musical charisma as praiseworthy, Friedrich Nietzsche, half a century later, attributed the balefully seductive appeal of Wagner's music entirely to its ability to "stimulat[e] tired nerves" (1911, p. 5), a shift in attitude reflecting the increasing currency after 1870 of the psychopathological diagnosis of neurasthenia and its impact on the understanding of music's effects on the nerves (Kennaway, 2012).

To summarize, the popular reception of eighteenth-century vibrating nerve theory seemingly depended to a large extent on two attributes that were closely associated with music. First and foremost was the well-known fact that music—regarded as consisting, physically, of nothing more than sonorous vibrations—can fundamentally alter our emotions. The idea that the mental changes brought about by music might somehow relate to material changes in the nerves in response to sound thus seemed highly intuitive.[2] The second attribute was the phenomenon of sympathetic resonance, which—since at least the Renaissance—was intimately linked with, and often paradigmatically demonstrated by, sound (Blanchard, 1941, p. 415; Savary, 1827). Vernacular understandings of vibrating nerve theory thus conflated music's powerful effects on the mind with an acoustic phenomenon familiar from the domain of music to account for a diverse range of cognitive and affective experiences.

Charcot's English contemporary Joseph Mortimer Granville—remembered today as the inventor of the *percuteur*, a mechanical vibrating device intended to treat muscle pain—provides a late-nineteenth-century example of the unexpectedly long afterlife of vibrating nerve theory and its musical entanglements. Discussing hysteria in the introduction of his *Nerve-Vibration and Excitation*, he proposed that the predisposition to hysteria was simply the nerves' overactive tendency to vibrate in sympathy with external or disruptive elements:

> If musical burners, supplied from different reservoirs of gas, will presently vibrate in concord; if strings or reeds vibrating at the same time, though a short distance apart, will fall into harmony, why is it unlikely that nervous organisms, possessing the same qualities of physical structure, should exhibit a corresponding affinity? (Granville, 1883, p. 27)

This passage, which could have been written a century earlier, if not for the reference to gas, invokes sympathetic vibration alongside musical instruments and terminology to account for the suggestibility of the nerves of patients suffering from hysterical afflictions. Granville goes on to remark that this condition was common chiefly in females, whose "organism is characterized not inaccurately, though popularly, by the phrase 'finely strung nerves,'" but also in men of a "feminine character" (1883, p. 28). The English physician, it should be emphasized, did not study hysteria—indeed, he averred that he had "never yet percussed a female patient"—but rather focused his research almost exclusively on the amelioration of pain (Granville, 1883, p. 57).[3]

Granville also offers a hypothesis about how the effects of vibration affect the nervous system:

> The first effect of nerve-vibration, therefore, is awakening or interrupting; the second is more like tuning a violin string, or the wire of a pianoforte. Nerve stretching acts in one of two ways for a time. If much force be used it disorganizes the nerve and prevents any vibration taking place in its elements, with the result, in successful cases, of giving it a new starting-point when the integrity of the nerve-fibre is restored in the process of natural repair. When less force is used, the nerve is acted upon precisely as screwing and stretching act on a violin string, altering its physical capacity for vibration, and either reducing or increasing the amplitude of the waves of movement into which agitation will throw it. . . . The manner in which vibration acts . . . is, I believe, explained by the law of musical concords and discords or harmonies. (1883, pp. 52–53)

Granville was renowned for his experiments with vibrating medical devices starting in the late 1870s, and his ideas were well known at the Salpêtrière; indeed, Charcot's student Gilles de la Tourette (1878) explicitly modeled a vibrating helmet, designed to treat neuralgia, on a similar invention by the Englishman. In parallel with his investigations into the effects of vibrations on hysterics, Charcot employed vibratory devices to treat neurological conditions we would recognize today: the giant tuning fork set on a resonating bench was used to alleviate locomotor ataxia, and upon learning that patients with Parkinson's disease obtained temporary relief from symptoms after a bumpy train ride, he developed a vibrating chair to mimic these results (Goetz, 2009). The apparent failure of these late-nineteenth-century physicians to acknowledge a clear distinction between electrical and acoustical vibrations reflected the broader scientific landscape in which their investigations were situated.

Evidence from Charcot's clinic suggests that submerged conceptions of the neurophysiological potential of acoustic vibrations nonetheless continued to play a role in pioneering psychiatric and physiological research long after substantial progress had been made by scientists such as Johannes Peter Müller (1835), Emil Du Bois-Reymond (1848), and Eduard Hitzig and Gustav Fritsch (1870) with regard to the actual electrical mechanism of nervous transmission. Thus, theories about the nerves' special susceptibility to certain kinds of sounds, which had been widespread in the preceding century, continued to exert authority well after their baseline assumptions had been either fully discredited or substantially refined. Understanding "*La médecine vibratoire*," to use Charcot's term, thus requires us to recognize the broader social and scientific context in which it was embedded, one forged in part—probably unbeknownst to practitioners themselves—by concepts borrowed from music's acoustic and expressive properties. This unexpected coexistence of different regimes of knowledge is particularly evident in experiments involving psychic symptoms, such as Charcot's attempts to provoke catalepsy by the sound of a tuning fork, which implicitly departed from the assumption that the hysterical patient's nervous system would somehow be uniquely responsive to vibratory stimuli via some form of harmonic resonance (Charcot experimented with a range of stimuli, including bright electric lights and gongs).

The idea that sonic vibrations have a direct and powerful effect on the nerves via sympathy—an idea whose origins can be traced directly to the eighteenth century—is still powerfully attractive today. As a quick Google search makes abundantly clear, tuning forks, alongside better-known therapeutic mainstays such as gongs and Tibetan singing bowls, have experienced a surprisingly long afterlife as healing devices in the alternative medicine community (on Charcot's use of gongs in his Salpêtrière experiments, see Pesic, 2020). An article on tuning fork therapy describes the practice as entailing "the 'energetic activations' of specific parts of the body" (Masala & Merolle, 2017, p. 368). Though it would be easy to simply dismiss such ideas as quackery, the modern music therapy literature includes a number of accounts of the successful application of tuning forks in a therapeutic context, even if only to achieve a placebo effect, suggesting that this form of treatment can be productively employed by patients and healers who accept the notion that certain frequencies can induce material changes in the body (Amir, 1993; Crowe & Scovel, 1996; Alio-Warr, 2012).

From an entirely different perspective, a twenty-first-century version of Charcot's attempt to treat Parkinson's disease with vibrations can be found in the technique of deep brain stimulation, in which a surgically implanted electrode transmits electrical impulses directly into the patient's basal ganglia (Lozano et al., 2019). The past few decades have also seen the emergence of new noninvasive techniques such as transcranial magnetic stimulation, which delivers an electrical current to the scalp and skull by means of electromagnetic induction (Hallett, 2007). The exact mechanism by which these interventions relieve symptoms is not entirely understood, but the devices provide periodic stimulation—that is, regular electric vibrations—and have been used by researchers and therapists for a variety of purposes, from interfering with or facilitating perception to the treatment of major depressive disorder.

Finally, within the contemporary neuroscientific community, research has sprung up around the idea that brain rhythms, or "distinct patterns of massed [electrical] neuronal activity associated with specific behaviors," exist endogenously and that these rhythms somehow align with both the external world and internal biological structures and mental activity (Frank, 2009, p. 37; see also Buzsáki, 2006). Some researchers have argued that if two connected cells, connected cell populations, or connected areas display similar neuronal oscillatory activity—measured by imaging technologies such as electroencephalography (EEG) or magnetoencephalography (MEG)—so that they are mathematically coherent, they may be understood as communicating, that is, as jointly processing information (Dikker et al., 2021). Similarly, the hypothesis that endogenous brain rhythms, that is, neural oscillations, are entrained by some external stimulus and that, consequently, oscillation-based entrainment has some kind of function for perception has been well demonstrated in the case of the rhythmic structures characteristic of music and speech (Doelling & Poeppel, 2015).

The striking parallels between these ideas and Hartley's observations on the electric and vibratory nature of brain function remain an intellectual curiosity, yet they invite us to speculate whether the historical preexistence of a paradigm of vibratory influence on the nervous system made researchers more open to the possibility of a phenomenon such as interneural coherent or "sympathetic" communication. More broadly, an awareness of how earlier epistemes can continue to exert a hidden attraction for decades if not centuries invites us to consider the historical contingency of our current conceptions of the mind and body and to ponder how the intertwined histories of music, neuroscience, and neurology might indirectly continue to affect the assumptions we make about our brains and minds.[4]

Acknowledgments

This chapter draws in part on material that appeared in my dissertation "Reverberating Nerves: Physiology, Perception, and Early Romantic Auditory Cultures" (Yale University, 2015), as well as a blog post I wrote for *AMSNow* in 2015. I would like to thank David Poeppel and David E. Cohen for their help with this piece.

Notes

1. For example, Charles Ferdinand Pohl writes in a pamphlet about "the objections made against the instrument, as having a tendency to affect the nerves indeed, so much so as to cause it to be forbidden in several countries by the police (in the Museum at Salzburg it is still shown to the visitors as such)" (1862, p. 8).

2. The eighteenth and nineteenth centuries saw the emergence of several detailed physiologically based accounts of musical affect—notably the music-therapeutic writings of Louis Roger (1758), Peter Lichtenthal (1807), Peter Joseph Schneider (1835), and Hector Chomet (1874). On the use of music therapy, specifically opera, in early-nineteenth-century insane asylums, see Raz (2019).

3. The context of these lines is worth reading: "I should here explain that, with a view to eliminate possible sources of error in the study of these phenomena, I have never yet percussed a female patient, and have not founded any of my conclusions on the treatment of hysterical males. This is a matter of much moment in my judgment, and I am, therefore, careful to place the fact on record. I have avoided, and shall continue to avoid, the treatment of women by

percussion, simply because I do not want to be hoodwinked, and help to mislead others, by the vagaries of the hysterical state or the characteristic phenomena of mimetic disease" (Granville, 1883, p. 57).

4. Some of the historical ideologies behind such assumptions are explored by Jonathan De Souza (chapter 6) and Alexander Cowan (chapter 16) in this volume.

References

Alio-Warr, E. F. (2012). Alternative medicine approach to integrative music therapy. *FULTENA. Natural Therapies Latin American Foundation*, 1–21.

Amir, D. (1993). Moments of insight in the music therapy experience. *Music Therapy, 12*(1), 85–100.

Beattie, J. (1776). *Essays: On poetry and music, as they affect the mind*. W. Creech.

Berlioz, H. (2003). *Grand traité d'instrumentation et d'orchestration modernes* (P. Bloom, Ed.). Bärenreiter.

Blanchard, J. (1941). The history of electrical resonance. *Bell System Technical Journal, 20*(4), 415–433.

Bonnet, C. (1782). Essai analytique sur les facultés de l'âme. In *Collection complete des oeuvres de Charles Bonnet* 13. Samuel Fauche.

Bourneville, D.-M., & Régnard, P. (1876–1880). *Iconographie photographique de la Salpêtrière* 1. Bureau du Progrès Médical.

Brittan, F. (2020). The electrician, the magician and the nervous conductor. *Nineteenth-Century Music Review*. https://doi.org/10.1017/S1479409820000099.

Buzsáki, G. (2006). *Rhythms of the brain*. Oxford University Press.

Cannon, J. T., & Dostrovsky, S. (1981). *The evolution of dynamics: Vibration theory from 1687 to 1742*. Springer.

Charcot, J. M. (1879). Induced hysterical catalepsy and somnambulism: Report of a lecture delivered at the hospital of the Salpêtrière by Professor J. M. Charcot. *St. Louis Clinical Record: A Monthly Journal of Medicine and Surgery, 5*(1), 280.

Charcot, J. M. (1890). Catalepsie et somnambulisme hystériques provoqués. In *Oeuvres complètes de JM Charcot* (vol. 9, pp. 262–265). Lecrosnier et Babé.

Cheselden, W. (1740). *Anatomy of the human body*. William Bowyer.

Chisolm, H. (Ed.). (1876). *Encyclopedia Britannica: A dictionary of arts, sciences, literature and general information*, vol. 5 (9th ed.). Adam and Charles Black.

Chomet, H. (1874). *Effets et influence de la musique sur la santé et sur la maladie*. Germer-Baillière.

Coleridge, S. T. (2001). The eolian harp. In *The collected works of Samuel Taylor Coleridge*, vol. 75. Princeton University Press.

Crowe, B. J., & Scovel, M. (1996). An overview of sound healing practices: Implications for the profession of music therapy. *Music Therapy Perspectives, 14*(1), 21–29.

de Chambaud, M. (1765). Musique, effets de la. In R. Morrissey (Ed.), *Encyclopédie*, vol. 10. University of Chicago, ARTFL Encyclopédie Project. https://portail.atilf.fr/cgi-bin/getobject_?p.79:55./var/artfla/encyclopedie/textdata/IMAGE/.

de la Tourette, G. (1878). "Considérations sur la médecine vibratoire ses applications et sa technique. *Nouvelle iconographie de la Salpêtrière, 5*, 265–275.

Descartes, R. (1998). *Descartes: The world and other writings* (S. Gaukroger, Ed. & Trans.). Cambridge University Press.

Diderot, D. (1979). Conversation between D'Alembert and Diderot (1769). In J. Kemp (Ed. & Trans.), *Diderot, interpreter of nature: Selected writings* (p. 56). Hyperion Press.

Dikker, S., Michalareas, G., Oostrik, M., Serafimaki, A., Kahraman, H. M., Struiksma, M. E., & Poeppel, D. (2021). Crowdsourcing neuroscience: Inter-brain coupling during face-to-face interactions outside the laboratory. *NeuroImage, 227*, 117436.

Doelling, K. B., & Poeppel, D. (2015). Cortical entrainment to music and its modulation by expertise. *Proceedings of the National Academy of Sciences, 112*(45), E6233–E6242.

Du Bois-Reymond, E. H. (1848). *Untersuchungen über thierische Elektricität.* G. Reimer.

Frank, M. G. (2009). Brain rhythms. In M. D. Binder, N. Hirokawa, & U. Windhorst (Eds.), *Encyclopedia of neuroscience* (Vol. 3166). Springer. https://doi.org/10.1007/978-3-540-29678-2_727.

Goetz, C. G. (2009). Jean-Martin Charcot and his vibratory chair for Parkinson's disease. *Neurology, 73*(6), 475–478.

Granville, J. M. (1883). *Nerve-vibration and excitation.* Churchill.

Hallett, M. (2007). Transcranial magnetic stimulation: A primer. *Neuron, 55*(2), 187–199.

Hartley, D. (1843). *Observations on man, his frame, his duty, and his expectations.* Thomas Tegg & Son.

Herder, J. G. (1985). Von Nachahmung der lateinischen Elegien. In U. Gaier (Ed.), *Frühe Schriften 1764–1772* (p. 488). Deutsche Klassiker Verlag.

Hitzig, E., & Fritsch, G. (1870). Ueber die elektrische Erregbarkeit des Grosshirns. *Archiv für Anatomie, Physiologie und wissenschaftliche Medicin, 37*, 300–332.

Hustvedt, A. (2011). *Medical muses: Hysteria in nineteenth-century Paris.* W. W. Norton.

Kassler, J. C. (1984). Man—a musical instrument: Models of the brain and mental functioning before the computer. *History of Science, 22*(1), 59–92.

Kassler, J. C. (1995). *Inner music: Hobbes, Hooke, and North on internal character.* Athlone.

Kennaway, J. (2012). *Bad vibrations: The history of the idea of music as a cause of disease.* Ashgate.

Krüger, J. G. (1748). *Naturlehre: Welcher die Physiologie, oder Lehre von dem Leben und der Gesundheit der Menschen in sich fasset*. Carl Herrmann Hemmerde.

Kumar, D. R., Aslinia, F., Yale, S. H., & Mazza, J. J. (2011). Jean-Martin Charcot: The father of neurology. *Clinical Medicine and Research, 9*(1), 46–49. https://doi.org/10.3121/cmr.2009.883.

Lichtenthal, P. (1807). *Der musikalische Arzt, oder: Abhandlung von dem Einflusse der Musik auf den Körper*. Wappler und Beck.

Locke, J. (1812). *Complete works* 10. Otrige & Sons.

Lozano, A. M., Lipsman, N., Bergman, H., Brown, P., Chabardes, S., Chang, J. W., & Krauss, J. K. (2019). Deep brain stimulation: Current challenges and future directions. *Nature Reviews Neurology, 15*(3), 148–160.

MacDonald, H. (Ed. & Trans.). (2002). *Berlioz's orchestration treatise: A translation and commentary*. Cambridge University Press.

Masala, D., & Merolle, V. (2017). The tuning fork and the "soundtherapy." *Senses and Sciences, 4*(2), 365–370. https://doi.org/10.14616/sands-2017-2-365370.

Müller, J. P. (1835). *Handbuch der Physiologie des Menschen*. Hölscher.

Newton, I. (1718). *Opticks: Or, a treatise of the reflections, refractions, inflections, and colors of light*. W. & J. Innys.

Newton, I. (1819). *The mathematical principles of natural philosophy* (A. Motte, Trans.). Sherwood and Neely.

Nietzsche, F. (1911). *The case of Wagner, Nietzsche contra Wagner and selected aphorisms* (A. M. Ludovici, Trans.). T. N. Foulis.

Pesic, P. (2020). Composing the crisis: From Mesmer's harmonica to Charcot's tam-tam. *Nineteenth-Century Music Review*. https://doi.org/1017/s1479409820000087.

Pohl, C. F. (1862). *Cursory notices of the origin and history of the glass harmonica*. Petter & Galpin.

Raz, C. (2014). 'The expressive organ within us': Ether, ethereality, and early romantic ideas about music and the nerves. *19th-Century Music, 38*(2), 115–144.

Raz, C. (2019). Operatic fantasies in early nineteenth-century psychiatry. In D. J. Trippet & B. Walton (Eds.), *Nineteenth-century opera and the scientific imagination* (pp. 63–83). Cambridge University Press.

Raz, C., & Finger, S. (2018). Musical glasses, metal reeds, and broken hearts: Two cases of melancholia treated by new musical instruments. In P. Gouk, J. Prins, W. Thormaehlen, & J. Kennaway (Eds.), *The Routledge companion to music, mind and wellbeing: Historical and scientific perspectives* (pp. 77–92). Routledge.

Richardson, A. (2001). *British romanticism and the science of the mind*. Cambridge University Press.

Roger, L. (1758). *Tentamen de vi soni et musices in corpus humanum*. J. Garrigan.

Savary, F. (1827). Memoire sur l'aimantation. *Annales de Chimie et de Physique, 34*, 5–57.

Schneider, P. J. (1835). *System einer medizinischen Musik: Ein unentbehrliches Handbuch für Medizin-Beflissene, Vorsteher der Irren-Heilanstalten, praktische Aerzte und unmusikalische Lehrer verschiedener Disziplinen*. Carl Giorgi.

Scudo, P. (1841). Esquisses biographiques et musicales: M. Franz Liszt. *Revue de Paris, 38*, 31–33.

Sulzer, J. G. (1773). Erklärung eines psychologischen paradoxen Satzes. In *Vermischte philosophische Schriften* (pp. 99–121). Wiedemann.

Trower, S. (2012). *Senses of vibration: A history of the pleasure and pain of sound*. Continuum.

Wardhaugh, B. (2008). Formal causes and immediate causes: The analogy of the musical instrument in late seventeenth-century natural philosophy. In *Philosophies of technology: Francis Bacon and his contemporaries* (pp. 411–428) (C. Zittel, Ed.). Brill.

Willis, T. (1965). *The anatomy of the brain* (S. Pordage, Trans., W. Feindel, Ed.). McGill University Press.

II Beyond Music as a Window into the Mind

Volume Editors

Cognitive science aims to understand the workings of the mind. Following efforts at the beginning of the twentieth century to establish psychology as "a purely objective experimental branch of natural science" (Watson, 1913), a period dominated by behaviorism ensued, during which psychologists studied measurable behavior responses to external stimuli while "blackboxing" interior states. With the cognitive revolution in the 1950s onward, cognitive scientists reclaimed mental life as a valid object of study. According to the historian of science Hunter Crowther-Heyck (1999), the idea that the brain is a computer was instrumental to the cognitive revolution, enabling a scientific theory of the mind that linked cognition with behavior, mind with body, and knowledge with action through the construction of mental models informed by notions of feedback loops and hierarchical organization. Associated with this central framework was a focus on information processing and computational models that constituted the mind as an object of scientific study.

It is within this context that contemporary music cognition took shape. During the 1980s and 1990s, research focused on cognitive representations of the syntactic formations that constitute musical structure, adopting the framework of Chomskyan linguistics. Multiple models were constructed that aimed to explain how listeners derived syntactic representations from sequences of pitches over time (Lerdahl & Jackendoff, 1983; Narmour, 1990). Emotional, affiliative responses to music were understood to be choreographed by the way musical patterns played with this syntax, building expectations and then thwarting them for expressive effect.

Advances in neuroimaging during the early 2000s brought a new emphasis on mapping the brain regions that respond to various aspects of music, contributing to the notion that music can reveal how the mind works. This idea was well suited to the heart-to-heart, soul-to-soul, mind-to-mind listening paradigms of Western art music, which envisioned musical listening as a solitary, stationary, contemplative activity and musical creativity as emanating from a composer's soul or disembodied mind. As musicologists and ethnomusicologists drew attention to bodily dimensions of music making and experience, however, the narrowness of such a disembodied conception of music became increasingly apparent (Holsinger, 2001; Le Guin, 2006; Becker 2004). At the same time, cognitive scientists and philosophers have offered alternatives to the mind-body dualism on which the passive-listening paradigm and its scientific implementations relied (Varela, Thompson, & Rosch, 2016; Noë, 2004).

De Souza and Witek's contributions to this volume (chapters 6 and 7, respectively) center on the increased explanatory power of a turn toward embodied cognition, in which the body is understood not as an arbitrary vessel in which the mind-soul is housed but as deeply co-constitutive of the mind itself. Listening as an embodied cultural practice is a central interest among musicologists, ethnomusicologists, and sound researchers, and engagement between these fields has the potential to offer new framings for research questions and empirical methods. Kassabian's 2013 analysis of background music, for example, emphasizes that it is not generally heard with high levels of focused attention and produces a kind of distributed subjectivity that makes analysis at the level of the individual insufficient. Robinson's 2020 work on listening, positionality, and intersubjectivity implies that studies need to account for variables that go beyond which set of notes is present in the stimuli. Clarke's 2005 work on perceptual ecology demonstrates that musical environments, instruments, and practices scaffold and shape musical thinking in ways that make analysis purely at the unit of the brain less illuminative.

To further complicate the picture provided by neuroimaging studies that reveal the parts of the brain engaged in particular music-related tasks, studies of degeneracy and redundant neuronal systems suggest that the brain enables cognition in more than one way (Price & Friston, 2002); the brain can and often does find multiple ways to accomplish the same task. Similarly, other researchers argue that examining brain and behavioral responses to isolated stimuli may never provide sufficient meaningful insight into real-world cognition (Lindquist et al., 2012; Siegel et al., 2018), a challenge that Faber and McIntosh (chapter 12) take up with the tools of complex systems research, and that Williams and Sachs (chapter 11) address by using more naturalistic stimuli and experimental contexts.

Cognitive processes do not operate in isolation from the surrounding world. There is an iterative relationship between environments and cognitive processes, both at the timescale of evolution (as discussed in Tomlinson's chapter 2) and at the level of the individual (as described in De Souza's chapter 6). How can research frameworks in music cognition take this into account? Drawing from his work on the social contexts of Sri Lankan drumming, Sykes's chapter 9 suggests that reconceptualizing the science of music as a "science of sound-as-relations will generate a perspective on the universal that presumes its radical diversity." Kragness, Hannon, and Cirelli's chapter 8 explores the way infant-environment interactions shape the development of musical behaviors. It is not simply a case of the mind's independent constraints meeting the circumstances of an environment, reactions to which reveal the structure of those constraints. Rather, interactions between the baby and the environment influence musical development in ways that are not reducible to either the mind or the environment considered as independent factors.

Taken together, these chapters suggest that empirical work on music should start from a broader understanding of what the mind encompasses and pursue a wider set of goals. In particular, the field might consider how research in music cognition can connect to artistic practice and to goals related to health and well-being. There's exciting potential here for another conception of science that complements the modern paradigm of experimentation as hypothesis testing, with experiments oriented toward interaction and emergence, where the ability to produce results may exceed the ability to explain them (Pickering, 2010, 2016). Other chapters explore some of these possibilities. Miranda (chapter 10) focuses on the way technologies developed for neuroscience can be harnessed to generate new compositional worlds. Leslie (chapter 13) examines how thinking about artistic creation as a kind of experiment in cognition can yield new insights and practices.

Throughout this section and this volume runs a thread about the importance of the historical perspective to scientific practices. Sean Silver points out the iterative nature of discovery in cognitive science, where

> a set of tools for thinking offers a model for thinking, which suggests in turn new refinements to the tools, which prompt new twists to the model, and so on. . . . The standard cognitive model is about the present moment, interested in how we think *right now*. . . . Historicity is, however, built into cognition understood in its ecological sense; elements of any thinking ecology evolve with reference to other elements in the ecology . . . the relationships between elements of the system (between us and our objects, objects and their caretakers) bear historical baggage. (2015, pp. 272, 273)

When scientists engage deeply with this history, in sustained communication with musicologists and historians, it can result in work that revolutionizes both fields, as vividly demonstrated by Raz's exploration in chapter 5 of how notions of sonic vibration shaped early theories of neural transmission and the subsequent development of modern neuroscience.

References

Becker, J. (2004). *Deep listeners: Music, emotion, and trancing*. Indiana University Press.

Clarke, E. F. (2005). *Ways of listening: An ecological approach to the perception of musical meaning*. Oxford University Press.

Crowther-Heyck, H. (1999). George A. Miller, language, and the computer metaphor of mind. *History of Psychology, 2*(1), 37.

Holsinger, B. W. (2001). *Music, body, and desire in medieval culture: Hildegard of Bingen to Chaucer*. Stanford University Press.

Kassabian, A. (2013). *Ubiquitous listening: Affect, attention, and distributed subjectivity*. University of California Press.

Le Guin, E. (2006). *Boccherini's body: An essay in carnal musicology*. University of California Press.

Lerdahl, F., & Jackendoff, R. S. (1983). *A generative theory of tonal music*. MIT Press.

Lindquist, K. A., Wager, T. D., Kober, H., Bliss-Moreau, E., & Barrett, L. F. (2012). The brain basis of emotion: A meta-analytic review. *Behavioral and Brain Sciences, 35*(3), 121–143.

Narmour, E. (1990). *The analysis and cognition of basic melodic structures: The implication-realization model*. University of Chicago Press.

Noë, A. (2004). *Action in perception*. MIT Press.

Pickering, A. (2010). *The cybernetic brain: Sketches of another future*. University of Chicago Press.

Pickering, A. (2016). Art, science and experiment. *MaHKUscript: Journal of Fine Art Research, 1*(1), 1–6.

Price, C. J., & Friston, K. J. (2002). Degeneracy and cognitive anatomy. *Trends in Cognitive Science, 6*(10), 416–421.

Robinson, D. (2020). *Hungry listening: Resonant theory for indigenous sound studies*. University of Minnesota Press.

Siegel, E. H., Sands, M. K., Van den Noortgate, W., Condon, P., Chang, Y., Dy, J., Quigley, K.S., & Barrett, L. F. (2018). Emotion fingerprints or emotion populations? A meta-analytic investigation of autonomic features of emotion categories. *Psychological Bulletin, 144*(4), 343–393.

Silver, S. (2015). *The mind is a collection: Case studies in eighteenth-century thought*. University of Pennsylvania Press.

Varela, F. J., Thompson, E., & Rosch, E. (2016). *The embodied mind: Cognitive science and human experience*. MIT Press.

Watson, J. B. (1913). Psychology as the behaviorist views it. *Psychological Review, 20*(2), 158.

6 Music, Mind, Body, and World

Jonathan De Souza

Sirens Behave so Strangely

What is the difference between song and speech? Between musical and nonmusical sound? These differences are commonly explained in terms of sonic features: for example, singing has more sustained pitches than speaking, more temporal regularity, and more exact repetition. In this view, a sound's musicality depends on its inherent, objective properties. But certain sounds can blur the boundary between these categories or even move from one to the other.

In 1857, Hermann von Helmholtz gave a public lecture in Bonn titled "On the Physiological Causes of Harmony in Music." It examined "musical sounds and sensations" and introduced recent research in "physical and physiological acoustics" (Helmholtz, 1995, pp. 46–47). Such lectures were a kind of popular science. Helmholtz was teaching general audiences about specific findings but also about science itself—how it was done and what it could do (Cahan, 1993; Steege, 2012). So, as he discussed sound waves and the structure of the ear, Helmholtz also described devices for producing and studying tone, including tuning forks, resonators, and the mechanical siren (for a discussion of tuning forks and nineteenth-century psychology, see Raz's chapter 5 in this volume). The siren blows air through holes punched in a rotating disc. At a slow speed, the resulting puffs of air create a rhythm; at higher speeds, they coalesce into a tone. This demonstrates a surprising continuity between rhythm and pitch (Rehding, 2016). Yet for Helmholtz, it also revealed a paradox:

> When the siren is turned slowly, and hence the puffs of air succeed each other slowly, you hear no musical sound. By the continually increasing rapidity of its revolution, no essential change is produced in the kind of vibration of the air. Nothing new happens externally to the ear. The only new result is the sensation experienced by the ear. . . . If you admire paradoxes, you may say that aerial vibrations do not become sound until they fall upon a hearing ear. (1995, p. 52)

Musical sound, then, does not simply exist in the physical world; it also involves human physiology and psychology.

The siren's tone recalls other pitch phenomena that are partially constituted by a listener's mind. In the speech-to-song illusion, a spoken phrase—most famously, "sometimes behave so strangely"—starts to sound like singing when it is looped (Deutsch, 2019, p. 152). This complicates the objective distinction between speech and song because the shift involves no external sonic change. Other examples include the missing fundamental effect, in which listeners sense a low tone that is not physically present, and Shepard tones, which can create an impossible never-ending ascent or descent because they have a clear pitch class but an indeterminate register. In each case, music emerges in an interaction between sound and mind.

This relation suggests two complementary possibilities. First, if the mind shapes music, then psychology might elucidate questions about musical organization. Insights about perception and cognition might guide new theories of harmony, as Helmholtz supposed, or of other musical elements. Conversely, analyses of musical organization by music theorists, musicologists, and ethnomusicologists might offer distinctive insights about the human mind.

Yet the association between music and mind can sometimes be problematic. In the popular imagination, music is often thought to improve intelligence (despite the debunking of the so-called Mozart effect), and the "genius" composers of Western art music such as Johann Sebastian Bach are considered exemplary minds. For example, one online article states that Bach had an IQ of 165 (Rizzi, 2018). This kind of claim is not new. Almost a hundred years ago, Catharine Morris Cox (1926, pp. 309–310) estimated Bach's IQ at 140, based on biographies by Forkel and Spitta. Even earlier, the nineteenth-century eugenicist Francis Galton (1892)—who designed the first standardized intelligence tests—used Bach and other musicians to support racist theories of "hereditary genius" (for a discussion of music and eugenics, see Cowan's chapter 16 in this volume). The 2018 article does not invoke biography or genetics, though. It seeks evidence of Bach's intelligence in musical patterns from his compositions, particularly numerological devices and the use of the B-A-C-H motif, which translates the letters of his surname into pitches. This implicitly compares Bach's contrapuntal achievements to pattern-based psychometric tests used to measure IQ. It reveals common assumptions about music and mind, approaching both in terms of abstract reasoning.

This chapter pursues a richer, more critical view. The human mind, as revealed in music, is not simply an innate or inherited capacity for abstract reasoning. Rather, it is multifaceted; it is embodied and situated in a world alongside objects and others. Like the tone of Helmholtz's siren, the mind emerges at the boundaries between different levels, shaped by physiology but also experience, culture, and technology. Like music, the mind is inherently relational.

Although this chapter could examine various musical elements (e.g., rhythm, meter, timbre), it focuses on pitch. And although it mentions styles from rock to Balinese gamelan, most of the illustrations are drawn from Bach. This facilitates a discussion of previous literature, where arguably Bach's music is overrepresented. More importantly, while it is imperative to decenter Western art music via cross-cultural research (Jacoby et al., 2020), it is also important to demonstrate that common assumptions are inadequate for this canonical example, which has long been used to support them. Bodies, technologies, or social relations are not only relevant to popular and non-Western musics, for example, and Bach is not some kind of transcendent "disembodied spirit" (Forkel, quoted in De Souza, 2017, p. 109). Even here, music is intertwined with mind, body, and world—and the chapter takes up each of these themes in turn.

Music, Computation, and Language

Cognitive science emerged as an interdisciplinary project in the mid-twentieth century. Inspired by recent technological advances, scholars started to imagine the mind as a computational system (McCulloch & Pitts, 1943). In this view, cognition is a kind of symbolic processing, with mathematics, logic, and language as paradigmatic examples. Linguistic syntax, for instance, was explained through systematic rules, tree-like hierarchies, and deep hidden structures (Chomsky, 1957). How could music fit into this model of the mind?

One computational approach to music cognition was developed by the composer-theorist Fred Lerdahl and the linguist Ray Jackendoff. Their *Generative Theory of Tonal Music* built on Chomskyan linguistics, using preference rules, hierarchies, and abstract deep structure to account for musical features such as harmony and meter (Lerdahl & Jackendoff, 1983). They argue that the brain's "music module, constructing the structure of the music in real time, unconsciously computes its moment-to-moment tensions and attractions" (Jackendoff & Lerdahl, 2006, p. 57). For example, Lerdahl's theory of tonal pitch space offers an algebraic model that quantifies the distance between any two pitches (Lerdahl, 2001). Such distances are hypothetically calculated in a listener's mind. Music, then, takes its place alongside language, revealing the computational mind at work (Jackendoff, 1987).

Figure 6.1 illustrates their approach in Bach's setting of the Lutheran chorale "Christus, der ist mein Leben" (Lerdahl & Krumhansl, 2007). The opening phrase starts from and returns to its tonic F-major chord. Each chord's relation to this tonic is indicated

Figure 6.1

Analysis of Bach's "Christus, der ist mein Leben," BWV 281 (Lerdahl & Krumhansl, 2007, fig. 25). "Diss." represents surface dissonance values, "Thier" hierarchical tension values, and "Attr." attraction values.

by a roman-numeral label. But this standard harmonic analysis is supplemented by numeric data for surface dissonance, hierarchical tension, and attraction, generated by the tonal-pitch-space model. A tree diagram above the notation represents each event as a branch. When a branch attaches at a higher position, the corresponding event is more stable and appears at a more abstract level; when it attaches at a lower position, the event is less stable and more ornamental. Event 2 (the tonic on the initial downbeat) anchors the entire phrase, whereas the tense passing motion of event 7 carries little weight. This tree can also be read from left to right: when a longer branch is followed by a shorter one, tonal tension increases; conversely, when a shorter branch is followed by a longer one, tension decreases. For example, tension decreases as event

1's pickup leads to event 2, or at a larger level when the dominant of event 6 resolves to the tonic of event 10. Overall, this analysis connects adjacent and nonadjacent events, suggesting a multilayered process of tonal departure and return.

Empirical studies have collected listeners' ratings of musical tension via stop-tension tasks (where the music stops at an event and participants rate its degree of tension) and continuous-tension tasks (rating the music as it unfolds). With both, ratings from musically trained participants correlated significantly with the model's predictions (Lerdahl & Krumhansl, 2007; see also Bigand et al., 1996; Krumhansl, 1996; Smith & Cuddy, 2003). This suggests that its quantification is consistent with listeners' cognition of tonal syntax. Yet, while responses validate the computational model, they are also used to *refine* it. Its parameters can be adjusted to improve the fit with the empirical data and to compare hypotheses or theoretical interpretations (e.g., different analytical methods provide different metrics for distance in chromatic music). So, the computational model is not static. It proceeds in dialogue with experimental research.

This approach draws on a hierarchical theory of mind but also on an earlier hierarchical theory of music. Heinrich Schenker (1935) developed similar methods of pitch reduction along with a distinctive analytical use of music notation. Broadly speaking, Lerdahl's tree diagram translates a Schenkerian analysis into a Chomskyan form. Schenker himself believed that music needs rules to be comprehensible, and he subscribed to a Kantian epistemology in which the listening mind represents discrete events and synthesizes them into an experiential whole (Korsyn, 1988). For Schenker, tonal hierarchy was natural, produced by the structure of the overtone series. He advocated supremacy of the Germanic canon, and his antidemocratic worldview linked musical, social, and racial hierarchies (Cook, 2007; Ewell, 2020).

By comparison, Lerdahl suggests that music should reflect the "nature" of the mind and that "the best music utilizes the full potential of our cognitive resources" (1992, p. 118; see also Cook, 1999, p. 241). While he considers much Western art music, jazz, Indian classical music, and Japanese koto music to meet these criteria, he dismisses other traditions: "Balinese gamelan falls short with respect to its primitive pitch space. Rock music fails on grounds of insufficient complexity" (1992, p. 119). Much as an IQ test ranks people according to a culturally specific definition of intelligence, this approach ranks musical genres according to culturally specific aesthetic values. Note that these exclusionary claims are offered without evidence. On one level, they are brought into question by scholarship on gamelan (e.g., Perlman, 2004; Tenzer, 2000) and rock (e.g., Temperley, 2018). Yet on another level, these claims are not properly testable. No computational model, no empirical results can prove that certain music or musicians or minds are best.

Naturalizing a particular style should also be avoided, as enculturation is fundamental to music cognition. The participants who rated tension in "Christus, der ist mein Leben" had formal training in Western music (Lerdahl & Krumhansl, 2007). A group with substantially different musical experience would provide different ratings. In an experiment investigating melodic expectations in a traditional Balinese melody, predictions by American musicians were better than chance, but they were substantially less accurate and less confident than those of Balinese musicians (Huron, 2006, pp. 53–55). In another study with Balinese musical stimuli, Westerners' responses reflected the statistical frequency of the melody's tones, but Balinese listeners were also sensitive to tonal relations in the slendro scale (Kessler et al., 1984). Moreover, cross-cultural research with the Tsimane', an indigeneous Amazonian people, shows that a preference for consonance and the sense of octave equivalence are not universal but rely on experience with particular musical systems (Jacoby et al., 2019; McDermott et al., 2016). So, although humans share certain perceptual and cognitive resources, these are inflected by cultural learning.

Like a computational model that is updated to better fit experimental data, listeners' musical expectations update to better fit the music around them. Through experience, they tune in to probabilities that define a musical style (Byros, 2012; Meyer, 1956). Tonal perception reflects the distribution of pitches in a key (Huron, 2006; Krumhansl, 1990), and even listeners without absolute pitch are sensitive to differences between keys (Eitan et al., 2017; see also Quinn & White, 2017). This does not require formal training and seems driven by statistical learning, a general way people pick up on regularities in complex stimuli. Statistical learning is relevant to many domains, including language, motor learning, and music (Perruchet & Pacton, 2006; Rohrmeier & Rebuschat, 2012). In music, listeners respond to the statistical regularities in melodic pitch structure in Western tonal music (Endress, 2010; Morgan et al., 2019; Pearce et al., 2010), North Indian rāgas (Rohrmeier & Widdess, 2017), and systems based on artificial musical scales (Loui, 2012; Loui & Wessel, 2008; Loui et al., 2010). Listeners also learn about probabilities in chord sequences (Huron, 2006; Jonaitis & Saffran, 2009; Loui et al., 2009). For example, at the end of every phrase in "Christus, der ist mein Leben," the dominant chord (V) resolves to the tonic (I). The sense of closure here relies in part on the ubiquity and predictability of that cadential formula. In the chorale's third phrase, Bach instead takes the dominant to the submediant (vi). In his chorale harmonizations, the V–vi progression is approximately fifty times less likely than V–I (Huron, 2006, p. 226; see also White & Quinn, 2018, p. 326), and it sounds more open-ended. Expectations for unfamiliar pieces, then, are shaped by prior experience.

This can be modeled via Bayesian methods, an approach to probability that balances current observations and prior beliefs (Temperley, 2007), and predictive processing theory, which offers a neurocomputational perspective on the interplay of top-down prediction and bottom-up sensory input (Clark, 2016; see also De Souza, 2020, sec. 3; Witek's chapter 7 in this volume). This theory distinguishes between uncertainty (no clear expectation) and surprise (expectation is not confirmed), and empirical research suggests that chord progressions are most pleasurable when they evoke low uncertainty with high surprise or high uncertainty with low surprise (Cheung et al., 2019). The dominant in the third phrase of Bach's chorale setting would engender low uncertainty, and the subsequent vi, relatively high surprise. According to predictive processing theory, this submediant should be especially salient for listeners, insofar as it is an unpredicted input that follows a relatively confident (but unsuccessful) prediction.

Each listener, however, can have different sets of expectations. Certain chords, such as the subtonic (♭VII), or harmonic progressions are rare in baroque music but common in rock. One computational study compared Bach's chorale settings and songs from *Billboard*'s "Hot 100" list, arguing that these repertoires have different tonal functions (White & Quinn, 2018). Accordingly, listeners prefer V–I over ♭VII–I in a classical context, but they find the latter acceptable in rock (Vuvan & Hughes, 2019). Similarly, studies of bimusicality have investigated listeners who are enculturated to more than one musical system (e.g., Western and Indian classical music; Wong et al., 2011). The ability to switch ears might seem to be attributable to contemporary technology and cultural exchange, to a world where listeners have access to diverse recordings and music from distant times and places. Yet arguably, Bach and his contemporaries already had multiple sets of expectations: the composer wrote several organ preludes based on modal chorale tunes, which can sound unresolved to listeners who are not familiar with seventeenth-century modal "church keys" (Barnett, 1998; Fitzpatrick, 2015). For example, "Nun komm, der Heiden Heiland" ends on the final of the A Dorian mode, which can easily be mistaken for the dominant of D minor (figure 6.2). So, tonal perception responds to statistical regularities but is also sensitive to context.

There is no single, universal harmonic syntax, then. Similarly, some cognitive linguists criticize Chomsky's universal grammar, highlighting the structural diversity of human languages. In this view, "language is a bio-cultural hybrid, a product of intensive gene-culture coevolution" (Evans & Levinson, 2009, p. 431; on gene-culture coevolution and music, see Patel's chapter 1 in this volume). Some argue, somewhat controversially, that syntactic recursion is not available in languages such as the Amazonian language Pirahã (Everett, 2005). This represents a different approach to language and mind.

Nonhierarchical linguistics have also been applied to music. Instead of focusing on discrete words (arranged according to meaning-neutral grammatical rules), construction

Figure 6.2
Bach's "Nun komm, der Heiden Heiland," BWV 599, mm. 9–10

grammar highlights extended linguistic units involving conventional pairing of syntax and semantics, form and function (Goldberg, 2006). Construction grammar–inspired approaches offer a different interpretation of the harmonic phenomena discussed earlier (Gjerdingen & Bourne, 2015; Zbikowski, 2017). In this view, listeners are not simply attuned to the distribution of individual notes or chords. Instead, they learn larger schemas, and a cadence is more meaningful than the parts that constitute it. Cadences involve processes in multiple syntactic layers (e.g., harmony, melody, rhythm), and this syntactic organization is tied to their meaning or function (providing a sense of closure). A listening mind, then, would not parse the music chord by chord, like a novice theory student working through a roman-numeral analysis. Instead, it might resemble a student at an eighteenth-century Italian conservatory, where pupils learned how to recognize, improvise, and compose with stock contrapuntal patterns (Gjerdingen, 2007). Bach's teaching was similar: it did not define a set of chords and rules for arranging them in functional progressions; instead, his students harmonized chorale melodies at the keyboard (Leaver & Remeš, 2018). Their training focused on constructions, on meaningful musical utterances.

While debates about musical grammar continue, it seems clear that music has often been considered cognitive insofar as it resembles language. Linguistic comparisons have shaped understanding of the musical mind, partially because language has long been considered the essence of thought. Certainly, language and music are both distinctively human cognitive phenomena, and they are processed in similar ways (Patel, 2003; Koelsch et al., 2005). And computational, statistical, or probabilistic approaches have much to offer interdisciplinary research on music cognition: they can help clarify links between musical organization and listeners' responses, especially when they account for individual and cultural learning. Yet these approaches often present music as text more than performance, as abstract idea more than embodied practice, excluding motion, dance, and other modes of physical engagement. To quote Suzanne Cusick (1994, p. 16), they treat music as a "*mind–mind* game," an art that is produced by one mind and received by other minds. But musicians and listeners are neither disembodied spirits nor sophisticated robots; they are living, breathing, feeling organisms who move and are moved, who feel pleasure and pain. And this, too, has implications for the musical mind.

Musical Minds, Musical Bodies

How are mind and body related? According to various forms of dualism, mind and body are more or less separate. This thinking is often attributed to the early modern philosopher René Descartes, though it has older roots. In a sixth-century treatise that was influential throughout medieval Europe, the late-Roman polymath Boethius argues that music theory is nobler than performance, just "as the mind is superior to the body" (1989, p. 50). Boethius, Descartes, or twentieth-century computationalists do not deny that humans have bodies. But many see the body as an auxiliary to cognition, not an active participant in it. In this view, the body provides sensory input to the mind and receives its commands, but the body itself does not think or speak or listen. As Descartes wrote, "it is the mind which senses, not the body" (2001, p. 87).

Does the body contribute to music cognition, then? Leonard Meyer considered this problem in the 1950s:

> On the one hand, it seems clear that almost all motor behavior is basically a product of mental activity rather than a kind of direct response made to the stimulus as such. . . . On the other hand, the facts indicate that somehow motor behavior does play an important part in facilitating and enforcing the musical aesthetic experience. (1956, p. 81)

Meyer argues that all motor activity corresponds to mental activity—but that mental activity, by contrast, is not reducible to motor activity. As such, he concludes that bodies do not require a separate analysis, despite their experiential significance, and "music is best examined in terms of mental behavior" (Meyer, 1956, p. 82). This conclusion has clear Cartesian implications. Meyer even quotes a line that paraphrases Descartes (from the French musicologist, pianist, and philosopher Gisèle Brelet): "Rhythm comes from the mind not the body" (quoted in Meyer, 1956, p. 81).

Yet consider a series of experiments by Jessica Phillips-Silver and Laurel Trainor (2005, 2007). Participants heard metrically ambiguous rhythmic patterns and bounced on every second or third beat. Later, they heard versions of the rhythmic pattern with accented strong beats in duple or triple meter. Infant participants chose to listen for a longer time to the accented version that matched their earlier movement, and adults matched the accented pattern to their bouncing pattern. Bodily movement was essential for this effect: it persisted when participants were blindfolded but not when they watched someone else bouncing. In this case, bodily movement influences rhythm perception, and listeners hear what the body feels. Moreover, beat perception involves brain regions that are implicated in motor activity, including the basal ganglia, premotor cortex, and supplementary motor area (Cannon & Patel, 2021; Grahn, 2009). Brain regions do have multiple functions (Anderson, 2014), and this does not mean that the experience of meter always requires bodily movement. Still, moving to music can affect metric perception, much as gesturing affects reasoning and problem solving (see Alibali et al., 2011; Goldin-Meadow & Beilock, 2010). So, despite Meyer's efforts, it is not easy to explain away bodily aspects of music perception and cognition.

The dualism that drives Meyer's dilemma is a theoretical commitment. It is not directly testable. Searching for a clean boundary between mind and body—or entirely rejecting the mind-body distinction—seems unproductive. Arguably, it is better to accept continuities between minds and bodies and ways in which they are intertwined without being identical (see De Souza, 2017, pp. 11–12). This approach can draw on established frameworks from phenomenology (Merleau-Ponty, 2012; Montague, 2011), embodied cognitive linguistics (Lakoff & Johnson, 1980; Johnson, 1987), or biology (Maturana & Varela, 1980). From the latter perspective, life itself is fundamentally cognitive: when a single-celled organism seeks nourishment or a plant turns its leaves toward sunlight, it is already engaged in a process of sense making; even though these organisms are not conscious, their behavior can be understood as a precursor to human cognition. This would expand the definition of mind and resolve Meyer's opposition by framing motor activity as a particular kind of cognitive activity.

Admittedly, this conceptual shift can be a source of controversy. Some proponents of embodied cognition reject tenets of standard cognitive science, such as computation and mental representation. However, these viewpoints can also be combined. Some cognitive scientists would accept the physical and neural foundations of computation (Shapiro, 2011, sec. 1.3) or reinterpret earlier findings in embodied terms, so that the benefits of computational research are not lost (Barsalou, 2010). Where standard cognitive science might view cognition as computation with abstract, amodal symbols, a more embodied paradigm would emphasize perceptual symbols grounded in the brain's systems for sensation, perception, and action (Barsalou, 1999, 2008). It can also open up new avenues for empirical research, including tasks with bodily movement or action, and data collection involving motion capture, physiological measures (e.g., heart rate), and so forth. And because the body has long been a topic of inquiry in musicology, phenomenology, anthropology, and related fields, cross-disciplinary research can investigate shared concerns: for example, certain phenomenological concepts related to embodiment have been operationalized and tested experimentally (Dotov et al., 2010; Maravita & Iriki, 2004). As such, an embodied viewpoint can facilitate collaboration between cognitive science and the humanities.

Rhythm offers clear examples of embodied music cognition, partially because physical movement and sound can share the same temporal patterning. Yet theories of embodied cognition also address aspects of pitch perception. In the West, pitch space is typically imagined in terms of vertical position. For example, the second melodic note in "Christus, der ist mein Leben" is "higher" than the preceding pickup, and in the chorale's first measure, Bach's bass line steadily "descends." This terminology matches visual aspects of the music notation, but obviously no literal, physical ascent or descent occurs here. These pitch changes instead involve variations in the sounds' frequency. Nonetheless, this understanding of pitch height affects perception. In representative experiments, audiovisual stimuli were presented to participants with and without formal musical training. When higher pitches were paired with higher visual stimuli, their responses were faster and more accurate (Lidji et al., 2007; Rusconi et al., 2006).

To explain such phenomena, music theorists have again taken inspiration from cognitive linguistics, suggesting that conceptualization of pitch height involves what Mark Johnson (1987) calls "image schemas" (see Brower, 1997; Cox, 2016, ch. 4; Johnson & Larson, 2003; Mead, 1999; Zbikowski, 2002). Image schemas are gestalts abstracted from sensorimotor perception (Rohrer, 2005), which can ground conceptual metaphors (Lakoff & Johnson, 1980). The verticality schema, derived from countless experiences of up-and-downness, supports conceptual metaphors ascribing locations to musical pitches, emotional states (in statements such as "Cheer up" or "I'm feeling down"), or even religious entities (e.g., God and Heaven above, Hell below). In music, this can support the sense of tonal "gravity" (Larson, 2012)—as in the end of "Christus, der ist mein Leben," where the melody steps down and comes to rest on the tonic. It can also support musical imagery—as in Bach's organ prelude "Durch Adams Fall," where dissonant descending leaps represent a spiritual fall from grace.

Once again, this is not universal but learned. Across cultures, pitch is mapped onto many different domains (Eitan & Timmers, 2010). For example, Javanese musicians describe pitch relationships in terms of tension and size: pitches are not low or high but loose or tight, big or small (Zanten, 1986, p. 85). Though they do not rely on verticality, these theoretical metaphors also relate music to sensorimotor experience. When Lerdahl and Krumhansl discuss tonal "tension," they draw on the same tension metaphor as Javanese theory: "Everyone experiences physical tension and relaxation," they note, "and it is common to extend the terms to mental and emotional terrains as well" (2007, p. 330). Image schema–based conceptual metaphors are common in various kinds of music theory (Saslaw, 1996; Zbikowski, 2002). Language for discussing and thinking about music builds on other areas, including bodily experience.

Listening to music does not involve only metaphorical movement. Foot tapping, dancing, head bobbing, and other movements often emerge spontaneously. They reflect embodied ways of engaging with the music (Kozak, 2015; Witek, 2017). When music evokes a desire to move, it is said to have high "groove" (Janata et al., 2012), and this pleasurable sensation involves motor and reward networks in the brain (Matthews et al., 2020). Although groove is related to moderate rhythmic complexity (particularly syncopation), it can also be affected by pitch. One study found that moderate harmonic complexity, combined with moderate rhythmic complexity, enhanced musical pleasure and the impulse to move (Matthews et al., 2019). Other experiments suggest registral differences: stronger bass parts contribute to groove (Stupacher et al., 2016), and when these low frequencies could be *felt* as well as heard, participants gave higher ratings of groove and aesthetic pleasure and displayed more spontaneous movement (Hove et al., 2020). This involves a multisensory experience of music, combining auditory and tactile stimulation.

That said, music's effects on the body are as varied as human bodies themselves. Drawing on disability studies, Joseph Straus critiqued embodied music theory for "the blithe assumption that we all inhabit the same kind of body, a normatively abled body, and thus all experience our bodies in pretty much the same way" (2006, p. 123). For example, tactile and visual aspects may be particularly central for musicians with impaired hearing, such as percussionist Evelyn Glennie or members of the deaf rock band Beethoven's Nightmare—although as Jessica Holmes (2017) discusses, deaf listening and deaf musicality are diverse and often misunderstood. Such research emphasizes that disability is cultural as well as biological, and it demonstrates the importance of individual and group differences.

Performance expertise points to another set of bodily differences in which action and perception are combined. For singers, tension is not simply metaphorical: vocal tension and muscular tension are palpable. Instrumentalists cultivate bodily skills that require an external tool, modes of music cognition that are impossible without technical mediation. For pianists, pitch space goes from left to right as well as from low to high, and this is reflected in their responses in the aforementioned experiments (Lidji et al., 2007; Rusconi et al., 2006). Their model of pitch space has been shaped by the technical interface of the keyboard. So, merely including the body in cognition does not go far enough. If embodied cognition involves learning, with cultural and individual differences, the mind-body is not fixed but dynamic; it is situated in and conditioned by the world. Our purview must expand to reveal the mind-body interacting with objects and others.

Music, Mind, and World

Every mind requires a body that supports it, and likewise, every body is situated in a world. This is central to ecological psychology, founded by James J. Gibson (1966, 1979a), which emphasizes that perception always involves an organism in an environment. And as Gibson notes, tools can complicate the boundary that separates them:

> When in use, a tool is a sort of extension of the hand, almost an attachment to it or a part of the user's own body, and thus is no longer a part of the environment of the user. . . . This capacity to attach something to the body suggests that the boundary between the animal and the environment is not fixed at the surface of the skin but can shift. (1979a, 41; see also Maravita & Iriki, 2004)

Tool use is fundamentally relational, involving an interaction between agent and object. The object has certain affordances (i.e., possibilities for action), but these are available only to an organism with appropriate abilities (Gibson, 1979b). For example, apples can be eaten by many animals and can be thrown by many humans (though not by horses), but apples can be juggled only by practiced performers. Similarly, musical instruments are playable in a general sense—after all, toddlers and professionals alike can press piano keys—but more affordances become available with training. And as new skills open up possibilities for action, they also change what is perceptible to the organism (Dreyfus, 2002).

As I argued in my book *Music at Hand*, learning to play an instrument affects players' perception, cognition, and imagination (De Souza, 2017). The instrument consistently converts action into sound. With practice, this establishes a two-way auditory-motor connection in a player's brain (Bangert et al., 2006; Chen et al., 2011; Drost et al., 2005; Lahav et al., 2005; Margulis et al., 2009). I describe this as a link between the hand and the ear. The auditory-motor coactivation can be stimulated in various ways: when instrumentalists listen to music they can play, there is corresponding activity in the primary motor cortex, even though they are not moving (Haueisen & Knösche, 2001); when they move their fingers to virtually play in the absence of the instrument, there is activation in the primary auditory cortex (Lotze et al., 2003). Just imagining the performance does not evoke this coactivation. Similarly, the auditory-motor coupling never emerges when beginners practice on a keyboard where the pitch-to-key mapping is random and changeable (Bangert & Altenmüller, 2003). This indicates that the instrument's stable affordances are essential for this distinctive form of multimodal integration.

Because this training responds to the instrumental interface, different kinds of instrumentalists tend to have particular ways of imagining and experiencing musical space. For example, keyboard instruments set up an opposition between the white notes' C-major scale and the sharp or flat black notes. This structure seems to influence many musicians with absolute pitch, who respond more quickly and accurately to white notes (Miyazaki, 1988, 1990). Pitch labeling is also more accurate with notes with piano timbre, relative to sine tones (Reis et al., 2021), and some musicians may have a form of instrument-specific absolute pitch (Reymore & Hansen, 2020; see also Hedger et al., 2013). Instrumental features can also ground musicians' theoretical models: for example, the big-small pitch mapping in Java, mentioned earlier, corresponds to the size of metal bars in tuned percussion instruments that are central to gamelan ensembles. In such cases, the instrument serves as a cognitive reference point.

Performance experience also varies with different instruments. The organ's multikeyboard interface—with manuals, pedals, and stops—offers a particular kind of embodied thinking. The instrument allows for textures where three independent lines are realized on three different keyboards—two for the hands and one for the feet. As the organist and musicologist David Yearsley writes, "Anyone who has played a trio at the organ . . . knows what thinking this way feels like in the body" (2012, p. 50). At the same time, anyone who has never played the organ does *not* know how this feels. The organist's knowledge is not only declarative but also procedural, a kind of know-how.

Instrumental interfaces can shape distinctive idioms. For example, pedal points (long, held bass notes) are facilitated by the organ's capacity for endless sustain and its powerful low register (De Souza, 2017, p. 40). Yet there are also aspects of performance that are felt by performers but not directly heard by listeners. Cusick illustrates this by describing a physically challenging moment in Bach's chorale prelude on "Aus tiefer Not" (figure 6.3):

> Neither foot can rest long enough to balance the body, neither hand can rest long enough to balance the body. For these few terrifying measures (terrifying in the organist's experience), one might as well be floating in mid-air, so confused and constantly shifting is the body's center of gravity. (1994, p. 18)

This physical imbalance corresponds to the absence of grace in the chorale's lyrics, and balance is regained at the arrow in the figure, when grace arrives in the text.

When Bach's students realized chorale tunes at the keyboard, they used body-instrument interaction to create and think about harmony. Seemingly abstract voice-leading rules can take on a physical character here: for example, moving the hands in contrary motion is a performance strategy that often avoids parallel fifths and octaves.

Figure 6.3

Bach's "Aus tiefer Not," BWV 686 (Cusick, 1994, fig. 2). The arrow indicates where the performer's body (hands playing the upper two staves, feet playing the bottom stave) returns to a state of balance.

In a sense, the students' knowledge of tonal syntax involved both the instrument and their hands (Bianco et al., 2016; Sammler et al., 2013). An experiment by Giacomo Novembre and Peter Keller (2011) illustrates this well. Pianists imitated silent videos that showed one hand playing chord progressions on a keyboard. Imitation was fastest when a chord fit the established harmonic context, which suggests that the videos supported both tonal and physical expectations. Additionally, imitation was less accurate when a harmonically conventional chord was performed with unconventional fingering. Overall, this suggests that for expert instrumentalists, musical syntax may involve an embodied "grammar of action" (Novembre & Keller, 2011).

Musical instruments, then, can complicate the boundary between mind-body and world, and theorists of cognitive extension would argue that such objects become part of the mind. This is not to say that a pipe organ can think; rather, it can serve as a functional component in a larger cognitive system (Clark, 2008). It is possible to harmonize a chorale melody, solve a math problem, or remember a set of directions in one's head—and these are undeniably cognitive. According to the extended mind hypothesis, these routines remain cognitive when they incorporate external objects such as a keyboard, calculator, or notebook. In any case, the instrument does make it possible to off-load certain aspects of the task. For example, the organ takes care of tuning, and it provides the energy that sustains the pitches. Such contributions are implicated in a comment on performance attributed to Bach by Johann Friedrich Köhler in 1776: "All one has to do is hit the right notes at the right time, and the instrument plays itself" (quoted in Marshall, 1999, p. 93).

Of course, body-instrument interaction occurs in a social context too. Bach's students watched and listened to him play, and imitation remains central to music lessons today. Neuroscientific studies have explored imitation in guitar learning, emphasizing how the brain's action observation network is engaged while watching a teacher and also when reproducing chords (Buccino et al., 2004; Gardner et al., 2017; see also Rizzolatti & Sinigaglia, 2008). This interpersonal coordination is also essential for ensemble performance, and it might be especially prominent in collective improvisation (as in the Balinese genres discussed by Tilley, 2019). For music theorist Arnie Cox, bodily imitation is important for audiences too. According to his mimetic hypothesis, listeners make sense of music by overtly or covertly imitating performers' actions and the sonic patterns they create (Cox, 2011, 2016). Interaction with others, then, would be central to music cognition.

Indeed, music supports social bonding across multiple human cultures (Savage et al., 2021). For example, choral singing increases participants' trust, cooperation, and feelings of closeness (Anshel & Kipper, 1988; Weinstein et al., 2016). It is possible that this kind of participatory music making affects neurochemistry, increasing oxytocin or reducing cortisol, although current results are inconclusive (Bullack et al., 2018; Chanda & Levitin, 2013; Keeler et al., 2015; Kreutz, 2014; Schladt et al., 2017). With religious ritual, dance, or collective work, this is one of music's central affordances: it facilitates the coordination of minds and bodies (Clarke, 2005; Kozak, 2019).

Emotion is also central to music's "social affordances" (Krueger, 2011a). Music can support emotional contagion, and it can be understood to represent a "super-expressive voice" (Juslin & Västfjäll, 2008). Empirical studies confirm parallels between acoustic cues for emotion in music and speech: both tend to sound "happier" when they are higher, faster, and louder (for a review, see Schutz, 2017). This points to the significance of register, an aspect of pitch that is often neglected by Western music theory focused on pitch class. Of course, musical simulations of emotional processes typically combine multiple syntactic layers, as Lawrence Zbikowski (2017, pp. 10–11, 79–91) demonstrates in an analysis of Bach's cantata *Ich habe genug*. If music functions as a kind of virtual other, then music cognition would overlap with aspects of social cognition (Wallmark et al., 2018). In music and other domains, the extended mind might include others as well as objects (Krueger, 2011b), with cognitive processes that integrate multiple minds and bodies.

Ultimately, the mind, as revealed in music, involves multiple levels—from mental computation to embodied action to cognitive extension. These interrelated levels can be studied on their own or in various blends. Each one mixes nature and culture, so each benefits from a combination of scientific and humanistic methods. Multiple disciplinary perspectives are needed, then, to understand musical minds, bodies, and worlds.

References

Alibali, M. W., Spencer, R. C., Knox, L., & Kita, S. (2011). Spontaneous gestures influence strategy choices in problem solving. *Psychological Science, 22,* 1138–1144.

Anderson, M. L. (2014). *After phrenology: Neural reuse and the interactive brain.* MIT Press.

Anshel, A., & Kipper, D. A. (1988). The influence of group singing on trust and cooperation. *Journal of Music Therapy, 25*(3), 145–155.

Bangert, M., & Altenmüller, E. O. (2003). Mapping perception to action in piano practice: A longitudinal DC-EEG study. *BMC Neuroscience, 4,* 26.

Bangert, M., Betzwieser, T., Schlaug, G., Rotte, M., Drescher, D., Hinrichs, H., Heinze, H.-J., & Altenmüller, E. (2006). Shared networks for auditory and motor processing in professional pianists: Evidence from fMRI conjunction. *Neuroimage, 30,* 917–926.

Barnett, G. (1998). Modal theory, church keys, and the sonata at the end of the seventeenth century. *Journal of the American Musicological Society, 51,* 245–281.

Barsalou, L. W. (1999). Perceptual symbol systems. *Behavioral and Brain Sciences, 22,* 577–660.

Barsalou, L. W. (2008). Grounded cognition. *Annual Review of Psychology, 59,* 617–645.

Barsalou, L. W. (2010). Grounded cognition: Past, present, and future. *Topics in Cognitive Science, 2,* 716–724.

Bianco, R., Novembre, G., Keller, P. E., Scharf, F., Friederici, A. D., Villringer, A., & Sammler, D. (2016). Syntax in action has priority over movement selection in piano playing: An ERP study. *Journal of Cognitive Neuroscience, 28*(1), 41–54.

Bigand, E., Parncutt, R., & Lerdahl, F. (1996). Perception of musical tension in short chord sequences: The influence of harmonic function, sensory dissonance, horizontal motion, and musical training. *Perception & Psychophysics, 58*(1), 125–141.

Boethius, A. M. S. (1989). *Fundamentals of music* (C. V. Palisca, Ed.; C. M. Bower, Trans.). Yale University Press.

Brower, C. (1997). Pathway, blockage, and containment in "Density 21.5." *Theory and Practice, 22/23,* 35–54.

Buccino, G., Vogt, S., Ritzl, A., Fink, G. R., Zilles, K., Freund, H.-J., & Rizzolatti, G. (2004). Neural circuits underlying imitation learning of hand actions: An event-related fMRI study. *Neuron, 42*(2), 323–334.

Bullack, A., Gass, C., Nater, U. M., & Kreutz, G. (2018). Psychobiological effects of choral singing on affective state, social connectedness, and stress: Influences of singing activity and time course. *Frontiers in Behavioral Neuroscience, 12*. https://www.frontiersin.org/articles/10.3389/fnbeh.2018.00223/full.

Byros, V. (2012). Meyer's "Anvil": Revisiting the schema concept. *Music Analysis, 31*, 273–346.

Cahan, D. (1993). Helmholtz and the civilizing power of science. In D. Cahan (Ed.), *Hermann von Helmholtz and the Foundations of Nineteenth-Century Science* (pp. 559–601). University of California Press.

Cannon, J. J., & Patel, A. D. (2021). How beat perception co-opts motor neurophysiology. *Trends in Cognitive Sciences, 25*(2), 137–150.

Chanda, M. L., & Levitin, D. J. (2013). The neurochemistry of music. *Trends in Cognitive Sciences, 17*(4), 179–193.

Chen, J. L., Rae, C., & Watkins, K. E. (2011). Learning to play a melody: An fMRI study examining the formation of auditory-motor associations. *Neuroimage, 59*(2), 1200–1208.

Cheung, V. K. M., Harrison, P. M. C., Meyer, L., Pearce, M. T., Haynes, J.-D., & Koelsch, S. (2019). Uncertainty and surprise jointly predict musical pleasure and amygdala, hippocampus, and auditory cortex activity. *Current Biology, 29*, 4084–4092.

Chomsky, N. (1957). *Syntactic structures*. Mouton.

Clark, A. (2008). *Supersizing the mind: Embodiment, action, and cognitive extension*. Oxford University Press.

Clark, A. (2016). *Surfing uncertainty: Prediction, action, and the embodied mind*. Oxford University Press.

Clarke, E. F. (2005). *Ways of listening: An ecological approach to the perception of musical meaning*. Oxford University Press.

Cook, N. (1999). Analysing performance and performing analysis. In N. Cook & M. Everist (Eds.), *Rethinking Music* (pp. 239–261). Oxford University Press.

Cook, N. (2007). *The Schenker project: Culture, race, and music theory in fin-de-siècle Vienna*. Oxford University Press.

Cox, A. (2011). Embodying music: Principles of the mimetic hypothesis. *Music Theory Online, 17*.

Cox, A. (2016). *Music and embodied cognition: Listening, moving, feeling, and thinking*. Indiana University Press.

Cox, C. M. (1926). *Genetic studies of genius II: The early mental traits of three hundred geniuses*. Stanford University Press.

Cusick, S. G. (1994). Feminist theory, music theory, and the mind/body problem. *Perspectives of New Music, 32*, 8–27.

Descartes, R. (2001). *Discourse on method, optics, geometry, and meteorology* (P. J. Olscamp, Trans.). Hackett.

De Souza, J. (2017). *Music at hand: Instruments, bodies, and cognition*. Oxford University Press.

De Souza, J. (2020). Instrumental transformations in Heinrich Biber's Mystery Sonatas. *Music Theory Online, 26*(4).

Deutsch, D. (2019). *Musical illusions and phantom words: How music and speech unlock mysteries of the brain*. Oxford University Press.

Dotov, D. G., Nie, L., & Chemero, A. (2010). A demonstration of the transition from ready-to-hand to unready-to-hand. *PLOS ONE, 5*(3), e9433.

Dreyfus, H. L. (2002). Intelligence without representation—Merleau-Ponty's critique of mental representation. *Phenomenology and the Cognitive Sciences, 1*, 367–383.

Drost, U. C., Rieger, M., Brass, M., Gunter, T. C., & Prinz, W. (2005). When hearing turns into playing: Movement induction by auditory stimuli in pianists. *Quarterly Journal of Experimental Psychology: Section A, 58*(8), 1376–1389.

Eitan, Z., Ben-Haim, M. S., & Margulis, E. H. (2017). Implicit absolute pitch representation affects basic tonal perception. *Music Perception, 34*(5), 569–584.

Eitan, Z., & Timmers, R. (2010). Beethoven's last piano sonata and those who follow crocodiles: Cross-domain mappings of auditory pitch in a musical context. *Cognition, 114*, 405–422.

Endress, A. D. (2010). Learning melodies from non-adjacent tones. *Acta Psychologica, 135*(2), 182–190.

Evans, N., & Levinson, S. C. (2009). The myth of language universals: Language diversity and its importance for cognitive science. *Behavioral and Brain Sciences, 32*(5), 429–448.

Everett, D. L. (2005). Cultural constraints on grammar and cognition in Pirahã: Another look at the design features of human language. *Current Anthropology, 46*(4), 621–646.

Ewell, P. A. (2020). Music theory and the white racial frame. *Music Theory Online, 26*(2).

Fitzpatrick, M. (2015). *J. S. Bach's modal compositional practice in the chorale preludes for solo organ: A Schenkerian perspective* [Unpublished doctoral dissertation]. University of Western Ontario.

Galton, F. (1892). *Hereditary genius: An inquiry into its laws and consequences*. Macmillan.

Gardner, T., Aglinskas, A., & Cross, E. S. (2017). Using guitar learning to probe the Action Observation Network's response to visuomotor familiarity. *NeuroImage, 156*, 174–189.

Gibson, J. J. (1966). *The senses considered as perceptual systems*. Houghton Mifflin.

Gibson, J. J. (1979a). *The ecological approach to visual perception*. Houghton Mifflin.

Gibson, J. J. (1979b). The theory of affordances. In R. Shaw & J. Bransford (Eds.), *Perceiving, acting, and knowing: Toward an ecological psychology* (pp. 67–82). Erlbaum.

Gjerdingen, R. O. (2007). Partimento, que me veux-tu? *Journal of Music Theory, 51*, 85–135.

Gjerdingen, R. O., & Bourne, J. (2015). Schema theory as a construction grammar. *Music Theory Online, 21*.

Goldberg, A. E. (2006). *Constructions at work: The nature of generalization in language*. Oxford University Press.

Goldin-Meadow, S., & Beilock, S. L. (2010). Action's influence on thought: The case of gesture. *Perspectives on Psychological Science, 5*, 664–674.

Grahn, J. A. (2009). The role of the basal ganglia in beat perception: Neuroimaging and neuropsychological investigations. *Annals of the New York Academy of Sciences, 1169*, 35–45.

Haueisen, J., & Knösche, T. R. (2001). Involuntary motor activity in pianists evoked by music perception. *Journal of Cognitive Neuroscience, 13*, 786–792.

Hedger, S. C., Heald, S. L. M., & Nusbaum, H. C. (2013). Absolute pitch may not be so absolute. *Psychological Science, 24*(8), 1496–1502.

Helmholtz, H. von. (1995). *Science and culture: Popular and philosophical essays* (D. Cahan, Ed.). University of Chicago Press.

Holmes, J. A. (2017). Expert listening beyond the limits of hearing: Music and deafness. *Journal of the American Musicological Society, 70*, 171–220.

Hove, M. J., Martinez, S. A., & Stupacher, J. (2020). Feel the bass: Music presented to tactile and auditory modalities increases aesthetic appreciation and body movement. *Journal of Experimental Psychology: General, 149*(6), 1137–1147.

Huron, D. (2006). *Sweet anticipation: Music and the psychology of expectation*. MIT Press.

Jackendoff, R. (1987). *Consciousness and the computational mind*. MIT Press.

Jackendoff, R., & Lerdahl, F. (2006). The capacity for music: What is it, and what's special about it? *Cognition, 100*, 33–72.

Jacoby, N., Margulis, E. H., Clayton, M., Hannon, E., Honing, H., Iversen, J., Klein, T. R., Mehr, S. A., Pearson, L., Peretz, I., Perlman, M., Polak, R., Ravignani, A., Savage, P. E., Steingo, G., Stevens, C. J., Trainor, L., Trehub, S., Veal, M., & Wald-Fuhrmann, M. (2020). Cross-cultural work in music cognition: Challenges, insights, and recommendations. *Music Perception, 37*(3), 185–195.

Jacoby, N., Undurraga, E. A., McPherson, M. J., Valdés, J., Ossandón, T., & McDermott, J. H. (2019). Universal and non-universal features of musical pitch perception revealed by singing. *Current Biology, 29*(19), 3229–3243.e12.

Janata, P., Tomic, S. T., & Haberman, J. M. (2012). Sensorimotor coupling in music and the psychology of the groove. *Journal of Experimental Psychology: General, 141*(1), 54–75.

Johnson, M. (1987). *The body in the mind: The bodily basis of meaning, imagination, and reason.* University of Chicago Press.

Johnson, M., & Larson, S. (2003). "Something in the Way She Moves"—metaphors of musical motion. *Metaphor and Symbol, 18*, 63–84.

Jonaitis, E., & Saffran, J. (2009). Learning harmony: The role of serial statistics. *Cognitive Science, 33*, 951–968.

Juslin, P. N., & Västfjäll, D. (2008). Emotional responses to music: The need to consider underlying mechanisms. *Behaviour and Brain Sciences, 31*(5), 559–575.

Keeler, J. R., Roth, E. A., Neuser, B. L., Spitsbergen, J. M., Waters, D. J. M., & Vianney, J.-M. (2015). The neurochemistry and social flow of singing: Bonding and oxytocin. *Frontiers in Human Neuroscience, 9*. https://www.frontiersin.org/articles/10.3389/fnhum.2015.00518/full.

Kessler, E. J., Hansen, C., & Shepard, R. N. (1984). Tonal schemata in the perception of music in Bali and in the West. *Music Perception, 2*(2), 131–165.

Koelsch, S., Gunter, T. C., Wittfoth, M., & Sammler, D. (2005). Interaction between syntax processing in language and in music: An ERP study. *Journal of Cognitive Neuroscience, 17*(10), 1565–1577.

Korsyn, K. (1988). Schenker and Kantian epistemology. *Theoria, 3*, 1–58.

Kozak, M. (2015). Listeners' bodies in music analysis: Gestures, motor intentionality, and models. *Music Theory Online, 21*.

Kozak, M. (2019). *Enacting musical time: The bodily experience of new music.* Oxford University Press.

Kreutz, G. (2014). Does singing facilitate social bonding? *Music and Medicine, 6*(2), 51–60.

Krueger, J. (2011a). Doing things with music. *Phenomenology and the Cognitive Sciences, 10*(1), 1–22.

Krueger, J. (2011b). Extended cognition and the space of social interaction. *Consciousness and Cognition, 20*, 643–657.

Krumhansl, C. L. (1990). *Cognitive foundations of musical pitch.* Oxford University Press.

Krumhansl, C. L. (1996). A perceptual analysis of Mozart's Piano Sonata K. 282: Segmentation, tension, and musical ideas. *Music Perception, 13*(3), 401–432.

Lahav, A., Boulanger, A., Schlaug, G., & Saltzman, E. (2005). The power of listening: Auditory-motor interactions in musical training. *Annals of the New York Academy of Sciences, 1060*(1), 189–194.

Lakoff, G., & Johnson, M. (1980). *Metaphors we live by.* University of Chicago Press.

Larson, S. (2012). *Musical forces: Motion, metaphor, and meaning in music.* Indiana University Press.

Leaver, R. A., & Remeš, D. (2018). J. S. Bach's chorale-based pedagogy: Origins and continuity. *Bach: Journal of the Riemenschneider Bach Institute, 48*(2), 116–150.

Lerdahl, F. (1992). Cognitive constraints on composition systems. *Contemporary Music Review, 6*(2), 97–121.

Lerdahl, F. (2001). *Tonal pitch space*. Oxford University Press.

Lerdahl, F., & Jackendoff, R. (1983). *A generative theory of tonal music*. MIT Press.

Lerdahl, F., & Krumhansl, C. (2007). Modeling tonal tension. *Music Perception, 24*, 329–366.

Lidji, P., Kolinsky, R., Lochy, A., & Morais, J. (2007). Spatial associations for musical stimuli: A piano in the head? *Journal of Experimental Psychology: Human Perception and Performance, 33*(5), 1189–1207.

Lotze, M., Scheler, G., Tan, H.-R. M., Braun, C., & Birbaumer, N. (2003). The musician's brain: Functional imaging of amateurs and professionals during performance and imagery. *NeuroImage, 20*, 1817–1829.

Loui, P. (2012). Learning and liking of melody and harmony: Further studies in artificial grammar learning. *Topics in Cognitive Science, 4*(4), 554–567.

Loui, P., & Wessel, D. (2008). Learning and liking an artificial musical system: Effects of set size and repeated exposure. *Musicae Scientiae, 12*(2), 207–230.

Loui, P., Wessel, D. L., & Hudson Kam, C. L. (2010). Humans rapidly learn grammatical structure in a new musical scale. *Music Perception, 27*(5), 377–388.

Loui, P., Wu, E. H., Wessel, D. L., & Knight, R. T. (2009). A generalized mechanism for perception of pitch patterns. *Journal of Neuroscience, 29*(2), 454–459.

Maravita, A., & Iriki, A. (2004). Tools for the body (schema). *Trends in Cognitive Sciences, 8*(2), 79–86.

Margulis, E. H., Mlsna, L. M., Uppunda, A. K., Parrish, T. B., & Wong, P. C. M. (2009). Selective neurophysiologic responses to music in instrumentalists with different listening biographies. *Human Brain Mapping, 30*, 267–275.

Marshall, K. (1999). The fundamentals of organ playing. In N. Thistlethwaite & G. Webber (Eds.), *The Cambridge companion to the organ* (pp. 93–112). Cambridge University Press.

Matthews, T. E., Witek, M. A. G., Heggli, O. A., Penhune, V. B., & Vuust, P. (2019). The sensation of groove is affected by the interaction of rhythmic and harmonic complexity. *PLOS ONE, 14*(1), e0204539.

Matthews, T. E., Witek, M. A. G., Lund, T., Vuust, P., & Penhune, V. B. (2020). The sensation of groove engages motor and reward networks. *NeuroImage, 214*, 116768.

Maturana, H. R., & Varela, F. J. (1980). *Autopoiesis and cognition: The realization of the living*. Reidel.

McCulloch, W. S., & Pitts, W. (1943). A logical calculus of the ideas immanent in nervous activity. *Bulletin of Mathematical Biophysics, 5*(4), 115–133.

McDermott, J. H., Schultz, A. F., Undurraga, E. A., & Godoy, R. A. (2016). Indifference to dissonance in native Amazonians reveals cultural variation in music perception. *Nature, 535*(7613), 547–550.

Mead, A. (1999). Bodily hearing: Physiological metaphors and musical understanding. *Journal of Music Theory, 43*, 1–19.

Merleau-Ponty, M. (2012). *Phenomenology of perception* (D. A. Landes, Trans.). Routledge.

Meyer, L. B. (1956). *Emotion and meaning in music.* University of Chicago Press.

Miyazaki, K. (1988). Musical pitch identification by absolute pitch possessors. *Perception & Psychophysics, 44*(6), 501–512.

Miyazaki, K. (1990). The speed of musical pitch identification by absolute-pitch possessors. *Music Perception, 8*(2), 177–188.

Montague, E. (2011). Phenomenology and the "hard problem" of consciousness and music. In D. Clarke & E. F. Clarke (Eds.), *Music and consciousness: Philosophical, psychological, and cultural perspectives* (pp. 29–46). Oxford University Press.

Morgan, E., Fogel, A., Nair, A., & Patel, A. D. (2019). Statistical learning and Gestalt-like principles predict melodic expectations. *Cognition, 189*, 23–34.

Novembre, G., & Keller, P. E. (2011). A grammar of action generates predictions in skilled musicians. *Consciousness and Cognition, 20*, 1232–1243.

Patel, A. D. (2003). Language, music, syntax and the brain. *Nature Neuroscience, 6*(7), 674–681.

Pearce, M. T., Ruiz, M. H., Kapasi, S., Wiggins, G. A., & Bhattacharya, J. (2010). Unsupervised statistical learning underpins computational, behavioural, and neural manifestations of musical expectation. *NeuroImage, 50*(1), 302–313.

Perlman, M. (2004). *Unplayed melodies: Javanese gamelan and the genesis of music theory.* University of California Press.

Perruchet, P., & Pacton, S. (2006). Implicit learning and statistical learning: One phenomenon, two approaches. *Trends in Cognitive Sciences, 10*(5), 233–238.

Phillips-Silver, J., & Trainor, L. J. (2005). Feeling the beat: Movement influences infants' rhythm perception. *Science, 308*, 1430.

Phillips-Silver, J., & Trainor, L. J. (2007). Hearing what the body feels: Auditory encoding of rhythmic movement. *Cognition, 105*, 533–546.

Quinn, I., & White, C. W. (2017). Corpus-derived key profiles are not transpositionally equivalent. *Music Perception, 34*(5), 531–540.

Rehding, A. (2016). Instruments of music theory. *Music Theory Online, 22*(4).

Reis, K. S., Heald, S. L. M., Veillette, J. P., Van Hedger, S. C., & Nusbaum, H. C. (2021). Individual differences in human frequency-following response predict pitch labeling ability. *Scientific Reports, 11*(1), 14290.

Reymore, L., & Hansen, N. C. (2020). A theory of instrument-specific absolute pitch. *Frontiers in Psychology, 11*. https://www.frontiersin.org/articles/10.3389/fpsyg.2020.560877/full.

Rizzi, S. (2018, November 8). *What was Bach's IQ and was he really a genius?* Classic FM. https://www.classicfm.com/composers/bach/bach-iq-genius-music/.

Rizzolatti, G., & Sinigaglia, C. (2008). *Mirrors in the brain: How our minds share actions and emotions* (F. Anderson, Trans.). Oxford University Press.

Rohrer, T. (2005). Image schemata in the brain. In B. Hampe & J. Grady (Eds.), *From perception to meaning: Image schemas in cognitive linguistics* (pp. 165–196). Mouton de Gruyter.

Rohrmeier, M., & Rebuschat, P. (2012). Implicit learning and acquisition of music. *Topics in Cognitive Science, 4*, 525–533.

Rohrmeier, M., & Widdess, R. (2017). Incidental learning of melodic structure of North Indian music. *Cognitive Science, 41*(5), 1299–1327.

Rusconi, E., Kwan, B., Giordano, B. L., Umiltà, C., & Butterworth, B. (2006). Spatial representation of pitch height: The SMARC effect. *Cognition, 99*, 113–129.

Sammler, D., Novembre, G., Koelsch, S., & Keller, P. E. (2013). Syntax in a pianist's hand: ERP signatures of "embodied" syntax processing in music. *Cortex, 49*(5), 1325–1339.

Saslaw, J. (1996). Forces, containers, and paths: The role of body-derived image schemas in the conceptualization of music. *Journal of Music Theory, 40*, 217–243.

Savage, P. E., Loui, P., Tarr, B., Schachner, A., Glowacki, L., Mithen, S., & Fitch, W. (2021). Music as a coevolved system for social bonding. *Behavioral and Brain Sciences, 44*, e59.

Schenker, H. (1935). *Free composition* (E. Oster, Trans.). Longman.

Schladt, T. M., Nordmann, G. C., Emilius, R., Kudielka, B. M., de Jong, T. R., & Neumann, I. D. (2017). Choir versus solo singing: Effects on mood, and salivary oxytocin and cortisol concentrations. *Frontiers in Human Neuroscience, 11*. https://www.frontiersin.org/article/10.3389/fnhum.2017.00430.

Schutz, M. (2017). Acoustic constraints and musical consequences: Exploring composers' use of cues for musical emotion. *Frontiers in Psychology, 10*. https://www.frontiersin.org/articles/10.3389/fpsyg.2017.01402/full.

Shapiro, L. (2011). *Embodied cognition*. Routledge.

Smith, N. A., & Cuddy, L. L. (2003). Perceptions of musical dimensions in Beethoven's Waldstein sonata: An application of tonal pitch space theory. *Musicae Scientiae, 7*(1), 7–34.

Steege, B. (2012). *Helmholtz and the modern listener*. Cambridge University Press.

Straus, J. N. (2006). Normalizing the abnormal: Disability in music and music theory. *Journal of the American Musicological Society, 59*, 113–184.

Stupacher, J., Hove, M. J., & Janata, P. (2016). Audio features underlying perceived groove and sensorimotor synchronization in music. *Music Perception, 33*(5), 571–589.

Temperley, D. (2007). *Music and probability*. MIT Press.

Temperley, D. (2018). *The musical language of rock*. Oxford University Press.

Tenzer, M. (2000). *Gamelan gong kebyar: The art of twentieth-century Balinese music*. University of Chicago Press.

Tilley, L. A. (2019). *Making it up together: The art of collective improvisation in Balinese music and beyond*. University of Chicago Press.

Vuvan, D. T., & Hughes, B. (2019). Musical style affects the strength of harmonic expectancy. *Music & Science, 2*, 1–9.

Wallmark, Z., Deblieck, C., & Iacoboni, M. (2018). Neurophysiological effects of trait empathy in music listening. *Frontiers in Behavioral Neuroscience, 12*. https://www.frontiersin.org/articles/10.3389/fnbeh.2018.00066/full.

Weinstein, D., Launay, J., Pearce, E., Dunbar, R. I. M., & Stewart, L. (2016). Singing and social bonding: Changes in connectivity and pain threshold as a function of group size. *Evolution and Human Behavior, 37*(2), 152–158.

White, C. W., & Quinn, I. (2018). Chord context and harmonic function in tonal music. *Music Theory Spectrum, 40*(2), 314–335O.

Witek, M. A. G. (2017). Filling in: Syncopation, pleasure and distributed embodiment in groove. *Music Analysis, 36*(1), 138–160.

Wong, P. C. M., Chan, A. H. D., Roy, A., & Margulis, E. H. (2011). The bimusical brain is not two monomusical brains in one: Evidence from musical affective processing. *Journal of Cognitive Neuroscience, 23*(12), 4082–4093.

Yearsley, D. (2012). *Bach's feet: The organ pedals in European culture*. Cambridge University Press.

Zanten, W. van. (1986). The tone material of the Kacapi in Tembang Sunda in West Java. *Ethnomusicology, 30*, 84–112.

Zbikowski, L. M. (2002). *Conceptualizing music: Cognitive structure, theory, and analysis*. Oxford University Press.

Zbikowski, L. M. (2017). *Foundations of musical grammar*. Oxford University Press.

7 Rhythmic Entrainment and Embodied Cognition

Maria A. G. Witek

Introduction

Musical activities—whether we are playing an instrument, listening through headphones, composing or producing, or humming along to an earworm—involve our bodies. Dancing to a rhythmic beat is perhaps one of the most obviously embodied ways of engaging with music. The way we synchronize our body movements to rhythmic structures—a process known as rhythmic entrainment—illustrates that the body plays a role in how music is perceived and experienced. There is now vast evidence from cognitive neuroscience that music activates all the major brain networks—the auditory, visual, somatosensory, motor, reward, memory, and executive function systems and more. But what is the relationship between the body and the brain in rhythmic entrainment? Is the body just a vessel through which musical information in the environment is transformed into neural signals in the brain, nothing more than the site of our senses (audition, vision, touch, proprioception, and so forth), which are then processed by the corresponding brain networks into internal representations? Or do the body and the environment have more central roles in music cognition, taking part in cognitive processing itself?

How brain, body, and environment relate has been one of the key questions in European philosophy since the seventeenth century, when René Descartes made an absolute separation between the spiritual and the material, concluding that human perceptual experience has to be explainable entirely in terms of processes inside the thinking subject and that these processes are the causes of the body's behavior (Descombes, 2001). This Cartesian dualism dominated research on musical experience most pointedly during the 1980s, when the brain was often likened to a computer programmed to decode information from the environment via the senses into mental representations of the world (e.g., Jackendoff, 1987) In this cognitivist view, synchronizing and dancing to a musical beat is nothing more than the brain using the body to extrapolate the rhythmic framework from the environment into an internal mental representation. An important counter to such dualist and representational positions emerged in the 1990s in response to Varela, Thompson, and Rosch's (1991) proposal that mind is embodied and that cognition should be understood as embodied action. In this view, the body and the environment play a more central role in mind, and mental representations are replaced with embodied know-how. According to embodied cognition, the movements we make when entraining to a beat are part of our perception and cognition of it.

Despite the popularity of embodiment in philosophy of mind, the idea that the brain is processing information from the environment into abstract mental representations continues to serve as a tacit, limiting assumption, especially in the cognitive neurosciences. However, a more nuanced debate has recently emerged, cutting across cognitive neuroscience and philosophy. In this debate, it is generally agreed that mind *is* embodied, but there is disagreement about the *degree* of embodiment. This chapter explores how different degrees of embodiment are reflected in some of the most influential theories of mind and brain, using rhythmic entrainment as a point of departure. After a brief introduction to rhythmic entrainment, I present some contemporary theories of embodied cognition and consider how they understand embodiment in entrainment. My two foci are (1) extended mind theory, with its related models in predictive processing, and (2) enactivism, with its related models in systems dynamics. Both frameworks presume that mind is embodied, but each differs in its understanding of how deeply embodied mind is understood to be. I consider two related implications of these frameworks: the effect on understanding the relationship among brain, body, and environment in music, and the effect on the status of internal representations in music cognition. I then consider how these embodied theories might appeal to other music disciplines, especially those concerned with humanistic, material, and social mechanisms of musical experiences. I address a recent concern in sound and music studies that antirepresentationalism favors the material over the social and the precognitive over the cognitive by showing how enactivism redefines mind as equally distributed among neural, corporeal, and environmental systems, leading to an understanding of musical and sonic experiences as always already cognitive, social, and material. I conclude by considering the implications of my arguments for the improved dialogue between music cognition researchers and scholars of music from humanistic and social-scientific perspectives.

Entraining to the Beat

Moving or dancing to a musical beat is perhaps one of the most overt expressions of the body's importance in music. This activity requires the perception of temporal regularity in a rhythmic pattern, an ability known as beat perception. This basic ability is considered to be largely innate and universal (Honing, 2012), although it can be improved through musical training (Repp & Su, 2013). Entrainment is widely recognized as explaining how humans (and some nonhuman animals; see, e.g., Cook et al., 2013) are able to perceive and synchronize with a beat, such as when they tap along, clap, bob their head, or dance to music. Entrainment is a form of coupling, where two or more entities become connected and begin to behave in an interactive and coordinated way. The dynamic attending theory (Jones & Boltz, 1989; Large & Jones, 1999) defines entrainment as the process by which an independent and self-sustaining oscillator comes into contact with and becomes driven by another independent and self-sustaining oscillator (oscillator here being understood as a periodic rhythm or beat pattern). The driving force of these oscillators on each other causes them to adapt their phase and period so that they eventually synchronize. We can think of the beat as one oscillator and the attentional process in the listener, musician, or dancer as another oscillator, and through entrainment, the attentional oscillator becomes driven by the beat oscillator, leading to beat-synchronized attention (Large & Kolen, 1994). The synchronization of the attentional oscillator depends on humans' ability to form temporal expectations. When an individual perceives a periodically recurring beat, the attentional oscillator forms expectations about when future beats will occur (Barnes & Jones, 2000), and it is these expectations that drive the phase and period adaptations toward synchrony. There are several types of entrainment (Phillips-Silver et al., 2010), such as self-entrainment, which describes a process whereby an individual entrains to a self-generated or imagined rhythm. Motor entrainment is said to occur when physical body movements are entrained to a stimulus. Social or interpersonal entrainment occurs when two or more individuals become mutually entrained to each other, such as when playing instruments together in an ensemble or dancing in a crowd. Entrainment also occurs naturally in various physical (e.g., pendulum clocks) and biological (e.g., circadian sleep-wake cycles) systems (Clayton et al., 2004).

Neural Entrainment, Rhythm, and Motor Networks in the Brain

The entrainment that happens in the brain can be thought of as involving both physical and biological entrainment—it is often simply referred to as neural entrainment. The brain behaves in a fundamentally rhythmic and oscillatory way. A neuron's principal activity is inherently periodic, and oscillatory activity is generated from both spontaneous neuronal firing and connections to other neurons (Buzsáki, 2006). Put simply, the change in a neuron's electric membrane potential, called an action potential, is oscillatory because it rhythmically fluctuates between a low- and high-energy state. Through short- and long-range synaptic connections between neurons, one oscillating action potential entrains other postsynaptic action potentials. Within the cortex, brain regions are connected via pathways that project in both directions, affording bidirectional coupling between areas and coordinated firing patterns. When groups of neurons interact, the group-level firing frequency can be different from that at the individual neuronal level. It is at the group level that neuronal activity can be measured with neurophysiological recording techniques such as electroencephalography (EEG) and magnetoencephalography (MEG).

The oscillatory nature of rhythmic entrainment has led some researchers to investigate the relationship between behavioral oscillations and neural oscillations. Some studies have shown that rhythm perception is associated with oscillations in higher-frequency bands, specifically beta (13–30 Hz) and gamma (>30 Hz) (Grahn, 2012). Beta oscillations are known to be prominent during various motor cognitive tasks, while gamma oscillations are associated with attention, memory, and anticipation. There is also evidence of more direct synchronization between auditory and neural rhythms. When listening to a musical beat, periodic EEG responses reflect not just the acoustic beats in the music but also the implied metric context (Nozaradan et al., 2011; Lenc et al., 2020). Therefore, the attentional oscillations that underpin our ability to perceive and entrain to a beat in music may have internal neurophysiological counterparts.

Using spatially sensitive functional magnetic resonance imaging (fMRI), it has been shown that temporal and rhythmic perception and production are associated with increased activity in both cortical and subcortical brain regions, most commonly the premotor cortex, supplementary motor area, parietal cortex, prefrontal cortex, cerebellum, and basal ganglia (Grahn & Rowe, 2009; Chapin et al., 2010). Most of these regions are also implicated in motor cognition more broadly; the basal ganglia are important for motor control and learning, the cerebellum allows fine motor control and supports coordination, and the secondary motor areas are involved in planning and imagining movements. Even during passive listening, especially when the rhythms have a strong metric beat, activations of motor-relevant areas increase (Chen et al., 2008).

These various studies converge to indicate that timing, beat, and rhythm perception are intimately linked with motor systems in the brain, suggesting that body movement is central to how we cognitively process rhythmic patterns in music. But they don't fully answer the questions posed at the beginning of this chapter. What are the roles of the body and the environment in cognitive processes such as entrainment? Are they equally important as the brain, or do mechanisms inside the head, such as mental representations, play a more constitutive part in the musical mind?

Embodiment in Cognitive Science, Philosophy, and Music

The answers to these questions depend on how we understand embodiment. The study of embodiment is an interdisciplinary endeavor, drawing on perspectives from cognitive science, psychology, philosophy of mind, computational modeling, and linguistics. Contemporary theories of embodiment emerged largely in response to a widespread belief that mind is reducible to computational processes happening inside the brain. Wallace lays out four main strands of this orthodoxy within cognitive science: "'cognitivism' (i.e. the belief that thought is and must be the following of rules 'stored' in the brain), 'information processing' (the idea that the brain 'processes information' via algorithms (i.e. the rules)), artificial intelligence (the faith that the brain is a digital computer, processing information via algorithms), and perhaps Chomskyan linguistics (which views the 'language acquiring device' as a 'computational system' using rules)" (2007, p. 20).

This neurocentrism and representationalism can be traced back to Cartesian and Platonic dualism; that is, human behavior in a material world is a mere copy of immaterial ideals that exist as a set of rules inside the thinking mind, and the brain represents the world through the application of these rules (Wallace, 2007). In music, the most well-known systematic application of such dualism is Lerdahl and Jackendoff's (1983) *Generative Theory of Tonal Music*, which applies a Chomskyan rule-based model to explain the cognition of basic tonal musical structures (Bach's four-part chorales) and attempts to account for tonality, harmony, and meter as internal representations of syntactically organized mental structures (see also De Souza's chapter 6 in this volume).

Embodied theories of cognition tend to disagree with this kind of modeling, holding that mind is not reducible to the brain and that the body and environment are central to mental processes. Much of the current debate among scholars is happening within the 4E framework of embodied cognition—4E standing for embodied, embedded, enacted, and extended. What is common to these theories is that cognition is seen as being "in some sense . . . dependent on the morphological, biological, and physiological details of an agent's body, an appropriately structured natural, technological, or social environment, and the agent's active and embodied interaction with this environment" (Newen et al., 2018, p. 5). They are distinguished by the aspects of embodiment on which they focus. In the context of 4E, *embodied* is often used as an umbrella term that assumes some form of bodily importance akin to that expressed in the foregoing quote, but it does not necessarily specify how this importance is actualized. *Embedded* usually refers to thinking in which mind and cognition are understood to be situated in the environment, which includes the body. When mind is said to be *enacted*, it is generally thought that cognition can be partly explained by the active engagement of agents and their environments. Finally, *extended* refers to perspectives in which cognitive processes are seen as including (extending into) the body and environment. As I describe below, the last two Es—enacted and extended—represent most explicitly the debate about how embodied mind is thought to be, precisely what roles the body and the environment take in constituting mind alongside the brain, and whether mind relies causally on internal mental representations to perceive and understand the world.

The embodied perspective has made its way into musical scholarship, primarily in music theory. Marc Leman's (2008) work represents a model in which the body is seen as the mediator through which information and representation are transferred between musical material and mental processes; musical instruments and technologies can take part in this mediatory work as well. Arnie Cox (2016) draws on linguistic research to argue that we experience music through embodied metaphors. Mariusz Kozak (2019) takes an embodied approach to the experience of musical time, arguing that it is constituted by the body as it engages in musical activities. Jonathan De Souza (2017) draws on phenomenology and transformation theory to argue for the embodied nature of playing an instrument. In affect theory, Friedlind Riedel (2015, 2020) emphasizes the embodied nature of affective atmospheres in musical cultures. Arguments for embodiment also appear in some experimental music cognition research, especially that which studies the body's responses to music (e.g., Burger et al., 2013).

Entrainment—or at least the more generalized concept of coupling—tends to be a central feature in embodied theories (Thompson, 2005; Krueger, 2014; Di Paolo et al., 2017), as it represents a relatively clear example of how the body, the brain, and the environment can be integrated in human behavior and cognition. Entrainment is especially prominent in theories of embodiment in music, as it plays a central part in one of the most obviously embodied musical activities: moving to a musical beat (Krueger, 2014; Kozak, 2019). Montague (2019) suggests that the nature of embodiment depends on the type of entrainment (e.g., biological versus attentional). Put another way, depending on which model of embodiment is applied, entrainment has a different function in mind and accords different roles to the body, the brain, and the environment. Different models also disagree in the extent of their rejection of internal mental representations as causes of rhythmic entrainment and of cognition more broadly. To illustrate these differences, the next sections consider the last two of the four Es and the neuroscientific models that frequently accompany them: extended mind and predictive processing, and enactivism and dynamic systems theory.

Extended Mind and Embodied Predictive Processing

The main premise of extended mind theory, first proposed by Andy Clark and David Chalmers (1998; see also Clark, 2008), is that mind is not an isolated organ inside the head but extends across the brain, the body, and the environment. In this version of embodiment, objects in the world that we use to perform cognitive tasks, such as writing a shopping list to remember which groceries to buy or consulting a map to find our way to a hotel, should not be thought of as mere mnemonic or navigational tools; rather, they are fully part of the cognitive processes of memory and navigation. For Clark and Chalmers, an external object is part of a cognitive process if it functions *just like* an internal cognitive process, thus conforming to what they term the parity principle. They illustrate this with a fictional story about Inga and Otto, who both live in New York City and want to visit the Museum of Modern Art (MOMA). Inga knows the directions by heart, but Otto, who has Alzheimer's disease, has to consult his notebook. Even though Otto's notebook is external, its function is indistinguishable from Inga's memory and should therefore be considered part of Otto's memory proper. The body is similarly considered part of mental processes in extended mind theory. If we use our fingers to count the number of times we have been to MOMA, the fingers participate in the cognitive process of remembering and counting because they function just like the arithmetic and mnemonic processes in our heads. Joel Krueger has proposed a model for extended musical minds, arguing that music acts as "affective scaffolding" onto which listeners can off-load some of the mental processes involved in regulating emotions. For Krueger, rhythmic entrainment is the mechanism through which emotional extensions occur: "We engage with music because, unlike most other non-musical sounds, it affords synchronously organizing our reactive behavior and felt responses; and we take pleasure in letting music assume some of these organizational and regulative functions that, in other contexts, normally fall within the scope of our own endogenous capacities" (2014, p. 3).

However, while cognitive processes are viewed as extending into the body and the environment, the *machinery* or *vehicles* of cognition and mental states are largely thought to remain within the brain (Clark, 2009, 2012). In extended mind theory, the feedback and feed-forward loops that cross between brain, body, and environment may be necessary to *generate* a given experience, but they are not themselves part of the circuitry of experience. To Clark, this largely rests on the fact that physical events in the body and in the world have a significantly slower timescale than those in the central nervous system. For Clark, at least, while extraneural materials play a crucial role in mind, neurons occupy a more significant position in the system.

The brain, according to Clark, has a special role in the embodied mind because it is responsible for the mechanisms that make embodied experience possible. Clark draws on the framework of predictive processing (Friston et al., 2011), also called predictive coding, to explain these mechanisms (Clark 2012, 2013). He suggests that the brain is a "guessing machine" that employs predictive processes to construct generative models of the external environment from which its ultimately hidden causes can be inferred (Clark, 2015). Rather than encoding sensory information directly, the brain procures, through a process that approximates Bayesian inference, a predictive model of the causes of its input. What the brain processes is not the information itself but the difference, or *prediction error*, between the predictive model and the information, making information processing highly efficient. This occurs in a hierarchical and looping fashion, where each layer of neural processing predicts the input from the levels below. The lower levels then feed forward the prediction error, which tells the levels above how accurate their predictive models are. The loops at the higher end of the hierarchy predict increasingly more complex and contextualized information. The brain's primary purpose is to minimize prediction error and maximize the accuracy of its predictive models and thus its perception of the world, and it does so via two closely related methods. It can choose the right predictive model, which amounts to *perception*, or it can signal the body to move in such a way that the input better matches the model, which amounts to *action*. Thus, the crux of the embodied argument in the embodied predictive processing framework is that the brain relies on both perceptual and motor systems to collectively predict sensory input, a mechanism referred to as *active inference* (Friston et al., 2011).

The notion of active inference is closely linked with another central principle in the prediction framework: the free energy principle, originally developed in statistical physics (Friston, 2010; Clark, 2013). According to this principle, an organism's primary goal is to maintain its organization and its boundaries, and it does this by minimizing free energy (in an information theoretical sense) in its interaction with the environment. Organisms can be understood as self-organizing systems to the extent that they maintain their internal structure by reducing free energy. As such, the interaction with the environment is instrumental in defining the organism itself. In the predictive processing view, free energy is minimized by reducing prediction error. When a body moves and physically interacts with its environment, it is actively shaping its sensory input and reducing free energy so as to minimize prediction error and feed the prediction machine that is the brain. Consistent with the extended mind view, this action-perception coupling allows cognitive processes to extend across the brain, the body, and the environment. However, the predictive mechanisms that give rise to these distributed processes are of an exclusively neural nature. And by viewing perception and cognition as fundamentally inferential, extended mind theory and the predictive processing framework must accept a minimally representational view of mind—even if the body and its acting in the world are crucial in giving rise to these inferences (Gallagher, 2017).

Due in large part to the fact that music is fundamentally temporal and thus implicates prediction in its cognition, predictive processing has become one of the most influential frameworks in music cognition research (e.g., Koelsch et al., 2019). And it is partly via the embodied aspects of predictive processing that some music cognition researchers have begun to argue for the embodied nature of mind in a more specific way (Schaefer, 2015; Koelsch et al., 2019). In particular, since beat perception requires temporal predictions, the framework has been widely embraced among rhythm and meter researchers (Vuust et al., 2018). Put simply, meter is a predictive model against which rhythmic input is compared, and depending on the degree of similarity between the metric model and the rhythmic input, stronger or weaker prediction errors are procured. Sometimes, such as when listening to polyrhythms, the prediction error might challenge the metric model to such an extent that a completely different meter provides an equally good or better model. In such cases, the active inference of moving to the beat reinforces a particular meter and helps minimize prediction error (Vuust et al., 2018).

By extension, extended mind theory and embodied predictive processing can understand rhythmic entrainment as an activity that gives the body a crucial role in the cognitive mechanism of beat perception. The beat can be actively inferred by moving the body in the environment, marking out the internal model in the external world in a way that reduces free energy and minimizes prediction error. The internal model then extends out to the external world through the body, and the active inferences made by the body extend back into the brain in the form of prediction error, creating a feedback loop of embodied information. But again, entrainment here serves the production of accurate predictive models for the brain. The physical movements of the body as it synchronizes to a beat are abstracted into prediction errors that are compared with prediction models, and the phase and period corrections are made possible primarily because of neural computation.

Enactivism and Systems Dynamics

In contrast, most theories of enactivism disagree that the brain should hold a more central role in the embodied mind than the body and the environment. In fact, the basic idea of embodiment according to enactivism is that mind is not an information-processing machine localized in the brain but an active process that cannot be localized at all. This active process is a form of bidirectional *relating*, or *coupling*, between an organism and its environment. When mind is seen as a relational activity, there is no need for abstract representations, not even those of a predictive or inferential nature. Perception and cognition are instead enacted in the engagement with the world. Since these ideas were first popularized by Francisco Varela, Evan Thompson, and Eleanor Rosch in 1991, they have been applied to a number of mental processes, such as emotion (Colombetti, 2014) and language (Di Paolo et al., 2018). Rather than appealing to the modules of an internal organ like the brain, enactivists seek to describe emotional and linguistic processes as active processes in the world. Thomas Fuchs compares mind to breathing: "Just as respiration cannot be restricted to the lungs but only functions in a systemic unity with the environment, so the individual mind cannot be restricted to the brain" (Fuchs, 2009, p. 222).

This kind of enactive and cyclic distribution is central across different levels of human life (Thompson & Varela, 2001). At the biological level, a process known as autopoiesis drives the metabolic self-production of a living system, where an organism's identity and autonomy are enacted through an interconnected network of environmental and homeostatic processes. At the personal level, perception and cognition are enacted by the sensorimotor coupling that occurs when an organism engages with its environment, giving rise to meaning as a kind of know-how or sense making (Di Paolo et al., 2017) and a sense of identity as an agent. When different autopoietic agents interact with each other at the social level, the sensorimotor coupling between them is the basis for how they perceive their intersubjective selves.

Together, these three levels produce a sense of self as an inextricable mixture of metabolic, sensorimotor, and intersubjective relationships, where mind is a distributed system that spans the body, the brain, and the environment, but the distribution is a consequence of mind being an active process driven by an organism rather than a reflection of where mind is. Another consequence is that neither mind nor the machinery of mind are restricted to neural activity inside the brain (Ward, 2012). Because the subpersonal (autopoietic) and the personal (sensorimotor and social) are intimately linked, restricting one to the internal or external necessarily restricts the other levels as well. Neither externalism nor internalism can fully reflect mind as an active process. In this view, internal neural factors play a crucial role in giving rise to mental processes, but no more so than external corporeal and material factors.

To explain the role of the brain in sensorimotor coupling, enactivists often apply principles from systems dynamics (Juarrero, 1999; Thompson & Varela, 2001; Di Paolo et al., 2017), which models physical, chemical, biological, and social systems such as fluids, fires, organisms, and economies in terms of how their elements dynamically interact with each other and with elements of other systems (Kelso, 2009). Here, patterns of behavior *emerge* from the coordination dynamics of the system elements rather than being caused by a central, inferential homunculus. In this way, human behavior can be understood as resulting from vertically and horizontally coupled subsystems in a self-organized system. In this model, the brain is just one subsystem among other equally important subsystems, like the body and the environment. The coordination of dynamic systems falls into three main categories: absolute coordination, relative coordination, and no coordination. In an oscillatory system, such as a brain, these categories represent degrees of synchrony between the elements, ranging from phase-locked synchrony to complete asynchrony and various mixed states in between.

The neuroscientist J. A. Scott Kelso is a central figure in the systems dynamics approach to the brain and to human behavior. In a famous experiment, Kelso and his colleagues tested the coordination dynamics of finger wagging in time to an external isochronous rhythm, either in phase or antiphase (Kelso et al., 1987; Schoner & Kelso, 1988). They observed that when the rhythm sped up, it became increasingly difficult for humans to maintain an antiphase relationship with the stimulus, eventually switching to in-phase synchrony. They then demonstrated the same switching in coordinated movements between different limbs, different humans, and different animal species (Fuchs & Kelso, 2018). In later experiments using EEG and MEG, investigators found that the switching was mirrored in the oscillatory components of the brain signal of individuals (Kelso et al., 2013) and in the coordination of brain signals of interacting subjects (Tognoli et al., 2007).

Kelso and his colleagues modeled this behavior as a system of coupled nonlinear oscillators (Haken et al., 1985) and demonstrated that at the point of criticality—where the system switches from antiphase to in phase—the sensorimotor and neural behavior split into two possible states, either stable (in phase) or unstable (antiphase). This is expressed mathematically as a pitchfork bifurcation. But near such critical bifurcation points, just before the switch, there is a third type of state known as *metastability* (Kelso & Tognoli, 2007). This dynamic coordination regime is a transient state of relative coordination in which previously stable attractor states are no longer stable but still exert a degree of attraction. In oscillatory systems, the previously phase-locked synchronizations are now loosely synchronized but not phase locked, leading to subtle "dwell and escape" patterns (Kelso, 2009). During such metastable states, there is a balance between integration and segregation, and the smallest change or perturbation to the system (such as new input) causes it to reorganize into a new spatiotemporal configuration. In sensorimotor systems such as the finger-wagging human, metastability is what makes the switch between different behavioral states possible. In brains, the tendency toward metastability explains how ensembles of neurons constantly couple and decouple, making it a highly flexible system that allows different areas to engage in different functions during different stages of processing (Bressler & Kelso, 2016).

What is important for enactivists is that metastability supports nonlinear reciprocal causality between a system and its subsystem, as well as between subsystems. There is no central cognitive computer causing the changes in the system; instead, the changes emerge through the responses of the system as a whole. Thus, mind can be seen as a system that is partly afforded by the dynamic coupling between its neural, sensorimotor, and environmental components and that partly determines the coupling of these subsystems. A change to any of these components will disrupt the stability of the system and, via metastable dynamics, can lead to any number of reconfigurations and phenomenological changes. Precisely which reconfiguration ensues depends on the system as a whole, rather than solely the neural component. Mind's embodiment and embeddedness are thus not added on higher up in the hierarchy of subsystems but are fundamental parts of the basic makeup of mind. Seeing the brain as a self-organizing system that is itself part of the self-organizing system of mind, incorporating the body and the environment, makes the separation of the brain unnecessary. It also removes the need for the brain to abstract mental representations from sensory information to explain how humans perceive the world and act on it.

In music, enactivist thinking has been applied to theoretical investigations of musical mind, skill, behavior, and human evolution (Witek, 2022; Doffman, 2009; Tomlinson, 2015; Schiavio et al., 2017), but it has rarely been tested in an experimental setting. However, as Kelso has demonstrated, important principles of dynamic systems *are* testable (Walton et al., 2015). In music, systems dynamics has been widely embraced among rhythm and beat perception researchers (Demos et al., 2012; Henry & Herrmann, 2014). In fact, dynamic attending theory's fundamental principles are built on systems dynamics (Large & Jones, 1999). A self-sustaining oscillation is itself a dynamic system that can form a subsystem in a coupled suprasystem of self-sustaining oscillations.

By extension, the relevance of systems dynamics to other forms of entrainment should be clear. Biological, motor, self-, and social entrainment are all forms of sensorimotor coupling, a kind of dynamic coordination or synchronization regime in which the patterns of behavior emerge from the interaction of the elements of the system as a whole, and no one element is more fundamental than another. By this logic, the mechanisms of rhythmic entrainment can be seen as physically enacted and thus fully distributed among the brain, the body, and the environment. The temporal corrections needed to synchronize the phase and period of our movements are not first extracted from the beat, then estimated and represented in the head, and finally performed by the body. Neither are they abstracted into prediction error that is compared with an internal predictive beat model. The temporal correction is enacted in the world (which includes the body and the brain). This does not mean that prediction is irrelevant to entrainment and, more broadly, to mind. Without prediction, neither entrainment nor mind would function as they do. Instead, it means that prediction is not the fundamental mechanism that explains the embodiment of entrainment and mind. It is the activeness, interactiveness, and distributedness of the process of entraining and the process of mind that render them embodied.[1]

Embodiment and Antirepresentationalism in the Humanities and Social Sciences

If we focus on representationalism, internalism, and neurocentrism, the degree of embodiment in enactivism can be seen as more extreme than in extended mind theory, since the former rejects these concepts while the latter retains them to a minimal degree (Gallagher, 2017). Antirepresentationalism, in particular, may have different implications for different disciplines seeking to integrate embodied models of mind into their frameworks. Abandoning representation entirely might pose problems for some humanistic accounts of musical entrainment, suggesting that extended mind theory and predictive processing might be most appropriate. Examples of such accounts may include formalist and generative music theories of rhythm, especially those relying on computational explanations for how rhythm and beat are perceived (e.g., Pressing, 2002; Rohrmeier, 2020). Other, more humanistic and social approaches might align better with enactive and systems dynamics, as they decentralize the brain and give more weight to bodily and environmental processes, thus being more open to explanations that depend on social, cultural, and historical embeddedness. Dance music studies, for example, may find this more equally distributed understanding of mind more compatible with its orientation toward holistic social accounts of music (e.g., Garcia, 2020; Alisch, 2020).

Recently, some humanities researchers investigating "concepts, practices, and technologies of sound and listening in different historical and cultural contexts" have moved toward antirepresentionalist paradigms (Thompson, 2020), while others have voiced skepticism of such a move. In particular, affect theorists of music and sound have contested the antirepresentationalism of the "ontological turn" in the humanities. Proponents of the ontological turn (i.e., as a shift away from paradigms like social constructivism and toward the "real") are in favor of understanding sound and sonic experiences as fluxes of material processes that sidestep representational processes (Cox, 2011), with representations understood here as both mental and textual (Kane, 2015).

Marie Thompson notes that the main theoretical movements within the ontological turn share the following themes: "The decentering of 'the human', the social subject and renunciation of anthropocentrism; a focus on the pre-, extra- or non-social 'real' and/or 'material' world; the utilization of 'scientific' approaches; and an interest in emergence, speculation, potentiality, the 'general' and the 'universal'" (2017, p. 267).

We can recognize some of these themes in enactivist thinking. Emergence, as we have seen, is key in systems dynamics, explaining how mind comes about without the need for a single central homunculus, such as the brain, to cause it. Enactivism can be seen as antianthropocentric due to the greater explanatory weight given to the material, both in the body and in the environment. However, its antianthropocentrism is a decentering not of the human but of the human *brain*. Other themes in Thompson's list stand in more direct opposition to enactivist thinking, most notably, the rejection of the social. The social is fundamental to enactivism, being a key level at which sense making and sensorimotor coupling occur. This suggests there are some important incompatibilities between the ontological turn and enactivism.

To illustrate the discrepancies between the ontological turn and enactivism, and clarify the possibilities for antirepresentationalism in music and sound studies, let us briefly consider Steve Goodman's (2010) analysis of sound as vibration and the affective and embodied nature of sonic and musical experiences (for a critique of Goodman, see Kane, 2015). Goodman's focus is on sounds that are weaponized (e.g., sonic bombs) and contribute to "an immersive atmosphere or ambience of fear and dread" (2010, p. xiv) and on music that taps into this fear response and subverts it (e.g., Afrofuturism and electronic dance music). Most relevant to this discussion is Goodman's understanding of how affect is transmitted between objects and agents. In his view, affect is a precognitive process that does not rely on representations in the brain but is instead transferred between bodies (of both objects and subjects) when they act on one another, via vibration. These vibrational properties of sounds have the greatest affective power on listeners, as the sonic is "emphasized in its sensory relation, in its intermodality, as rhythmic vibration, in excess and autonomous from the presence of a human, phenomenological subject or auditor" (Goodman, 2010, p. 9).

Brian Kane takes issue with Goodman's claim that precognitive affective vibration is antidualistic, claiming it "is betrayed by his rigid temporal and theoretical separation of affective from cognitive realms'" (2015, p. 8). Marie Thompson critiques ontological antirepresentationalism's decentering of the human and the social because it leads to a break from the political implications of analysis, ultimately ignoring the racialization of sonic experiences. To her, a move too far toward the material strays too far away from "lived experiences, social mediation and historicism" (Thompson, 2017, p. 269; see also

Valiquet, forthcoming). In the remainder of this chapter, I argue that the problems of antirepresentationalism in the ontological turn do not apply to the antirepresentationalism of enactivism, and that this difference is partly explained by the type of synchronization assumed in vibration compared with entrainment. Finally, I conclude that enactivist antirepresentationalism offers music scholars in the humanities and social sciences a way to understand music and sound cognition grounded in lived experience.

The antirepresentationalism of enactivism does not reject representations per se but rejects their causal role in cognitive processes. This means that while there are representations (semantic, pictorial, mathematical, and so forth) in the world (which involve humans and their bodies and brains), representations internal to the brain do not cause cognition, including entrainment. Instead, entrainment emerges from the sensorimotor coupling of an agent with its environment. These environments are made up of objects and other agents, and the coupling both embeds and is embedded by the social, cultural, and historical lives of the agent. Interpersonal motor entrainment between dancers in a nightclub enacts the culture of the dance scene and its communities. The way the dancers move their bodies, the way the music's rhythms are articulated, and the meanings the dancers experience are all particular to the dancers' culture and its history. Therefore, replacing mental representations with enactive sense making can account for the embodied experiences of oppression that Marie Thompson highlights, without negating culturally situated knowledge (see Di Paolo et al., 2018).

Enactivist antirepresentationalism also rejects the separation of precognitive affect from cognition that Goodman proposed and Kane critiqued. This is evident in the nature of the mechanism that Goodman claims is responsible for transmitting affect between the environment and the human perceiver: rhythmic vibration. On first glance, vibration as transmitter of affect may appear to translate to the notion of coupling and entrainment as transmitters of emotion in enactivism (Witek, 2022; Colombetti, 2014; Krueger, 2014). They both rely on the mechanism of synchronization to enable the transfer from one object or subject to another. However, there are crucial differences between the synchronization of coupling and vibration. In enactive coupling, synchronization causes the interacting entities to dynamically influence each other, either unidirectionally or bidirectionally, leading to an emergent, structurally coherent system while still retaining their autonomy (Thompson, 2005). The vibration of Goodman's theory, however, is understood more as resonance, which does not involve this dynamically interactive system of autonomous entities. Resonance involves a more passive form of synchronization, where the frequency of one vibrating entity enhances that same vibrating frequency in another, such as when a tuning fork resonates with a guitar string tuned to the same frequency. Furthermore, where

Goodman sees the vibrational force of sound as precognitive, enactivism understands coupling as cognitive in the enactive sense, that is, as sense making. Enactivism does not distinguish between pre- and postognitive processes in its antirepresentationalism; it instead redefines cognition as antirepresentational itself.

Conclusion and Future Directions

Enactivism's antirepresentational definition of cognition, which is always socially, culturally, and historically situated, might appease the critics of antirepresentationalism according to the ontological turn. It might also lead to more interdisciplinary collaboration between music cognition researchers and humanities and social science researchers. However, this will require music cognition researchers to take the embodied and situated aspects of the concepts they study more seriously, such as by frontloading them in behavioral and neuroscientific experiments rather than adding them as qualifiers post hoc (Gallagher, 2003). Researchers in the humanities and social sciences will need to accept that cognition as a system plays a role equal to that of social, cultural, and historical systems in music.

Compared with extended mind theory and predictive processing, the degree of embodiment and type of antirepresentationalism in enactivism and systems dynamics appear to mark clearer routes for bringing the humanities, social sciences, and cognitive neurosciences of music closer together. By rejecting the (albeit minimal) inferential representationalism of extended mind and predictive processing, enactivism can help us see musical activities as emerging from the coordination of systems in the body, brain, and environment, all of which play equally important roles in mind. I have shown how this applies to rhythmic entrainment in music, which illustrates the sensorimotor coupling in mind so explicitly, but enactivism and systems dynamics are frameworks that explain all types of activities and cognitive processes in living systems.

Note

1. See Gallagher and Allen (2018), Allen and Friston (2018), and Di Paolo et al. (2021) for discussions of the conceptual links between autopoiesis, enactivism, and the free energy principle.

References

Alisch, S. (2020). 'I opened the door to develop kuduro at JUPSON': Music studios as spaces of collective creativity in the context of electronic dance music in Angola. *Contemporary Music Review, 39*(6), 663–683.

Allen, M., & Friston, K. J. (2018). From cognitivism to autopoiesis: Towards a computational framework for the embodied mind. *Synthese, 195*(6), 2459–2482.

Barnes, R., & Jones, M. R. (2000). Expectancy, attention and time. *Cognitive Psychology, 41*(3), 254–311.

Bressler, S. L., & Kelso, J. A. S. (2016). Coordination dynamics in cognitive neuroscience. *Frontiers in Neuroscience, 10*, 397.

Burger, B., Saarikallio, S., Luck, G., Thompson, M. R., & Toiviainen, P. (2013). Relationships between perceived emotions in music and music-induced movement. *Music Perception, 30*(5), 517–533.

Buzsáki, G. (2006). *Rhythms of the brain*. Oxford University Press.

Chapin, H. L., Zanto, T., Jantzen, K. J., Kelso, J. A. S., Steinberg, F., & Large, E. W. (2010). Neural responses to complex auditory rhythms: The role of attending. *Frontiers in Psychology, 1*(224), 1–18.

Chen, J. L., Penhune, V. B., & Zatorre, R. J. (2008). Listening to musical rhythms recruits motor regions of the brain. *Cerebral Cortex, 18*, 2844–2854.

Clark, A. (2008). *Supersizing the mind: Embodiment, action, and cognitive extension*. Oxford University Press.

Clark, A. (2009). Spreading the joy? Why the machinery of consciousness is (probably) still in the head. *Mind, 118*(472), 963–993.

Clark, A. (2012). Dreaming the whole cat: Generative models, predictive processing, and the enactivist conception of perceptual experience. *Mind, 121*(483), 753–771.

Clark, A. (2013). Whatever next? Predictive brains, situated agents, and the future of cognitive science. *Behavioral and Brain Sciences, 36*(3), 181–204.

Clark, A. (2015). *Surfing uncertainty: Prediction, action, and the embodied mind*. Oxford University Press.

Clark, A., & Chalmers, D. (1998). The extended mind. *Analysis, 58*(1), 7–19.

Clayton, M., Sager, R., & Will, U. (2004). In time with the music: The concept of entrainment and its significance for ethnomusicology. *European Meetings in Ethnomusicology, 1*, 1–86.

Colombetti, G. (2014). *The feeling body: Affective science meets the enactive mind*. MIT Press.

Cook, P., Rouse, A., Wilson, M., & Reichmuth, C. (2013). A California sea lion (*Zalophus californianus*) can keep the beat: Motor entrainment to rhythmic auditory stimuli in a non-vocal mimic. *Journal of Comparative Psychology, 4*, 412–427.

Cox, A. (2016). *Music and embodied cogntion: Listening, moving, feeling, and thinking*. Indiana University Press.

Cox, C. (2011). Beyond representation and signification: Toward a sonic materialism. *Journal of Visual Culture, 10*(2), 145–161.

Demos, A. P., Chaffin, R., Begosh, K. T., Daniels, J. R., & March, K. L. (2012). Rocking to the beat: Effects of music and partner's movements on spontaneous interpersonal coordination. *Journal of Experimental Psychology, General, 141*(1), 49.

Descombes, V. (2001). *The mind's provisions: A critique of cognitivism*. Princeton University Press.

De Souza, J. (2017). *Music at hand: Instruments, bodies and cognition*. Oxford University Press.

Di Paolo, E., Buhrmann, T., & Barandiaran, X. (2017). *Sensorimotor life: An enactive proposal*. Oxford University Press.

Di Paolo, E., Thompson, E., & Beer, R. D. (2021). Laying down a forking path: Incompatibilities between enaction and the free energy principle. *Psyarxiv*. https://doi.org/10.31234/osf.io/d9v8f.

Di Paolo, E. A., Cuffari, A. C., & De Jaegher, H. (2018). *Linguistic bodies: The continuity between life and language*. MIT Press.

Doffman, M. (2009). Making it groove! Entrainment, participation and discrepancy in the "conversation" of a jazz trio. *Language & History, 52*(1), 130–147.

Friston, K. (2010). The free-energy principle: A unified brain theory? *Nature Reviews Neuroscience, 11*(2), 127–138.

Friston, K., Mattout, J., & Kilner, J. (2011). Action understanding and active inference. *Biological Cybernetics, 104*(1–2), 137–160.

Fuchs, A., & Kelso, J. A. S. (2018). Coordination dynamics and synergetics: From finger movements to brain patterns and ballet dancing. In S. Muller, P. Plath, G. Radons, & A. Fuchs (Eds.), *Complexity and synergetics* (pp. 301–316). Springer.

Fuchs, T. (2009). Embodied cognitive neuroscience and its consequences for psychiatry. *Poiesis und Praxis, 6*(3–4), 219–233.

Gallagher, S. (2003). Phenomenology and experimental design: Toward a phenomenologically enlightened experimental science. *Journal of Consciousness Studies, 10*(9–10), 85–99.

Gallagher, S. (2017). *Enactivist interventions: Rethinking the mind*. Oxford University Press.

Gallagher, S., & Allen, M. (2018). Active inference, enactivism and the hermeneutics of social cognition. *Synthese, 195*(6), 2627–2648.

Garcia, L. M. (2020). Feeling the vibe: Sound, vibration, and affective attunement in electronic dance music scenes. *Ethnomusicology Forum, 29*(1), 21–39.

Goodman, S. (2010). *Sonic warfare: Sound, affect and the ecology of fear*. MIT Press.

Grahn, J. A. (2012). Neural mechanisms of rhythm perception: Current findings and future perspectives. *Topics in Cognitive Science, 4*(4), 585–606.

Grahn, J. A., & Rowe, J. B. (2009). Feeling the beat: Premotor and striatal interactions in musicians and nonmusicians during beat perception. *Journal of Neuroscience, 29*(23), 7540–7548.

Haken, H., Kelso, J. A. S., & Bunz, H. (1985). A theoretical model of phase transitions in human hand movements. *Biological Cybernetics, 51*(5), 347–356.

Henry, M. J., & Herrmann, B. (2014). Low-frequency neural oscillations support dynamic attending in temporal context. *Timing and Time Perception, 2*(1), 62–86.

Honing, H. (2012). Without it no music: Beat induction as a fundamental musical trait. *Annals of the New York Academy of Sciences, 1252*(1), 85–91.

Jackendoff, R. (1987). *Consciousness and the computational mind.* MIT Press.

Jones, M. R., & Boltz, M. (1989). Dynamic attending and responses to time. *Psychological Review, 96*(3), 459–491.

Juarrero, A. (1999). *Dynamics in action: Intentional behavior as a complex system.* MIT Press.

Kane, B. (2015). Sound studies without auditory culture: A critique of the ontological turn. *Sound Studies, 1*(1), 2–15.

Kelso, J. A. S. (2009). Coordination dynamics. In *Encyclopedia of complexity and system science* (pp. 1537–1564). Springer.

Kelso, J. A. S., Dumas, G., & Tognoli, E. (2013). Outline of a general theory of behavior and brain coordination. *Neural Networks, 37*, 120–131.

Kelso, J. A. S., Schöner, Z., Barnea, Z., & Haken, H. (1987). Phase-locked modes, phase transitions and component oscillators in biological motion. *Physica Scripta, 35*(1), 79–87.

Kelso, J. A. S., & Tognoli, E. (2007). Towards a complementary neuroscience: Metastable coordination dynamics of the brain. In L. Kozma & L. Perlovsky (Eds.), *Neurodynamics of cognition and consciousness* (pp. 39–60). Springer.

Koelsch, S., Vuust, P., & Friston, K. (2019). Predictive processes and the peculiar case of music. *Trends in Cognitive Sciences, 23*(1), 63–77.

Kozak, M. (2019). *Enacting musical time: The bodily experience of new music.* Oxford University Press.

Krueger, J. (2014). Affordances and the musically extended mind. *Frontiers in Psychology, 4*, 1003.

Large, E. W., & Jones, M. R. (1999). The dynamics of attending: How people track time-varying events. *Psychological Review, 106*(1), 119–159.

Large, E. W., & Kolen, J. (1994). Resonance and the perception of musical meter. *Connection Science, 6*(1), 177–208.

Leman, M. (2008). *Embodied music cognition and mediation technology.* MIT Press.

Lenc, T., Keller, P. E., Varlet, M., & Nozaradan, S. (2020). Neural and behavioral evidence for frequency-selective context effects in rhythm processing in humans. *Cerebral Cortex Communications, 1*(1), tgaa037.

Lerdahl, F., & Jackendoff, R. (1983). *A generative theory of tonal music.* MIT Press.

Montague, E. (2019). Entrainment and embodiment in musical performance. In Y. Kim & S. L. Gilman (Eds.), *The Oxford handbook of music and the body.* Oxford University Press.

Newen, A., De Bruin, L., & Gallagher, S. (2018). 4E cognition: Historical roots, key concepts and central issues. In A. Newen, L. De Bruin, & S. Gallagher (Eds.), *The Oxford handbook of 4E cognition* (pp. 3–18). Oxford University Press.

Nozaradan, S., Peretz, I., Missal, M., & Mouraux, A. (2011). Tagging the neuronal entrainment to beat and meter. *Journal of Neuroscience, 31*(28), 10234–10240.

Phillips-Silver, J., Aktipis, C. A., & Bryant, G. A. (2010). The ecology of entrainment: Foundations of coordinated rhythmic movement. *Music Perception, 28*(1), 3–14.

Pressing, J. (2002). Black Atlantic rhythm: Its computational and transcultural foundations. *Music Perception, 19*(3), 285–310.

Repp, B. H., & Su, Y.-H. (2013). Sensorimotor synchronization: A review of recent research (2006–2012). *Psychonomic Bulletin & Review, 20*(3), 403–452.

Riedel, F. (2015). Music as atmosphere: Lines of becoming in congregational worship. *Lebenswelt: Aesthetics and Philosophy of Experience, 6*.

Riedel, F. (2020). Atmospheric relations: Theorising music and sound as atmosphere. In F. Riedel & J. Torvinen (Eds.), *Music as atmosphere: Collective feelings and affective sounds* (pp. 1–42). Routledge.

Rohrmeier, M. (2020). Towards a formalisation of musical rhythm. In *International Society for Music Information Retrieval Conference*. Bangalore, India.

Schaefer, R. S. (2015). Mental representations in musical processing and their role in action-perception loops. *Empirical Musicology Review, 9*(3–4), 161.

Schiavio, A., van der Schyff, D., Cespedes-Guevara, J., & Reybrouck, M. (2017). Enacting musical emotions: Sense-making, dynamic systems, and the embodied mind. *Phenomenology and the Cognitive Sciences, 16*(5), 785–809.

Schoner, G., & Kelso, J. A. S. (1988). Dynamic pattern generation in behavioral and neural systems. *Science, 239*(4847), 1513–1520.

Thompson, E. (2005). Sensorimotor subjectivity and the enactive approach to experience. *Phenomenology and the Cognitive Sciences, 4*, 407–427.

Thompson, E., & Varela, F. J. (2001). Radical embodiment: Neural dynamics and consciousness. *Trends in Cognitive Sciences, 5*(10), 418–425.

Thompson, M. (2017). Whiteness and the ontological turn in sound studies. *Parallax, 23*(3), 266–282.

Thompson, M. (2020). *Sound studies*. Oxford Bibliographies.

Tognoli, E., Lagarde, J., DeGuzman, G. C., & Kelso, J. A. S. (2007). The phi complex as a neuromarker of human social coordination. *Proceedings of the National Academy of Sciences, 104*(19), 8190–8195.

Tomlinson, G. (2015). *A million years of music: The emergence of human modernity*. MIT Press.

Valiquet, P. (forthcoming). Grimes' hymn to technocracy, insolvent affordances, and the need for reparative organology. *Journal of Popular Music Studies*.

Varela, F. J., Thompson, E., & Rosch, E. (1991). *The embodied mind: Cognitive science and human experience*. MIT Press.

Vuust, P., Dietz, M., Witek, M. A. G., & Kringelbach, M. L. (2018). Now you hear it: A novel predictive coding model for understanding rhythmic incongruity. *Annals of the New York Academy of Sciences, 1423*(1), 19–29.

Wallace, B. (2007). Introduction. In B. Wallace, A. Ross, J. Davies, & T. Anderson (Eds.), *The mind, the body and the world: Psychology after cognitivism?* (pp. 1–29). Imprint Academic.

Walton, A. E., Richardson, M. J., Langland-Hassan, P., & Chemero, A. (2015). Improvisation and the self-organization of multiple musical bodies. *Frontiers in Psychology, 6*, 313.

Ward, D. (2012). Enjoying the spread: Conscious externalism reconsidered. *Mind, 121*(483), 731–751.

Witek, M. A. G. (2022). The mind is a DJ: Rhythmic entrainment in beatmatching and embodied neuroscience. In M. Doffman, T. Young, & E. Pain (Eds.), *The Oxford handbook of time in music* (pp. 235–252). Oxford University Press.

8 The Musical Mind: Perspectives from Developmental Science

Haley E. Kragness, Erin E. Hannon, and Laura K. Cirelli

As developmental scientists who study music, our goals are to describe and track musical experiences and behaviors in the broader context of other developing capacities, characterize the nature of musical experiences at different points in the life span, and understand how individuals both respond to and shape their own unique histories of musical development. Yet to participate in a truly interdisciplinary discussion of the musical mind, we may need to scrutinize and reconsider some of our most fundamental concepts. As noted elsewhere in this volume (e.g., Mundy, chapter 4; Savage et al., chapter 18), basic concepts such as *music*, *musicality*, and *culture* are rooted in a network of ideas and categories of difference that may constrain the kinds of questions we ask and the conclusions we draw in ways that are limiting and perhaps even harmful. In this chapter we argue that a developmental science approach may offer a helpful perspective, given its emphasis on complex interactions of interlocking systems operating at multiple levels (Bronfenbrenner, 1999); its assumption of overlap between perceptual, cognitive, motor, and social processes; and its focus on the unique history of the individual.

First, when attempting to understand the concept of *music*, developmental science considers how an individual's conceptualization of music shifts with accrued experience and changing cognitive, social, and affective mechanisms. Research on infant auditory perception reminds us that we cannot assume that infants parse musical sounds like adults. For example, the perceptual integration of a sound wave and its harmonics (or overtones) as a single auditory event does not occur until four months of age (Folland et al., 2015). And while the definition of music is heavily influenced by culture, even for an individual adult, the point at which patterns of sound are interpreted as music is sometimes extremely unclear. This is illustrated by the popular "sometimes behave so strangely" illusion (Deutsch et al., 2008), in which the repetition of a spoken phrase suddenly sounds "musical" to the listener. When do infants and children form conceptual categories for music and nonmusic? Much of the time, when we say that infants *respond to music* in a certain way, what we really mean is that they respond to *something most Western adults consider to be music*.

Second, what does it mean to be musical? How do these definitions change as development unfolds? As noted elsewhere in this volume (e.g., Ilari & Habibi, chapter 17; Cowan, chapter 16), *musicality* is often conflated with specialized musical skills that typically require music training or formal experience with music. This is problematic for a number of reasons. First, musically trained individuals in Western societies often differ a priori from their peers on a number of factors, including music perception abilities (e.g., Swaminathan et al., 2017; Kragness, Swaminathan, et al., 2021). Moreover, notions of music as performative and reserved for the expert are largely a product of the Western art music culture (Turino, 2008). Everyday musical interactions, however, appear to be universal (in function, if not form), frequent, and salient in early life (Ilari, 2005). Evolutionary psychologists have reframed the notion of *musicality* to refer to capacities that enable musical engagement without training (e.g., Patel, chapter 1 in this volume; Duengen et al., chapter 3 in this volume). Indeed, well before infants and young children can engage in formal music training, they are highly musical creatures who can sing, dance, and respond in socially and emotionally meaningful ways to musical stimuli.

Third, how can we define *culture* without relying on categories that "other" and overgeneralize groups of diverse individuals? As acknowledged here and elsewhere (e.g., Savage et al., chapter 18; Henrich, 2020), the study of music cognition—much like the study of human nature in general—has relied on an oversampling of Western, educated, industrialized, rich, and democratic (WEIRD) populations, which has created an incomplete, nonrepresentative, and fundamentally noninclusive picture of human musicality. At the same time, criticizing the weirdness of WEIRD runs the risk of overessentializing and exoticizing the non-WEIRD and thus reinforcing the very problems the concept is meant to critique (Savage et al., chapter 18; Clancy & Davis, 2019). Instead of labeling a person's musical culture or grouping individuals based on their culture, a developmental perspective might emphasize what that individual "grew up with"—the individual's unique life history of musical experiences and how those past experiences shape that listener's current and future musical experiences. Given that context and age contribute to each and every musical experience we have, conceptualizing culture in terms of an individual's experience with particular musical activities or traditions (e.g., dancing, specific song genres) may enhance precision as well as better capture the wide range of musical cultural experiences that can exist within a particular society or geographic region.

In this chapter, we apply perspectives from developmental science to explore how the so-called musical mind emerges and changes from infancy into adulthood. Given the preceding conceptualizations of music, musicality, and culture, we expand on three ideas from developmental science that may be helpful when thinking about the musical mind. First, we assert that the musical mind does not develop in isolation from other domains. The way individuals perceive and engage with music must be considered in the context of their developing sensory and perceptual systems, cognitive abilities, motor milestones, and social-emotional capacities across the life span. Second, we highlight the need for a more nuanced appreciation of developmental processes for understanding the dynamic nature of musical experience and capacities at different ages. We use musical entrainment as an example of a musical behavior that is often assumed to be "available" from birth but appears to have a strikingly protracted developmental trajectory, depending on the tasks and methods of investigation. Finally, we note that infants and children are not simply passive recipients of musical input; their early and active responses to music influence how parents integrate music into their lives. Beginning early in life, musical experience is driven by a dynamic interaction between the environment and the individual.

Musical Development and Development in Other Domains

Development in any one domain is deeply intertwined with development in others. Consider the case of visual and motor development. Learning to locomote involves the complex coordination of numerous muscle groups and postural control. Once self-propulsion is achieved, infants' interactions and affordances in the world are fundamentally different. Visual information can be harnessed in a new, more sophisticated way and can be used to plan and self-correct infants' own movements in the world. Higgins, Campos, and Kermoian (1996) measured the impact of this motor milestone using a clever "moving walls" contraption that simulates visual input associated with locomoting forward or backward. While prelocomotor infants sit happily and enjoy the visual "show," crawling and walking infants of the same age shift their center of gravity forward or backward to account for the apparent feeling of motion. This is a powerful illustration of the idea that visual perception is affected by motor abilities and that the two systems do not develop independently. Likewise, the musical mind does not have an independent, isolated trajectory but develops in tandem with other perceptual, cognitive, and socioemotional abilities.

Cognitive abilities: Put simply, cognitive abilities are the mental processes that enable us to hold information in memory, as well as recall and manipulate it. Some cognitive abilities are already observable in fetuses—for instance, newborns recognize musical excerpts they heard prenatally (Panneton, 1985; Partanen et al., 2013; Polverini-Rey, 1992; Satt, 1984), implying that a fetus can encode and retain auditory input. With maturation comes increasingly sophisticated mental operations and processing abilities. The speed of mental operations, the amount of information that can be maintained in memory at one time, and the ability to focus attention all increase substantially over childhood as neural function matures and experience is accumulated.

Cognitive abilities are clearly implicated in many tasks that measure musical abilities—for instance, "same or different?" tasks rely heavily on the ability to retain auditory information in short-term memory for comparison. The capacity to perform such tasks is necessarily tied to advances in cognitive development (see the next section for a discussion of this in the context of rhythm perception). Perhaps more importantly, numerous day-to-day responses to musical phenomena presumably rely on cognitive mechanisms. For example, melodic enculturation may depend on a statistical learning mechanism by which cumulative exposure to musical structures shapes musical expectations (e.g., Pearce, 2018). Musical reward is hypothesized to be associated with cognitive evaluation of expectation violation and nonviolation (e.g., Cheung et al., 2019; Gold et al., 2019). Similarly, it has been suggested that the sensation of "groovy" syncopation emerges as a consequence of active inference instantiated at intermediate levels of rhythmic ambiguity (Witek et al., 2014; Matthews et al., 2020; Koelsch et al., 2018). Changes in cognitive ability are therefore potential targets for understanding developmental change in responses to music.

To the extent that changing cognitive abilities account for changes in music cognition, they could have implications for other aspects of musical engagement. The ability to recognize familiar sequences of syllables, rhythms, and pitches enables the emergence of social preferences based on songs (e.g., Cirelli & Trehub, 2018; Mehr et al., 2016; Soley & Spelke, 2016). Similarly, sharing musical expectations with others enables increasingly sophisticated musical engagement—the ability to clap, sing, or dance along and to share emotional responses to music with the people around us.

Language and communication abilities: Comparisons between language (particularly speech) and music are frequently made in the literature, which is not surprising, given that they (usually) share the auditory modality and are communicative in nature. This connection is especially prominent in early life, given that song appears to be an effective form of preverbal communication between parents and infants: caregivers frequently sing to soothe tired infants to sleep or engage them in playful interactions.

The musical qualities of speech, too, seem to be highly salient components of infants' early language development. Rhythm appears to be one of the earliest properties of language (Nazzi et al., 1998). Language and music share parallel rhythmic structures—for instance, the relative frequency of dotted rhythms in English versus French folk songs parallels the contrast in the two languages' spoken rhythms (Huron & Ollen, 2003;

Patel & Daniele, 2003), and this pattern is enhanced in children's songs (Hannon et al., 2016). Aside from communicating affect, early song may provide temporal frameworks on which language abilities are scaffolded. Indeed, rhythm perception has consistently been linked to language ability in children (Gordon et al., 2015; Ladányi et al., 2020).

Exposure to different speech accent patterns also shapes infants' perceptual grouping tendencies. For instance, infants who are surrounded by the Japanese language tend to perceptually group accented tones differently than infants surrounded by English, which is presumably attributable to differences in accents and grouping in those languages (Yoshida et al., 2010). Language experience continues to shape music perception into childhood and adulthood. Tone language speakers have superior performance on melody discrimination tasks (but not other music perception tasks) as early as preschool (Creel et al., 2018), and this advantage persists through adulthood (Swaminathan & Schellenberg, 2020; Swaminathan et al., 2021; Zhang et al., 2020).

Yet an open question is the extent to which music and language are experienced as separate domains, especially in the context of infancy. Though laboratory studies consistently show enhanced responses to song versus speech (e.g., Cirelli & Trehub, 2019; Corbeil et al., 2016), these results do not necessarily imply that infants are able to make a categorical distinction between the two. Instead, they may be responding to a difference in *degree* (of musicality, emotionality, or rhythmicity, for example) conferred by song versus speech. At the same time, nonspeech sounds, such as sine tones or birdsong, fail to facilitate infant learning as readily as speech does (Ferguson & Waxman, 2017; Woodruff-Carr et al., 2021). The contexts in which these domains are distinct have yet to be elucidated.

Motor abilities: When listening to a song, we sometimes find ourselves tapping our toes without even realizing we are doing so. The ability to control and coordinate beat-aligned movements intentionally, let alone without conscious awareness, is not straightforward. In its earliest stages, infants face constant motor challenges as they continuously recalibrate sensory (and multisensory) information as their bodies grow and change. It is no surprise, then, that moving to music also has an extended developmental trajectory. Improvement in beat alignment across childhood is slow, with improvements apparent into adolescence (McAuley et al., 2006).

Importantly, however, movement is not simply a consequence of hearing music. The act of *moving* to music can also shape perception. Tapping along to a beat improves beat perception (Manning & Schutz, 2013), and different effectors are involved, depending on their natural oscillatory frequency, which changes with growth. Research on infants being moved *by an experimenter* shows that different movement influences the perception of a metrically ambiguous rhythmic sequence (Phillips-Silver & Trainor, 2005).

Given the influence of movement on perception, it follows that changes in motor development may lead to changes in music perception. For instance, differences in effector size might differentially affect the easiest metrical level to move to, which could influence tempo judgments. The extent to which syncopation engages the motor system could be affected by motor affordances. And widening the perspective to other implications, being able to move oneself with music enables self-generated social bonding experiences in the context of music (e.g., Cirelli, 2018).

Musical Functions and Capacities Change Throughout Development

Perhaps in support of the idea that musicality is a universal, core feature of human nature, a common approach to characterizing the development of musicality has been to identify or inventory the capacities that appear earliest in development and assume that this evidence supports the existence of evolved, biologically based adaptations for music. Although it is clearly important to examine potential starting points of human musicality, this approach is problematic and may arise from preformationist assumptions that ignore the critical role of developmental processes in shaping the musical mind.

First, this approach reflects a common but unsupported assumption that if something is an adaptation, it should be observed at birth or shortly thereafter, and anything that emerges later must be acquired through learning. This assumption is undermined by examples such as breasts and bipedalism, which are clearly products of natural selection that are not present at birth but emerge later, when they are needed (Al-Shawaf et al., 2018). Likewise, as noted by Patel (chapter 1), developmental plasticity can play a crucial role in the emergence of traits designed by natural selection, such as when a particular capacity or specialization (e.g., motion perception or tonotopic cortical maps) depends on optimally timed exposure to environmental structure (experience-expectant plasticity).

A disproportionate focus on early emerging musical capacities may also promote a more static view of musical development that assumes younger listeners simply have less mature versions of the experiences and musical concepts of (Western) adults. But in fact, the musical experiences of younger individuals may be different from those of older individuals in important ways. Just as others have promoted a less anthropocentric approach to music cognition in nonhuman animals (see Duengen et al., chapter 3), we urge readers to consider embracing a more "ecological" approach to infants and children that doesn't take adults' musical concepts and definitions as the only reference point.

A second and perhaps bigger problem is that any inference about the developing musical mind is inherently limited by the paradigms, tasks, and stimuli used to examine it, and we cannot assume these paradigms always give rise to the same musical experiences, regardless of age. Even if children and adults respond in similar ways to a particular stimulus or task, this does not necessarily mean they rely on the same underlying processes, especially when the task examines only one aspect of a much more complex and multimodal musical behavior or experience. Musical entrainment provides a good example of a fundamental but complex musical capacity that appears to emerge very early or very late, depending on the tasks and methods used to measure it.

Without any specialized training, most adults are capable of dancing, clapping, marching, and other forms of entrainment, and these are arguably some of the most ubiquitous and directly observable human musical behaviors (Begel et al., 2017; Sowiński & Dalla Bella, 2013). Even though these behaviors are widespread, they are highly multimodal and complex, dependent on a host of underlying perceptual, cognitive, and motor processes that are still not fully understood. To be able to dance to music, a listener must segment multimodal musical input into discrete units or events, interpret the rhythmic patterning of those events, infer the beats within the metrical framework, establish and maintain endogenous rhythms that correspond to the beat and meter, and precisely coordinate body movements with that structure.

An extensive body of empirical work has examined adult entrainment behaviors, either by asking listeners to tap in synchrony with metronomes, rhythms, or music (Repp & Su, 2013; Snyder & Krumhansl, 2001) or, more recently, by using motion capture technology to observe dancing directly (Burger et al., 2014; Naveda & Leman, 2010; Toiviainen et al., 2010). When similar technologies are applied to infants and children, however, it is clear that entrainment behaviors develop rather gradually. For instance, although infants exhibit some rhythmic movements in response to music (Zentner & Eerola, 2010; Fujii et al., 2014; Rocha & Mareschal, 2017), and some preschoolers can drum or move periodically (rather than randomly) to music (Eerola, Luck, & Toiviainen, 2006; Kirschner & Ilari, 2014; Kirschner & Tomasello, 2009; Woodruff-Carr et al., 2014), even adolescents as old as sixteen or seventeen are not as precise in their musical entrainment as young adults (Braun Janzen et al., 2014; Drake et al., 2000; McAuley et al., 2006; Thompson et al., 2015). For this reason, much developmental research has focused on more indirect measures of the perceptual and cognitive underpinnings of entrainment behavior.

Studies of children typically use rhythm discrimination or change detection to indirectly measure beat perception during development. These paradigms have provided evidence that newborns and young infants can discriminate simple rhythmic patterns (Chang & Trehub, 1977; Demany et al., 1977; Hannon & Johnson, 2005) and that they are better at doing this for strongly metrical than weakly metrical rhythmic patterns (Trehub & Hannon, 2009; Hannon et al., 2011). Studies have also shown that the newborn brain exhibits larger mismatch negativity responses to deviants occurring on strong rather than weak beat positions (Winkler et al., 2009) and that beat frequencies can be observed in the steady-state evoked brain potentials of infants listening to rhythms (Cirelli et al., 2016). This indirect perceptual and neural evidence is often cited to support the claim that the basic capacity for entrainment to a musical beat is present at or shortly after birth and is continuously available throughout development.

These findings are at odds, however, with a growing body of evidence that rhythm and beat perception undergo significant and gradual changes over the course of childhood, becoming increasingly specific to the child's cultural musical listening experiences (Gerry et al., 2010; Einarson & Trainor, 2016; Hannon & Trehub, 2005a, 2005b; Soley & Hannon, 2010; Hannon, Vanden Bosch der Nederlanden, & Tichko, 2012) and eventually giving rise to robust cross-cultural differences in rhythm and beat perception and production in adulthood (Hannon, Soley, & Ullal, 2012; Kalendar et al., 2013; Jacoby & McDermott, 2017; Ullal et al., 2014). Recent evidence suggests that even for North American children listening to Western music, the capacity to match music to the correct metronome at both the beat and measure levels does not become adult-like until adolescence (Nave-Blodgett, Hannon, & Snyder, 2021; Nave-Blodgett, Snyder, & Hannon, 2021). Likewise, the capacity to sustain an endogenous beat while listening to an ambiguous rhythm appears to have a protracted course of development (Nave et al., in preparation).

It is possible to reconcile these seemingly conflicting findings by proposing that the capacity for beat perception is essentially present at birth but becomes refined in a culture-specific manner over the course of development in parallel with or as a result of improved motor coordination. However, it is also possible that the observed developmental changes represent fundamental differences in the perceptual, cognitive, and motor processes available to listeners of different ages. The fact that these abilities take so long to become adult-like is consistent with evidence of protracted development in other areas such as language, working memory, disgust, and crystallized intelligence (Hartshorne & Germine, 2015; Hartshorne et al., 2018; Rottman, 2014). Assuming that evolved traits should be available at an age when they are most beneficial for survival and reproduction, such developmental evidence can actually be used to evaluate claims or theories about their evolved functions (Rottman, 2014). In this case, although the capacity to entrain to music could be part of an evolved adaptation for music making, the fact that it continues to change throughout childhood and is not fully available until adolescence or early adulthood may have implications for understanding its functional role (perhaps to forge bonds with adolescent peers rather than between infants and caregivers). Thus, rather than assuming that indirectly observed, early-emerging capacities are essentially equivalent to the more complex and directly observed musical behaviors of adulthood, we emphasize the importance of understanding developmental processes and timing and the ways musical functions might change throughout life.

We Play an Active Role in Our Musical Development

The idea that humans start as blank slates ready to be written on by their environment dates back to Aristotle. This perspective on development not only downplays the important role of genetics and biology but also depicts the developing child as a passive recipient of input. Instead, research in the past century reveals that children actively construct knowledge about the world around them, select which inputs to attend to, and influence the behavior of those around them. For example, when infants lock eyes with a parent, their responsive smiles and coos encourage continued parental attention.

We continue to play an active role in our own development across the life span. In this section, we use examples from early childhood to highlight how active engagement and feedback loops affect the musical mind and shape future musical experiences.

Caregiver song provides a compelling example of such a feedback loop. It is one of the most common forms of musical exposure in early life (Ilari, 2005; Mendoza & Fausey, 2021) and is common all over the world (Trehub & Gudmundsdottir, 2015). When parents (as well as nonparents and older children) sing to infants, they adopt an infant-directed singing style. Compared with adult- or self-directed song, infant-directed song is more emotional and often higher in pitch, with slower and more regular rhythms (Nakata & Trehub, 2011; Trainor, 1996). This increased emotionality and modification of acoustic properties are similar to what occurs when we speak to infants (Fernald, 1989).

From birth through infancy, infant attention is captured and emotions are regulated by infant-directed song (Masataka, 1999; Trainor, 1996), often more so than infant-directed speech. For example, infants happily listen to recordings of infant-directed song for twice as long as infant-directed speech before becoming distressed (Corbeil et al., 2016). If infants are already distressed, caregiver song is better than caregiver speech at mitigating distress, reducing physiological arousal, and capturing infant attention (Cirelli & Trehub, 2020). In short, infant-directed song captures infants' attention, regulates their emotional responses, and influences their behavior.

These infant responses reinforce and shape caregivers' emotional responses and continued musical behaviors (Cirelli, Jurewicz, & Trehub, 2020). For example, if an infant responds with interest and joy to a caregiver's song, this may encourage the caregiver to continue singing or to use song in a similar context in the future. If it effectively calms infants before nap time or cheers them up when they are distressed, caregivers may be more likely to build these strategies into daily routines. Guided by infant feedback, even the musically inexperienced caregiver may rapidly develop a repertoire of songs or vocal behaviors. In general, developmental research has demonstrated that infant feedback has a dynamic impact on caregiver behavior. For example, when infants respond with joy and interest to higher-pitched infant-directed speech, caregivers continue to modulate their pitch upward (Smith & Trainor, 2008). This feedback loop may explain why it is so natural for humans to adopt infant-directed vocalization styles when interacting with infants, encouraged by their interest and responsiveness.

Infant musical preferences may also shape musical exposure in the home. For example, familiar songs are more likely to capture infant attention and generate rhythmic movements than rarely heard songs. This preference for familiar songs is robust, whether they are sung by a parent or a stranger, and it remains robust even when the stranger's and parent's renditions differ substantially in terms of tempo and pitch (Kragness, Johnson, & Cirelli, 2021). And although song is generally more effective than speech at cheering up distressed infants, an infant's favorite song is the most effective and is more likely to generate infant smiles (Cirelli & Trehub, 2020). Infants and children also show social preferences for singers of familiar songs (Cirelli & Trehub, 2018; Mehr et al., 2016; Soley & Spelke, 2016). If an infant responds more positively to one particular song than others, this likely encourages the parent to keep that song in the rotation. But how do these preferences emerge? Certainly, caregivers play a significant role in introducing particular songs to infants. However, it is likely that the songs that generate the largest responses from the infant will become part of the daily routine. Adult research highlights that musical preferences can be influenced by personality characteristics (Vella & Mills, 2017). Whether early temperament also influences early musical preferences is yet to be explored.

The preceding examples from infancy highlight that early responses to music actively shape how we experience music in our everyday lives, but these active processes continue throughout life. For example, most of us become active music makers by engaging informally with music, such as by humming while washing the dishes or tapping our feet while listening to a favorite song. Precursors to music making, including rhythmic movements in response to music and early song-like vocalizations, emerge before the first birthday (Reigado & Rodrigues, 2017; Zentner & Eerola, 2010) and become more frequent in toddlerhood (Sole, 2017; Cirelli & Trehub, 2019).

These early music-making behaviors are generally spontaneous and embedded in social-emotional contexts. For example, toddlers' song-like vocalizations are more likely to emerge in response to an experimenter's song (Reigado & Rodrigues, 2017). Music also becomes integrated into early everyday play. A naturalistic study of spontaneous sibling interactions in the home reported common musical play (singing, dancing, or playing real or imaginary instruments) between two- and four-year-old siblings. Musical play became substantially more frequent between these same sibling pairs two years later, and musical play was positively correlated to measures of sibling prosociality (Cirelli, Peiris, Tavassoli, et al., 2020). These observations in a naturalistic context support laboratory work linking musical movements to expressions of joy in infancy (Kragness, Johnson, & Cirelli, 2021; Zentner & Eerola, 2010) and work exploring how musical movement with others encourages prosociality in infants and children (Cirelli, 2018).

The musical mind is not the product of passive exposure to musical input. Individual differences in perception, attention, and personality, combined with genetic and sociocultural factors, lead some infants to respond more favorably to music than others. Responsive caregivers find this favorable reaction reinforcing and continue to integrate music into their infants' lives. Infants also become active participants in their own musical development when they begin to produce musical behaviors (sing, dance) and spontaneously integrate music into play. Early musical experiences are often associated with feelings of joy. In short, infants actively shape and contribute to their musical worlds. Infancy provides powerful examples of how we play an active role in our own musical development, but these feedback loops continue to refine our musical identities across the life span.

The Future of Developmental Science in Understanding Musicality

In this chapter, we conceptualized the musical mind from a developmental perspective and applied three major ideas from modern developmental science: that development occurs across multiple domains, that development is not simply the unfolding of continuous and static capacities, and that individuals shape their own musical experiences.

These ideas point to the importance of combining reductionist and ecological approaches. Reducing musical phenomena to highly controlled experimental settings necessarily omits elements of the musical experience. But without using a reductionist approach to understand basic questions about auditory pattern perception, acuity, and cognitive abilities, we run the risk of imposing unwarranted assumptions and generalizations about any given musical experience. At the same time, laboratory work cannot tell us about an individual's music listening history, the social and emotional contexts in which that person hears music, or the societal contexts in which these experiences are embedded. In the past decade, researchers have used unobtrusive home recording devices to collect daylong samples of infants' auditory home environments (e.g., Benetti & Costa-Giomi, 2019; Mendoza & Fausey, 2021). Combining such approaches with traditional laboratory work offers a promising avenue to better understand how environmental input (versus presumed "group" differences) shapes music perception (see Wojcik et al., 2022, for a related discussion on music-language links).

A focus on *experience-based perception* also implies a role for studying *experience-based change*. Rather than concentrating on the age at which certain musical abilities emerge, we argue that a more productive approach would be to focus on the accumulation of experiences, as well as the neural and behavioral preparedness that might drive the development of these abilities. Although longitudinal studies present logistical challenges, they are crucial for understanding the shape and rate of developmental change. Likewise, studies that emphasize the time periods in which rapid change is most likely to occur ("microgenetic" studies) can be particularly informative in terms of explaining how different elements of music perception emerge and their association with developmental change in other domains.

In conclusion, we return to our emphasis on questioning and redefining our most basic concepts: What is music? What does it mean to be musical? What is culture? These questions are addressed elsewhere from various perspectives. Iyer (2012) describes music as "the sound of bodies in motion." Patel (chapter 1) discusses how music is a socially constructed category and focuses instead on musicality. Ilari and Habibi (chapter 17) point out that the definition of musicianship varies across disciplines and cultures and often ignores musical skills acquired informally during childhood. Building on these ideas, we propose that any claims about music and musicality should be considered in light of development in other domains and an individual's lifetime of musical experiences. We cannot ignore individual differences in how musical abilities mature, interact with other facets of development, and become driven by feedback loops if we want to understand the complex and idiosyncratic musical mind. Just as we should not assume that infants experience music the same way as adults, we should not assume that every adult's experience of music is similar. This entails better measures of prior musical experiences and the sociocultural contexts in which these experiences took place. It also requires careful consideration of the tasks and stimuli we use so that the diversity of individual experiences can be more purposefully integrated in the research process.

References

Al-Shawaf, L., Zreik, K.A., & Buss, D. M. (2018). 13 misunderstandings about natural selection. In T. K. Shackelford & V. A. Weekes-Shackelford (Eds.), *Encyclopedia of evolutionary science*. Springer.

Begel, V., Benoit, C.-E., Correa, A., Cutanda, D., Kotz, S. A., & Dalla Bella, S. (2017). "Lost in time" but still moving to the beat. *Neuropsychologia*, *94*, 129–138.

Benetti, L., & Costa-Giomi, E. (2019). Infant vocal imitation of music. *Journal of Research in Music Education, 67*(4), 381–398. https://doi.org/10.1177/0022429419890328.

Braun Janzen, T., Thompson, W. F., & Ranvaud, R. (2014). A developmental study of the effect of music training on timed movements. *Frontiers in Human Neuroscience, 8*, 801.

Bronfenbrenner, U. (1999). Environments in developmental perspective: Theoretical and operational models. In S. L. Friedman & T. D. Wachs (Eds.), *Measuring environment across the life span: Emerging methods and concepts* (pp. 3–28). American Psychological Association.

Burger, B., Thompson, M. R., Luck, G., Saarikallio, S. H., & Toiviainen, P. (2014). Hunting for the beat in the body: On period and phase in music-induced movement. *Frontiers in Human Neuroscience, 8*, 903. https://doi.org/10.3389/fnhum.2014.00903.

Chang, H., & Trehub, S. E. (1977). Infants' perception of temporal grouping in auditory patterns. *Child Development, 48*(4), 1666–1670.

Cheung, V. K. M., Harrison, P. M. C., Meyer, L., Pearce, M. T., Haynes, J.-D., & Koelsch, S. (2019). Uncertainty and surprise jointly predict musical pleasure and amygdala, hippocampus, and auditory cortex activity. *Current Biology, 29*(23), 4084–4092.e4.

Cirelli, L. K. (2018). How interpersonal synchrony facilitates early prosocial behavior. *Current Opinion in Psychology, 20*, 35–39.

Cirelli, L. K., Jurewicz, Z. B., & Trehub, S. E. (2020). Effects of maternal singing style on mother-infant arousal and behavior. *Journal of Cognitive Neuroscience, 32*(7), 1213–1220.

Cirelli, L. K., Peiris, R., Tavassoli, N., Recchia, H., & Ross, H. (2020). It takes two to tango: Preschool siblings' musical play and prosociality in the home. *Social Development, 29*, 964–975.

Cirelli, L. K., Spinelli, C., Nozaradan, S., & Trainor, L. J. (2016). Measuring neural entrainment to beat and meter in infants: Effects of music background. *Frontiers in Neuroscience, 10*, 229.

Cirelli, L. K., & Trehub, S. E. (2018). Infants help singers of familiar songs. *Music & Science, 1*, 205920431876162.

Cirelli, L. K., & Trehub, S. E. (2019). Dancing to Metallica and Dora: Case study of a 19-month-old. *Frontiers in Psychology, 10*, 1073.

Cirelli, L. K., & Trehub, S. E. (2020). Familiar songs reduce infant distress. *Developmental Psychology, 56*(5), 861–868.

Clancy, K. B. H., & Davis, J. L. (2019). Soylent is people, and WEIRD is white: Biological anthropology, whiteness, and the limits of the WEIRD. *Annual Review of Anthropology, 48*, 169–186.

Corbeil, M., Trehub, S. E., & Peretz, I. (2016). Singing delays the onset of infant distress. *Infancy, 21*(3), 373–391.

Creel, S. C., Weng, M., Fu, G., Heyman, G. D., & Lee, K. (2018). Speaking a tone language enhances musical pitch perception in 3–5-year-olds. *Developmental Science, 21*(1), e12503.

Demany, L., McKenzie, B., & Vurpillot, E. (1977). Rhythm perception in early infancy. *Nature, 266*, 718–719.

Deutsch, D., Henthorn, T., & Lapidis, R. (2008). The speech-to-song illusion. *Journal of the Acoustical Society of America, 124*(4), 2471.

Drake, C., Jones, M. R., & Baruch, C. (2000). The development of rhythmic attending in auditory sequences: Attunement, referent period, focal attending. *Cognition, 77*(3), 251–288.

Eerola, T., Himberg, T., Toiviainen, P., & Louhivuori, J. (2006). Perceived complexity of Western and African folk melodies by Western and African listeners. *Psychology of Music, 34*(3), 337–371.

Eerola, T., Luck, G., & Toiviainen, P. (2006). An investigation of preschoolers' corporeal synchronization with music. *Proceedings of the 9th International Conference on Music Perception and Cognition* (pp. 472–476). Bologna, Italy.

Einarson, K. M., & Trainor, L. J. (2016). Hearing the beat: Young children's perceptual sensitivity to beat alignment varies according to metric structure. *Music Perception, 34*, 56–70.

Ferguson, B., & Waxman, S. (2017). Linking language and categorization in infancy. *Journal of Child Language, 44*(3), 527–552.

Fernald, A. (1989). Intonation and communicative intent in mothers' speech to infants: Is the melody the message? *Child Development, 60*(6), 1497–1510.

Folland, N. A., Butler, B. E., Payne, J. E., and Trainor, L. J. (2015). Cortical representations sensitive to the number of perceived auditory objects emerge between 2 and 4 months of age: Electrophysiological evidence. *Journal of Cognitive Neuroscience, 27*(5), 1060–1067.

Fujii, S., Watanabe, H., Oohashi, H., Hirashima, M., Nozaki, D., & Taga, G. (2014). Precursors of dancing and singing to music in three- to four-month-old infants. *PLOS ONE, 9*, e97680.

Gerry, D. W., Faux, A. L., & Trainor, L. J. (2010). Effects of Kindermusik training on infants' rhythmic enculturation. *Developmental Science, 13*(3), 545–551.

Gold, B. P., Pearce, M. T., Mas-Herrero, E., Dagher, A., & Zatorre, R. J. (2019). Predictability and uncertainty in the pleasure of music: A reward for learning? *Journal of Neuroscience, 39*(47), 9397–9409.

Gordon, R. L., Shivers, C. M., Wieland, E. A., Kotz, S. A., Yoder, P. J., & McAuley, J. D. (2015). Musical rhythm discrimination explains individual differences in grammar skills in children. *Developmental Science, 18*(4), 635–644.

Hannon, E. E., & Johnson, S. P. (2005). Infants use meter to categorize rhythms and melodies: Implications for musical structure learning. *Cognitive Psychology, 50*, 354–377.

Hannon, E. E., Lévêque, Y., Nave, K. M., & Trehub, S. E. (2016). Exaggeration of language-specific rhythms in English and French children's songs. *Frontiers in Psychology, 7*, 939.

Hannon, E. E., Soley, G., & Levine, R. S. (2011). Constraints on infants' musical rhythm perception: Effects of interval ratio complexity and enculturation. *Developmental Science, 14*, 865–872.

Hannon, E. E., Soley, G., & Ullal, S. (2012). Familiarity overrides simplicity in rhythmic pattern perception: A cross-cultural examination of American and Turkish listeners. *Journal of Experimental Psychology: Human Perception and Performance, 38*, 543–548.

Hannon, E. E., & Trehub, S. E. (2005a). Metrical categories in infancy and adulthood. *Psychological Science, 16*, 48–55.

Hannon, E. E., & Trehub, S. E. (2005b). Tuning in to musical rhythms: Infants learn more readily than adults. *Proceedings of the National Academy of Sciences, 102*, 12639–12643.

Hannon, E. E., Vanden Bosch der Nederlanden, C. M., & Tichko, P. (2012). Effects of perceptual experience on children's and adults' perception of unfamiliar rhythms. *Annals of the New York Academy of Sciences, 1252*(1), 92–99.

Hartshorne, J. K., & Germine, L. T. (2015). When does cognitive functioning peak? The asynchronous rise and fall of different cognitive abilities across the life span. *Psychological Science, 26*, 433–443.

Hartshorne, J. K., Tenenbaum, J. B., & Pinter, S. (2018). A critical period for second language acquisition: Evidence from ⅔ million English speakers. *Cognition, 177*, 263–277.

Henrich, J. (2020). *The WEIRDest people in the world: How the West became psychological peculiar and particularly prosperous*. Farrar, Straus and Giroux.

Higgins, C. I., Campos, J. J., & Kermoian, R. (1996). Effect of self-produced locomotion on infant postural compensation to optic flow. *Developmental Psychology, 32*(5), 836–841.

Huron, D., & Ollen, J. (2003). Agogic contrast in French and English themes: Further support for Patel and Daniele (2003). *Music Perception, 21*(2), 267–271.

Ilari, B. (2005). On musical parenting of young children: Musical beliefs and behaviors of mothers and infants. *Early Child Development and Care, 175*(7–8), 647–660.

Iyer, V. (2012). Liner notes to *Accelerando*. ACT Music.

Jacoby, N., & McDermott, J. H. (2017). Integer ratio priors on musical rhythm revealed cross-culturally by iterated reproduction. *Current Biology, 27*, 359–370.

Kalendar, B., Trehub, S. E., & Schellenberg, E. G. (2013). Cross-cultural differences in meter perception. *Psychological Research, 77*, 196–203.

Kirschner, S., & Ilari, B. (2014). Joint drumming in Brazilian and German preschool children: Cultural differences in rhythmic entrainment, but no prosocial effects. *Journal of Cross-Cultural Psychology, 45*(1), 137–166.

Kirschner, S., & Tomasello, M. (2009). Joint drumming: Social context facilitates synchronization in preschool children. *Journal of Experimental Child Psychology, 102*(3), 299–314.

Koelsch, S., Vuust, P., & Friston, K. (2018). Predictive processes and the peculiar case of music. *Trends in Cognitive Sciences, 23*(1), 63–77.

Kragness, H. E., Johnson, E. K., & Cirelli, L. K. (2021). The song, not the singer: Infants prefer to listen to familiar songs, regardless of singer identity. *Developmental Science*, e13147.

Kragness, H. E., Swaminathan, S., Cirelli, L. K., & Schellenberg, E. G. (2021). Individual differences in musical ability are stable over time in childhood. *Developmental Science*, e13081.

Ladányi, E., Persici, V., Fiveash, A., Tillmann, B., & Gordon, R. L. (2020). Is atypical rhythm a risk factor for developmental speech and language disorders? *WIREs Cognitive Science*, *11*(5).

Manning, F., & Schutz, M. (2013). "Moving to the beat" improves timing perception. *Psychonomic Bulletin & Review*, *20*(6), 1133–1139.

Masataka, N. (1999). Preference for infant-directed singing in 2-day-old hearing infants of deaf parents. *Developmental Psychology*, *35*(4), 1001–1005.

Matthews, T. E., Witek, M. A. G., Lund, T., Vuust, P., & Penhune, V. B. (2020). The sensation of groove engages motor and reward networks. *NeuroImage*, *214*, 116768.

McAuley, J. D., Jones, M. R., Holub, S., Johnston, H. M., & Miller, N. S. (2006). The time of our lives: Life span development of timing and event tracking. *Journal of Experimental Psychology: General*, *135*(3), 348–367.

Mehr, S. A., Song, L. A., & Spelke, E. S. (2016). For 5-month-old infants, melodies are social. *Psychological Science*, *27*(4), 486–501.

Mendoza, J. K., & Fausey, C. M. (2021). Everyday music in infancy. *Developmental Science*, e13122.

Nakata, T., & Trehub, S. E. (2011). Expressive timing and dynamics in infant-directed and non-infant-directed singing. *Psychomusicology: Music, Mind and Brain*, *21*(1–2), 45–53.

Nave, K. M., Snyder, J. S., & Hannon, E. E. (in preparation). Sustained subjective musical beat perception develops through adolescence, and is related to phonological processing.

Nave-Blodgett, J. E., Hannon, E. E., & Snyder, J. S. (2021). Auditory superiority for perceiving the beat level but not measure level in music. *Journal of Experimental Psychology: Human Perception and Performance*, *47*(11), 1516–1542.

Nave-Blodgett, J. E., Snyder, J. S., & Hannon, E. E. (2021). Hierarchical beat perception develops throughout childhood and adolescence and is enhanced in those with musical training. *Journal of Experimental Psychology: General*, *150*, 314–339.

Naveda, L., & Leman, M. (2010). The spatiotemporal representation of dance and music gestures using topological gesture analysis (TGA). *Music Perception*, *28*, 93–112.

Nazzi, T., Bertoncini, J., & Mehler, J. (1998). Language discrimination by newborns: Toward an understanding of the role of rhythm. *Journal of Experimental Psychology: Human Perception and Performance*, *24*(3), 756–766.

Panneton, R. K. (1985). *Prenatal experience with melodies: Effects on postnatal auditory preference in human newborns*. [Unpublished doctoral dissertation]. University of North Carolina–Greensboro.

Partanen, E., Kujala, T., Tervaniemi, M., & Huotilainen, M. (2013). Prenatal music exposure induces long-term neural effects. *PLOS ONE, 8*(10), e78946.

Patel, A. D., & Daniele, J. R. (2003). An empirical comparison of rhythm in language and music. *Cognition, 87*(1), B35–B45.

Pearce, M. T. (2018). Statistical learning and probabilistic prediction in music cognition: Mechanisms of stylistic enculturation. *Annals of the New York Academy of Sciences, 1423*(1), 378–395.

Phillips-Silver, J., & Trainor, L. J. (2005). Feeling the beat: Movement influences infant rhythm perception. *Science, 308*(5727), 1430.

Polverini-Rey, R. A. (1992). *Intrauterine musical learning: The soothing effect on newborns of a lullaby learned prenatally*. [Unpublished doctoral dissertation]. California School of Professional Psychology.

Reigado, J., & Rodrigues, H. (2017). Vocalizations produced in the second year of life in response to speaking and singing. *Psychology of Music, 46*(5), 626–637.

Repp, B. H., & Su, Y.-H. (2013). Sensorimotor synchronization: A review of recent research (2006–2012). *Psychonomic Bulletin & Review, 20*, 403–452.

Rocha, S., & Mareschal, D. (2017). Getting into the groove: The development of tempo-flexibility between 10 and 18 months of age. *Infancy, 22*(4), 540–551.

Rottman, J. (2014). Evolution, development, and the emergence of disgust. *Evolutionary Psychology, 12*, 417–433.

Satt, B. J. (1984). *An investigation into the acoustical induction of intrauterine learning*. [Unpublished doctoral dissertation]. California School of Professional Psychology.

Smith, N. A., & Trainor, L. J. (2008). Infant-directed speech is modulated by infant feedback. *Infancy, 13*(4), 410–420.

Snyder, J. S., & Krumhansl, C. L. (2001). Tapping to ragtime: Cues to pulse finding. *Music Perception, 18*, 455–489.

Sole, M. (2017). Crib song: Insights into functions of toddlers' private spontaneous singing. *Psychology of Music, 45*(2), 172–192.

Soley, G., & Hannon, E. E. (2010). Infants prefer the musical meter of their own culture: A cross-cultural comparison. *Developmental Psychology, 46*(1), 286.

Soley, G., & Spelke, E. S. (2016). Shared cultural knowledge: Effects of music on young children's social preferences. *Cognition, 148*, 106–116.

Sowiński, J., & Dalla Bella, S. (2013). Poor synchronization to the beat may result from deficient auditory-motor mapping. *Neuropsychologia, 51*, 1952–1863.

Swaminathan, S., Kragness, H. E., & Schellenberg, E. G. (2021). The musical ear test: Norms and correlates from a large sample of Canadian undergraduates. *Behavior Research Methods*. https://doi.org/10.3758/s13428-020-01528-8.

Swaminathan, S., & Schellenberg, E. G. (2020). Musical ability, music training, and language ability in childhood. *Journal of Experimental Psychology: Learning, Memory, and Cognition, 46*, 2340–2348.

Swaminathan, S., Schellenberg, E. G., & Khalil, S. (2017). Revisiting the association between music lessons and intelligence: Training effects or music aptitude? *Intelligence, 62*, 119–124.

Thompson, E. C., White-Schwoch, T., Tierney, A., & Kraus, N. (2015). Beat synchronization across the lifespan: Intersection of development and musical experience. *PLOS ONE, 10*, e0128839.

Toiviainen, P., Luck, G., & Thompson, M. R. (2010). Embodied meter: Hierarchical eigenmodes in music-induced movement. *Music Perception, 28*, 59–70.

Trainor, L. J. (1996). Infant preferences for infant-directed versus noninfant-directed playsongs and lullabies. *Infant Behavior & Development, 19*(1), 83–92.

Trehub, S. E., & Gudmundsdottir, H. R. (2015). Mothers as singing mentors for infants. In G. F. Welch, D. M. Howard, & J. Nix (Eds.), *The Oxford handbook of singing*, (pp. 455–469). Oxford University Press.

Trehub, S. E., & Hannon, E. E. (2009). Conventional rhythms enhance infants' and adults' perception of musical patterns. *Cortex, 45*, 110–118.

Turino, T. (2008). *Music as social life: The politics of participation*. University of Chicago Press.

Ullal, S., Hannon, E. E., & Snyder, J. S. (2014). Tapping to a slow tempo in the presence of simple and complex musical meters reveals experience-specific biases for processing music. *PLOS ONE, 9*(7), e102962.

Vella, E. J., & Mills, G. (2017). Personality, uses of music, and music preference: The influence of openness to experience and extraversion. *Psychology of Music, 45*(3), 338–354.

Winkler, I., Háden, G. P., Ladinig, O., Sziller, I., & Honing, H. (2009). Newborn infants detect the beat in music. *Proceedings of the National Academy of Sciences, 106*, 2468–2471.

Witek, M. A. G., Clarke, E. F., Wallentin, M., Kringelbach, M. L., & Vuust, P. (2014). Syncopation, body-movement and pleasure in groove music. *PLOS ONE, 9*(4), e94446. https://doi.org/10.1371/journal.pone.0094446.

Wojcik, E. H., Lassman, D. J., & Vuvan, D. T. (2022). Using a developmental-ecological approach to understand the relation between language and music. *Frontiers in Psychology, 13*, 762018. https://doi.org/10.3389/fpsycg.2022.762018.

Woodruff-Carr, K., Perszyk, D. R., & Waxman, S. R. (2021). Birdsong fails to support object categorization in human infants. *PLOS ONE, 16*(3), e0247430. https://doi.org/10.1371/journal.pone.0247430.

Woodruff-Carr, K., White-Schwoch, T., Tierney, A. T., Strait, D. L., & Kraus, N. (2014). Beat synchronization predicts neural speech encoding and reading readiness in preschoolers. *Proceedings of the National Academy of Sciences, 111*, 14559–14564.

Yoshida, K. A., Iversen, J. R., Patel, A. D., Mazuka, R., Nito, H., Gervain, J., & Werker, J. F. (2010). The development of perceptual grouping biases in infancy: A Japanese-English cross-linguistic study. *Cognition, 115*(2), 356–361.

Zentner, M., & Eerola, T. (2010). Rhythmic engagement with music in infancy. *Proceedings of the National Academy of Sciences, 107*(13), 5768–5773.

Zhang, L., Xie, S., Li, Y., Shu, H., & Zhang, Y. (2020). Perception of musical melody and rhythm as influenced by native language experience. *Journal of the Acoustical Society of America, 147*(5), EL385–EL390.

9 The Science of Music Is about Relations

Jim Sykes

An Ethnomusicologist in the Science-Music Borderlands

How should we conceptualize the perceived border between the science of music (in psychology, biology, neuroscience, and so on) and music's place in the humanities and social sciences (ethnomusicology, musicology, music theory)? Is it a sharp dividing line between distinct cultures or modes of inquiry? Based on the evidence in this volume, I think the answer is no. But if we speak of Science-Music Borderlandia not as a sharp partition but as a region that encompasses the science-humanities border, what kind of place is it? It will be useful to consider Alex Chavez's point that "although the scholarly field of border studies and the metaphorical use of the borderlands are often conflated, they are distinct" (2017, p. 11). Border studies tend to explore the material conditions of physical spaces, while the use of borderlands as a metaphor typically "speak[s] of a liminal state of in-betweenness in work in the humanities, largely cultural studies" (ibid.). In this volume, liminality bespeaks "the emotional residue of an unnatural boundary" or borderland between the music-sciences and music-humanities (Anzaldúa, quoted in ibid.).[1]

With this in mind, let us say that Science-Music Borderlandia (SMB) contains "districts" (subfields) that straddle the oft-politicized music-science and music-humanities borders. To those who immigrate to SMB, its outsider status lends it the romance of the underdog. There is also a culture of access in SMB that allows its residents to travel fairly easily across its own internal science-humanities border. But those who spend a long time in SMB feel the frustrations of a rural province neglected by the metropoles—with a concomitant lack of funding. SMB's underdog status, however, looks radically different as one moves farther away from it. The boundary between SMB and several regions of the music-humanities remains heavily policed: SMBers are usually happy to let outsiders in, but many in the music-humanities metropoles don't want SMBers visiting them. Ethnomusicologists, in particular, have long viewed SMB with suspicion: the specter of comparative musicology, its imposition of European musical values and conceptions of the human under the rubric of science, and the concomitant history of racism lurking in discussions of music as a biological phenomenon loom large. Ethnomusicologists fought their own border war with SMB a few generations back, and this remains a national trauma in Ethnomusicology-land.

In this chapter, my goal is to show how my own broader engagement with anthropology (not anthropological studies of music) helped me resolve problems in my own research that I now attribute to the tenacity of a certain musicological vocabulary that is widely shared across the music-humanities and music-sciences (and is, in fact, generative of them). In what follows, I explore the potential—for all of us—of perspectives that have been brewing for a long time in anthropology but recently gained steam in ethnomusicology as part and parcel of the "nonhuman turn" in the humanities. I will touch on a few other anthropological ideas along the way that are underused by music scholars of all stripes: the distributed person, the gift, sound-as-medium, and art-as-agency.[2] In this, my aim is to push forward vocabularies that can be useful across the music-humanities and music-sciences, regardless of whether these areas are brought into dialogue. In doing so, I strive to move beyond the typical references to ethnomusicology made within music science literature, which tend to hinge on old writings (Nettl, Blacking) and two major but well-worn ideas: that not every society has a word for music and that, in some places, everyone's a musician.

My starting point is the idea that loosening our attachment to music will allow researchers to place more attention on sound as a medium and agent, situating relations as central to music studies. This will require a rethinking of scale: for ethnomusicologists, this means stepping outside our area studies paradigm and reconceptualizing the global; for SMBers, it means better recognizing sound as a form of exchange in which the relations among sound, self, Others, place, and time may be normatively construed over the *longue durée* quite differently from how the history of scholarship in all disciplines of music studies normatively suggests. For ethnomusicologists, such turns should result in geographic innovations as a result of ontological innovations; for SMBers, the opposite. My aim is to do more than simply advocate for a "relational musicology" (Born, 2010); I suggest that turning to sound-as-medium and sound-as-agent, in tandem with attention to the porosity of human beings, will allow us to see *similarities* between the West and the rest, a division that is still common in the music-sciences and music-humanities, even when scholars attempt to overcome it. This means that ethnomusicology should *not* be positioned as relevant for scientific studies of music simply by virtue of its ability to provide examples of difference that can help refine scientists' notion of musical universals.[3] Nor do I think ethnomusicologists should set out simply to disprove scientists' arguments by providing evidence of musical differences from our own fieldwork—what Steven Feld (see chapter 19 of this volume) calls the "not among the bongo bongo" syndrome.[4] Rather, I suggest that a science of sound-as-relations will generate a perspective on the universal that presumes its radical diversity: "The single world is precisely the place where an unlimited set of differences exist . . . far from casting doubt on the unity of the world, these differences are its principle of existence" (Badiou, 2008, p. 39).[5] My broader claim—which I can only hint at here—is that the categories "music" and "musician" have been overdetermined in *both* the music-sciences and the music-humanities by *capitalism*. The perspective I build here is intended to demonstrate that displacing categories generated by capital (and Christianity, on which capitalist framings and uses of music initially drew) can allow a different understanding of how sound-as-relations functions broadly across human and nonhuman species.

What You Might Have Expected

Initially, I went to the field to study a drumming tradition performed by Sri Lanka's Sinhala Buddhist ethnic majority. I had a rather traditional ethnomusicological interest: I had heard a recording of a drumming genre and could not count its rhythms, which sounded wildly sentence-like and uncountable in duple or triple meter, even though the drummers played in unison and started and stopped at the same time. I wondered how the drumming was structured. What I was hearing was the "low-country drum" (*pahata rata beraya*), referred to in certain rituals as the "demon drum" (*yak beraya*), an instrument used in all-night rituals where offerings are made to the Buddha and deities to ask for protection from natural calamities or (in another set of rituals) to stave off illnesses caused by beings of low karmic standing (*yakku*). Though I didn't realize it initially, the genre has long been performed by a caste of ritualists called the Beravā (a word that means "drummer"—*bera* is the plural of *beraya*, or "drum"), a name that is quite stigmatized in Sri Lanka due to the historically downtrodden nature of this community. Some drummers call themselves *näeketi* (astrologer), and most are also dancers, Ayurvedic medicinal specialists, ritualists who know the required mantras, singers, and builders of complex and voluminous ritual objects and backdrops made largely out of bamboo (Kapferer, 1983). In precapitalist times, this community received agricultural land in exchange for services at Buddhist temples (a caste duty called *rajakariya*). Now, many have given up their ritual duties, or they perform in rituals all night long and then hop on a bus in the morning to arrive bleary-eyed at their nine-to five jobs.

Everything I've described so far probably resonates with many readers as typical ethnomusicology. It surely fits squarely with an image of ethnomusicology that many (including myself) decry as the study of "the music of the Other" (see Feld, chapter 19). As a white, male, cisgendered ethnomusicologist with a grant in hand and attached to a PhD program at a North American university, I traveled to a distant place and "discovered" musical difference. I set out to learn the low-country drum. I attended rituals with my *gurunnanse* (teacher), Herbert Dayasheela. Eventually, I "uncovered" the Beravā worldview about this ritual drumming. Let me continue in this vein before I unpack some of the roadblocks I faced and consider certain anthropological perspectives that helped me move beyond them.

First, as many ethnomusicologists discover during their field research (Wong, 2014), I learned that Sinhala Buddhist ritual drumming is not considered music. It is sacred speech. According to the standard tenets of Theravada Buddhism (the dominant type of Buddhism in Sri Lanka and Southeast Asia), serious practicing Buddhists should not revel sensuously in music and dance—one should never see a Buddhist monk dancing, for example. All Buddhists accept the Five Precepts (a list that includes dicta such as thou shall not steal, murder, or commit adultery), but a longer list taken up by monks and the laity on full-moon (*poya*) days and other important events includes refraining from dance and music. All beings are ordered cosmically according to their karma, with humans occupying a middle ground between deities, on the one hand, and ghosts and demons (*yakku*), on the other—the latter having low karma on account of deeds in past lives (Holt, 2004). The Buddha has left the realm of rebirth but has delegated authority to the "four warrant gods" (*hatara varam deviyo*). Although he is given offerings at rituals, these are generally considered symbolic. Two of the warrant gods (Natha or Maitreya, the next Buddha; and Vishnu, the protector of the island) receive offerings but are so karmically high that rituals are not typically oriented toward them—they are too removed from human affairs. Major rituals center on offerings to deities a bit lower on the karmic scale, particularly the two warrant gods Kataragama and Pattini (the latter being the sole goddess of the four; Obeyesekere, 1984). Because these gods are also Buddhist—that is, they are karmically high beings farther ahead on the journey to Buddhahood than you or I—it is inappropriate to offer them music. What the drummers offer is sacred speech. Even the pitches of the singing in the rituals are heavily constrained—which is to say, they do not have a large pitch range, perhaps to avoid sounding musical (a technique used also by Buddhist monks for chant or *pirith*—though singers do not consciously make this connection). To foreign ears, the singing sounds a bit like chanting, but Sinhala sung poetry is different from chanting and quite complex, as there are numerous variations that emerge at different points in the rituals.

Finally, in a set of rituals to ward off illnesses caused by beings of low karmic standing, demons (a term criticized for its Christian baggage; Scott, 1994) or *yakku* are celebrated as though they were gods. They are drawn to the ritual space, whereupon a ritualist utters mantras to get them to stop harming the patient. Drumming for *yakku* is not metered either; it sounds speech-like to foreign ears, but the patterns are shorter and more repetitive than the drum stanzas (*padas*) offered to deities.

The immediate question I faced in my research was how to situate this drumming in its social context. The drumming, which is almost always performed with dancers (traditionally, only men dance in rituals, although women commonly dance outside of ritual contexts), has played a role in Sinhala Buddhist nationalism, a movement that has tended to alienate minorities, sometimes violently. It seemed wrong to say that the tradition I was studying was a reflection or production of Sinhala identity, for the drum stanzas traditionally belong to the Beravā, not the entire Sinhala ethnic group. Buddhists don't conceive of the self as a stable, eternal, and unchanging soul; it is always in flux and ultimately a fiction. What actually exists, the Theravada scriptures (*Tipitaka*) state, is a not-self (*anatta*; Collins, 1982). Although the Beravā have a complex and sometimes ambiguous relationship to doctrinal Buddhism, I realized that I needed to ask how Beravā drum speech relates to a not-self. Ultimately, through discussions with drummers and explorations of the vocabulary surrounding the drumming, I discovered that drumming functions as a gift that emerged in a noncapitalist environment but now overlaps with it. Several occasions when drumming is performed are classified as "sound offerings" (*sabda pujava*); in all ritual contexts I studied, drumming was an offering to a nonhuman being, except when performed for visiting dignitaries or to support monks in procession. Gifts in Theravada Buddhism must be "disinterested," which means that gift giving is typically not recognized as such (Kapferer, 1983). So the idea of drumming as a gift was underrecognized by the drummers, but once I began to politely probe into it and examine the Beravā vocabulary, it became a core concept for me—in fact, the word *gift* appears in the title of my book about Sri Lankan musics (Sykes, 2018a). It bears emphasizing, too, that the drummers don't believe they composed this sacred speech—it was handed down to them. One ritual I studied (Sykes, 2018b), for example, consists of the drum speech uttered by the gods when they celebrated the Buddha's Awakening (what is traditionally referred to in Western scholarship as his Enlightenment).

To sum up: The music I studied is not music but speech; it is not reflective of a stable interior soul and isn't "expressive" in the normative Western sense; and it is the caste's duty to hoard and protect these gifts, with the Beravā traditionally hiding their knowledge from the broader Sinhala ethnic group. Much of the drumming was composed by the gods, not by people, and the entire tradition is passed down through the generations. My teacher believes this tradition is dying in part because it was situated in schools for the arts rather than in schools for Ayurvedic medicine—delegitimizing the efficacy of the rituals, transforming them into culture, and eliminating people's need for the rituals as medicine. I feared that describing the drumming as music and situating it as a Sinhala genre would continue the process of killing his tradition and play into the ethnicization process that has long been part of Sri Lanka's communal conflicts.

This is the point where readers, particularly the SMB community, might expect me to lean into my research as an example of "musical difference" with which I can chastise scientists' universal conceptions of music in which the Beravā don't fit. But that's not my intention. First, remember that the Beravā *do* have a conception of music—and Beravā drumming is structured to avoid being that. This notion of music historically came from neighboring communities, particularly the Tamil minority, so this notion of music is not equivalent to the normative Western version. What I suggest is that music and speech in and across diverse global contexts can likewise be conceived as a gift that connects (and alienates) people to (and from) Others, so long as we recognize that definitions of sound and its function and notions of personhood, community, space, and time might radically diverge. One problem, though, is that our normative language in the academy for discussing music undercuts this perspective at every turn: music is not a caste duty but reflects a broader communal identity (race, class, gender, nation); music says something about a self or soul, rather than being a tool that connects with Others or that protects or heals. It does not take much effort to see that this perceived relationship between music and the self (and, by extension, our standard framings of the relationship between traditional music genres and their communities) is—if not overtly Judeo-Christian in origin—certainly un-Buddhist. From here, we can grasp that our normative language for representing the music-self relation positions numerous Others, not just Buddhists, as though they are simply wrong about what a person is, what music is, and how they relate. But what if we should be viewing things the other way around? What if (for example) Buddhists are right in believing that the self is unstable and inherently changing, that sounds can have some efficacious power separate from us and affecting us, that sound is fundamentally a thing to be offered, and that sound acts as an agent in its own right? I'm issuing a reverse universalism here, so to speak, which I (playfully? I'm not sure) want to probe further. It is here that recent developments in anthropology—drawing on numerous older intellectual genealogies—can help situate my work and conjure possible avenues for research in SMB that might make a broader rapprochement between ethnomusicology and SMB possible.

Capitalism and Christianity

The category WEIRD (white, educated, industrialized, rich, and democratic; Henrich, 2019) is often used as shorthand for a type of society that is taken to be self-evident. This concept has been discussed and critiqued elsewhere in this volume, so here, I merely posit WIRED societies instead (white-dominated, [historically] imperialistic, rich, [former or current] empires [that] don't always export democracy).[6] The Western classical tradition, including certain perspectives on tonality, developed in WIRED societies, spread through the mass media, and were transformed by musicians in genres such as jazz, gospel, rock, and hip-hop. If we are to recognize such broad influences, though, we need to move beyond formalism and acknowledge that such musical concepts were diffused through musical labor, which includes selling concert tickets, performing onstage, and making recordings featuring certain normative technologies (from jazz horn sections to guitars, drum sets, and laptops). But now we're talking about capitalism, a phenomenon that is hardly reducible to WIRED societies, even if key aspects of music-in-capitalism (e.g., the first record labels) germinated from them. Put another way, when Beravā drum speech finds its way onto a recording, the ontology of the gift and its godly authorship is easily obscured by a cover (or thumbnail) depicting the name of a human who now appears as the artist. Rather than alienate workers from their product, the musical commodity circulates (even digitally) with their names on it. (By contrast, as agricultural laborers, the Beravā were experts in "keeping-while-giving" [Weiner, 1992]: they offered sounds to deities while retaining the ritual knowledge.) It is easy to show that the "proper object" of music in capitalism played a larger transformative role than any part of the WEIRD acronym—and this capitalist framing of musical labor now far transcends such WEIRD societies (which is a misnomer anyway).

I suggest that rather than focus on how music is transformed by capitalism, we need to attend to the conditions in which contexts for sonic exchange and notions of the music-person relation were transformed, promoting an ontology of music in which it is intimately related to a stable self or identity and expressive of it, a concept of musical personhood that became globally, historically normative ("in nature"). How often are precapitalist forms of musical labor (of which music-as-caste is just one example) and the relationships to land tenure that such forms imply taken as globally normative for what musical labor is or was? If one's answer is that such forms of life no longer exist, it would be easy to point to a small library's worth of books to show that this is not true—and the past few hundred years are a small sliver of human existence. But what I really mean to ask is, has capitalism (rather than Western classical music or WIRED epistemologies per se) overdetermined the musical object and person-music relations as these are explored experimentally in laboratory contexts and ethnography?

The typical narrative of transformation in Western music history has long been attributed in large part to Romantic notions of musical transcendence, an attitude that has been deemed uniquely Western (another surfacing of the WEIRD concept). But musical transcendence is not what makes the West unique—many societies conceptualize their music as transcendent, albeit in different ways. Consider that some Beravā drum stanzas come from the gods, have existed unchanged over the *longue durée*, and are conceived as being separate from society and unaffected by it. Elsewhere, I note that even a casual perusal of the textbook *Excursions in World Music* generates eight easy-to-find examples—from indigenous communities in the Americas to numerous communities in disparate parts of Asia—in which music is conceived as a gift from gods or spirits (Sykes, 2020a, p. 3). It is important to state that I am not trying to surface some notion of premodern or precolonial authenticity—which is the stuff that drives ethnonationalist movements—but rather noting similarities among myriad diverse traditions around the globe in which humans are not considered the composers of music and music is fundamentally about an exchange with nonhumans. The West was not the first to invent musical transcendence, but it did invest heavily in a Judeo-Christian notion of the soul that achieved broad political power in tandem with the decimation of "pagan" musical ontologies in which musical offerings to gods were fundamental—and I think this is what undergirds the capitalist ontology of music that spread globally. Its global reach—through colonialism, missionization, and the recording industry—influences researchers to accept the Judeo-Christian notion of the music-self relation as natural, since they see it reflected in the musical practices they study, which have been shaped by capitalism and its Christian ontology of music and personhood (elsewhere, I term this feedback loop "secular resonance"; Sykes, 2020b).

Scaling Up, Down, and Sideways

In a book about the continued relevance of anthropology in today's world, Anand Pandian (2019) explores at length how early-twentieth-century anthropologists argued for "the psychic unity of mankind" while issuing warnings about the relationship between "description" and "explanation." Malinowski emphasized "the unity between European habits of thought and the thinking of Trobriand Islanders," stating that "the native mind works according to the same rules as ours" (quoted in Pandian, 2019, p. 5). Margaret Mead noted that Franz Boas "saw the scientific task as one of progressive probing into a problem now of language, now of physical type, now of art style," but Boas also told his students that "no probe must go too far lest it lead to premature generalization—a development which he feared like the plague and against which he continually warned us" (quoted in ibid., p. 23). Participant observation, James Clifford wrote many years later, "serves as shorthand for a continuous tacking between the 'inside' and 'outside' of events: on the one hand grasping the sense of specific occurrences and gestures empathetically, on the other stepping back to situate these meanings in wider contexts" (quoted in ibid., p. 26). Pandian describes participant observation as

> the idea that the significance of things depends on the local contexts that give rise to them, that there is little value to work that does not enter into these local situations. At the same time, however, there is also the idea that this significance can only be grasped ultimately by stepping away from these particular contexts, by recasting them in relation to an analytical framework of greater scale. (2019, p. 26)

This is far different from the mere documentation of cultural difference. It's an emphasis on taking interlocutors' conceptual frameworks seriously as a base from which to ask broader questions, rather than developing the questions first and testing them out among various peoples. I'm sure many scientists have moved beyond this latter framing and might view my description as caricature, but it bears emphasizing that this is a significant reason why ethnomusicologists became distrustful of the "science of music." Why not, we ask, find some aspect of music or sound among specific peoples and then use that to ask broader questions that might have greater significance?

Interrogating the relationship between description and explanation remains important for anthropologists today. Tim Ingold (2019), for example, argues that art and anthropology are alike, in that they are both "future-oriented disciplines, united in the common task of fashioning a world fit for coming generations to inhabit." He believes that research is an "open-ended search for truth and a practice of correspondence" that "necessarily overflows the boundaries of objectivity" and is thus "a form of experience":

> In experience, things are with us in our thoughts, dreams, and imaginings, and we with them. It is here, I believe, that we can begin to see where science can align with art, and indeed with anthropology. It means calling into question the division between fact and fantasy, truth and illusion, which has underpinned the development of science ever since the days of Francis Bacon and Galileo Galilei. (Ingold, 2019, pp. 613–614)

My point here is not to advocate for Ingold's position per se but to provide a glimpse of an anthropological trend in which the turn toward subjectivity and flux has been removed from the Kantian-derived correlationism that dominated postmodern and poststructuralist thinking, where the dominant idea was that "we only ever have access to the correlation between thinking and being, and never to either term considered apart from the other" (Meillassoux, 2006, p. 5). Note that Ingold openly endorses the terms *truth* and *correspondence*, while also emphasizing the impossibility of complete objectivity. I'm not making this statement because I think it will be revelatory for scientists; rather, I'm showing that some points of convergence likely exist between scientists of music and anthropologists.[7]

Sticking with Ingold a bit longer, elsewhere he argues for understanding sound as a medium rather than a context in which sonic objects might be akin to images in a landscape: "When we look around on a fine day, we see a landscape bathed in sunlight, not a lightscape. Likewise, listening to our surroundings, we do not hear a soundscape. For sound, I would argue, is not the object but the medium of our perception. It is what we hear in. Similarly, we do not see light but see in it." He goes on to compare sound with wind, noting, "After all, the wind whistles, and people hum or murmur as they breathe. Sound, like breath, is experienced as a movement of coming and going, inspiration and expiration. If that is so, then we should say of the body, as it sings, hums, whistles or speaks, that it is *ensounded*" (Ingold, 2007, p. 12).

Sound has agency on human bodies, as anyone who has heard a loud fire alarm knows; but Ingold's point is revelatory for my purposes for pointing out not the agency of sound but the porosity of human beings (see also Kapchan, 2015). There is a long history of anthropological studies documenting the porous nature of humans, most famously writings on the "partible person" by Marilyn Strathern, Nancy Munn, and Annette Weiner. Strathern claims that for Melanesians, humans are "frequently construed as the plural and composite site of the relationships that produce them," and "it is at the point of interaction that a singular identity is established" (1988, pp. 13, 128). "From this perspective," Roger Sansi notes, "people are constantly being made and re-made through relations, and things are constantly being created not in contradistinction to persons but 'out of persons' (Strathern 1988: 172)" (2015, p. 11). Similarly, researchers have noticed in sub-Saharan Africa that "folk ideologies of the person" tend to "stipulate ontological interdependencies between persons and the material and spiritual domains" (McIntosh, 2009, p. 17; Comaroff, 1991). South Indianists, too, have noticed an "ontological consubstantiality" between humans and "the communities, places, and objects with which the person interacts" among Tamil people (McIntosh, 2009, p. 17). Many other examples could be given from across the globe.

Barad (2003) notes that "the belief that nature is mute and immutable and that all prospects for significance and change reside in culture is a reinscription of the nature/culture dualism that feminists have actively contested. Nor, similarly, can a human/nonhuman distinction be hardwired into any theory that claims to take account of matter in the fullness of its historicity" (quoted in Kapchan, 2015, p. 42).[8] Viveiros de

Castro (1998) notes that different animals understand the relations between nature and culture, including that between human and nonhuman animals, in different ways and proposes *multinaturalism*—many natures, many cultures. Similarly, the new materialist trend, drawing on now-classic works by Latour (2005), Deleuze and Guattari (1987), Bennett (2010), and others, aims to "foster an approach that operates in terms of human-nonhuman assemblages, tangles of heterogeneous things, and hybrid formations beyond the nature-culture divide" (Bräunlein, 2016, p. 378). As Bräunlein puts it, "Despite the divergent aims and ambitions of theorists of New Materialism, one assumption shared by all is that the 'fall of man' began with the erroneous anthropocentric logic of binaries, mainly the opposition between subject and object" (2016, p. 378).

Alfred Gell's (1998) work on the agency of the art object is famous in anthropology and art scholarship but has received surprisingly little attention in music studies (Born, 2010). I cite Sansi's summary of Gell's work at length:

> Gell proposed to look at works of art as indexes of agency. Indexes of agency are the result of intentions: "Whenever an event is believed to happen because of an 'intention' lodged in the person or thing which initiates the causal sequence, that is an instance of 'agency'" (Gell 1998: 17). To have intentions means to have a mind. The "life" we attribute to things, and works of art in particular, would be the result of a process of abduction or indirect inference of a "mind" in a thing. Artworks don't just index the agency of the artist, but of all the agents that have been "entrapped" by the artwork: they contain their distributed person, or distributed mind, which for Gell were the same thing. Gell argued that works of art can be seen as persons "because as social persons, we are present not just in our singular bodies, but in everything in our surroundings which bears witness to our existence, our attributes, and our agency" (1998: 103). For Gell this is not an exotic belief but on the contrary, he affirms that works of art are some of the more accomplished objectifications of human agency. Artworks can contain several different agencies from the artist, to the person represented or the person who commissioned it, to the person who bought it, to the curator that displays it. An artwork can be a "trap of agencies" sometimes contradictory, sometimes complementary (Gell 1998). (Sansi & Strathern, 2016, p. 427)

Sansi notes the similarity between anthropology and the art world, exploring the latter through one of anthropology's core concepts—the gift—but shifting the emphasis away from agency and toward a broader notion of art-as-relationality: "The form of the artwork is in the relations it establishes: to produce a form is to create the conditions for an exchange." Describing the artist as a mediator, he notes that contemporary artworks are often conceived as gifts that are "free, spontaneous, personal, and disinterested events," as opposed "to commodification and mass consumption" (Sansi & Strathern, 2016, p. 426). This conception moves past Mauss's famous definition of the gift as "*On se donne en donnant*" (Mauss, 2003, p. 227)—one gives oneself while giving—which hinges on a notion of the gift as obligation and necessitating reciprocity.

This latter perspective could make the relationality, processes of exchange, and distributed person in artworks seem to be a kind of false consciousness. As Steingo (2016) notes, since the advent of the New Musicology, the job of the critical music scholar has often seemed to require unmasking the social conditions behind the production of music. Steingo asks us to take his South African interlocutors' position toward their music seriously, which one of them describes as "the positive side of escapism." Though his interlocutors, who perform the electronic music genre Kwaito, do not engage social conditions directly, Steingo argues that they double reality to generate "a new sensory reality." More strongly, he argues that "there is nothing in 'music' that magically separates it from life. On the contrary, *music is the very name of this separation*—a separation that requires a very particular sensory apparatus and a very particular set of operations carried out through that apparatus" (Steingo, 2016, pp. 6, 9; emphasis in original). This relationality is not binary *in nature* but rather a psychological condition akin to the production of identity and commonly mistaken for it by the types of music scholars Steingo criticizes, in which the relational conditions of identity disappear. Consider Lacan's mirror stage—the proposition that one gains an awareness of self when one distinguishes oneself from the Other—which has been taken up in anthropology to explain what Willford calls the "ethnic fetish": "postcolonial demarcations of ethnic boundaries, and with them hierarchical assertions of an ethnosymbolic hierarchy . . . can produce uncanny doubles . . . [W]e might ask whether the disavowal or surmounting of the Other can produce uncanny doublings which, in turn, fuel overidentification with the ego ideal, or in Heidegger's words, 'ensnare' the subject" (2006, p. 2). The point is not to debunk the idea of musical traditions linked to ethnic heritage but to note that they emerge in dialogic or dialectical fashion—relationally—with what they are not. Music, the creation of a new sensory reality through processes of gift giving and exchange that ensounds bodies, is easily posited as the natural reflection of a stable self or community divorced from the relations that produced it. And it is precisely this notion, I suggest, that the music-humanities and music-sciences have tended to treat as what music just naturally *is*.

Conclusion

In sum, I suggest that studying canonical SMB concepts such as the musical mind and the musician-nonmusician dichotomy requires understanding how decisions about and experiences of ensoundment occur through social conditions that produce (and are produced by) relations of relations—such as between differently construed (and differently valued) humans in a sociopolitical context and between social constructions of musical labor that generate understandings of sonic labor in relation to other forms of labor. A science of music is a science of relations in which partible and distributed persons emerge in sounds that, while doubling reality and existing as "trapped agencies," act to ensound bodies. To my ears, such studies must take seriously non-Western sciences of music, not through the guise of a medical anthropology or an anthropological "music therapy" where difference from science is presumed and kept separate, but as a core part of a global "history of science" approach to music. Ayurvedic medicine, of which Beravā sonic exchange with nonhuman beings is a component, is one example. The point is not to substantiate their particular beliefs about what music does or what a person is but to better recognize the global ubiquity of sonic exchange and the partible person that has been obscured by the influence of Western rationalism, the Western concept of the individual, and the Romantic notion of music as expressive of an interior, enclosed self, all of which have come to seem natural and universal through their embedding in the commodification of music in capitalism.

In turn, I ask my fellow ethnomusicologists whether we might allow scientific studies that explore topics such as music perception and the evolution of music from a variety of perspectives that provincialize European-derived conceptions of sound, music, the human, and social relations. Clearly, what we have long resisted is the presumption by some scientists that European-derived notions of music and the human have some sort of global ubiquity (or, to put it another way: we cringe when such perspectives are unwittingly adopted to shape research questions and methods, as though they are naturally what music *is*). But what if twenty-first-century scientists are different? If we are, on principle, against all scientific studies of music, we need to ask why. If the answer is the legacy of a racist past or the hegemony of European-derived perspectives, we need to support a science of music that moves past these problems. If the response has to do with the impossibility of finding universal perspectives on music, this book shows that many scientists are not concerned with universals at all (and it bears emphasizing that my goal is not to argue in favor of musical universals but to offer new ways to generate questions of broad significance by taking seriously the world's many epistemologies). I have suggested that an array of underresearched vocabularies about music—pertaining to the distributed person, art-as-agency, sound-as-medium, and sound-as-gift—are historically and globally significant, even when (or perhaps *because*) they emerge through radical diversity. My hope is that recognizing and working with such concepts will generate new conversations about the perceived border between the scientific and humanistic studies of music.

Notes

1. It is also important to recognize the lack of *physical* engagement between music departments and those who study music in the "hard" sciences, as we are typically housed in separate buildings. However, this border is crossed surprisingly frequently in fleeting public events at universities (often well publicized) that purport to bring together the sciences, arts, and humanities.

2. Bear in mind that this chapter is not a review of this literature, which would require a very long essay.

3. Tim Ingold describes anthropologists in terms that also fit ethnomusicologists: "Anthropologists do their thinking, talking and writing in and with the world . . . we do our philosophy out of doors" (2011, pp. 241–242). I suggest that ethnomusicology should be viewed as a broad domain of contrasting opinions in which critical thinking about music and sound has developed "out of doors" through dialogic, embodied, engaged encounters with communities.

4. It's worth emphasizing to scientists that ethnomusicologists typically define their field by method rather than region—from this perspective, ethnomusicology involves conducting ethnography (fieldwork) and is not the study of non-Western musics (as it is characterized even at times in this volume). It is not by definition the study of "the music of the Other." One could do an ethnomusicological study of scientists in a lab at a North American university conducting experiments on music, so long as the anthropologist stuck around long enough to do participant observation. This definition of ethnomusicology-as-ethnography has its own problems, of course. For example, since ethnomusicologists are often positioned as the people who study non-Western musics, the latter are still routinely accessed through ethnography in music scholarship rather than history, generating a presentist view of the past.

5. See also Sykes (2020a). I recognize Feld's classic concept of "acoustemology," and while I feel it is essential, it is far better publicized than the perspectives I draw on here. Readers should think of my argument as being in dialogue with acoustemology (e.g., Feld, 2015, 2017), not challenging or supplanting it.

6. There are two main reasons (among many) why I do not support the use of WEIRD for understanding music (in history and in the present day): first, Western classical music emerged mainly in non-English-speaking countries and today is performed by diverse peoples (such as East Asians); and second, certain countries assumed to export Western classical music, such as Britain in the nineteenth century, were not democratic.

7. In a much earlier article, Ingold advocates for an "alternative biology" that "comprehends the social life of persons as an aspect of organic life in general," such that "an anthropology of persons" would be "encompassed within a biology of organisms whose focus is on processes rather than events, replacing the 'population thinking' of Darwinian evolutionary biology with a logic of relationships" (1990, p. 208). I don't pretend to know where biology is regarding this subject matter since Ingold published this essay in 1990.

8. Ana Maria Ochoa Gautier (2016) has criticized ecomusicology for reproducing the nature-culture distinction through studies that demarcate nature sounds and soundscapes as set apart from the human and culture.

References

Badiou, A. (2008). The communist hypothesis. *New Left Review, 49*, 29–42.

Bennett, J. (2010). *Vibrant matter: A political ecology of things*. Duke University Press.

Born, G. (2005). On musical mediation: Ontology, technology and creativity. *Twentieth-Century Music, 2*(1), 7–36.

Born, G. (2010). For a relational musicology: Music and interdisciplinarity, beyond the practice turn. *Journal of the Royal Musical Association, 135*(2), 205–243.

Bräunlein, P. (2016). Thinking religion through things: Reflections on the material turn in the scientific study of religion/s. *Method & Theory in the Study of Religion, 28*(4/5), 365–399.

Chavez, A. (2017). *Sounds of crossing: Music, migration, and the aural poetics of Huapango Arribeño*. Duke University Press.

Collins, S. (1982). *Selfless persons*. Cambridge University Press.

Comaroff, J. (1991). *Of revelation and revolution: Christianity, colonialism, and consciousness in South Africa* (Vol. 1). University of Chicago Press.

Deleuze, G., & Guattari, F. (1987). *A thousand plateaus: Capitalism and schizophrenia*. University of Minnesota Press.

Feld, S. (2015). Acoustemology. In D. Novak and M. Sakakeeny (Eds.), *Keywords in sound* (pp. 12–21). Duke University Press.

Feld, S. (2017). On post-ethnomusicology alternatives. In *Perspectives on a 21st century comparative musicology: Ethnomusicology or transcultural musicology*? Nota—Valter Colle/Udine.

Gell, A. (1998). *Art and agency*. New York: Oxford University Press.

Henrich, J. (2019). *The WEIRDest people in the world: How the West became psychologically peculiar and particularly prosperous*. Farrar, Straus and Giroux.

Holt, J. C. (2004). *The Buddhist Vishnu: Religious transformations, politics, and culture*. Columbia University Press.

Ingold, T. (1990). An anthropologist looks at biology. *Man (N.S.), 25*, 208–229.

Ingold, T. (2007). Against soundscape. In A. Carlyle (Ed.), *Autumn leaves: Sound and the environment in artistic practice* (pp. 10–13). Double Entendre.

Ingold, T. (2011). *Being alive: Essays on movement, knowledge, and description*. Routledge.

Ingold, T. (2019). Art and anthropology for a sustainable world. *Journal of the Royal Anthropological Institute, 25*, 659–675.

Kapchan, D. (2015). Body. In D. Novak and M. Sakakeeny (Eds.), *Keywords in sound*. Duke University Press.

Kapferer, B. (1983). *A celebration of demons: Exorcism and the aesthetics of healing in Sri Lanka*. Smithsonian Books.

Latour, B. (2005). *Reassembling the social: An introduction to actor-network-theory*. Oxford University Press.

Mauss, M. (2003) [1925]. *The gift: Forms and function of exchange in archaic societies*. W. W. Norton.

McIntosh, J. (2009). *The edge of Islam: Power, personhood, and ethnoreligious boundaries on the Kenya coast*. Duke University Press.

Meillassoux, Q. (2006). *After finitude: An essay on the necessity of contingency*. Bloomsbury.

Obeyesekere, G. (1984). *The cult of the goddess Pattini*. University of Chicago Press.

Ochoa Gautier, A. M. (2016). Acoustic multinaturalism, the value of nature, and the nature of music in ecomusicology. *Boundary 2*, *43*(1), 107–141.

Pandian, A. (2019). *A possible anthropology: Methods for uneasy times*. Duke University Press.

Sansi, R. (2015). *Art, anthropology and the gift*. Routledge.

Sansi, R., & Strathern, M. (2016). Art and anthropology after relations. *HAU Journal of Ethnographic Theory*, *6*(2), 425–439.

Scott, D. (1994). *Formations of ritual: Colonial and anthropological discourses on the Sinhala Yak Tovil*. University of Minneapolis Press.

Steingo, G. (2016). *Kwaito's promise: Music and the aesthetics of freedom in South Africa*. University of Chicago Press.

Strathern, M. (1988). *The gender of the gift*. University of California Press.

Sykes, J. (2018a). *The musical gift: Sonic generosity in post-war Sri Lanka*. Oxford University Press.

Sykes, J. (2018b). On the sonic materialization of Buddhist history: Drum speech in southern Sri Lanka. *Analytical Approaches to World Music*, *6*(2), 1–78.

Sykes, J. (2020a). The anthropocene and music studies. *Ethnomusicology Review*, *22*(1), 4–21.

Sykes, J. (2020b). The secularism of music studies. *Yale Journal of Music & Religion*, *6*(2), 119–143.

Viveiros de Castro, E. (1998). Cosmological deixis and Amerindian perspectivism. *Journal of the Royal Anthropological Institute*, *4*(3), 469–488.

Weiner, A. (1992). *Inalienable possessions: The paradox of keeping-while-giving*. University of California Press.

Willford, A. (2006). *Cage of freedom: Tamil identity and the ethnic fetish in Malaysia*. University of Michigan Press.

Wong, D. (2014). Sound, silence, power, music. *Ethnomusicology*, *58*(2), 347–353.

Interlude

10 Toward Neurotechnology for Musical Creativity

Eduardo Reck Miranda

[*Editors' note*: Moving toward a truly coevolutionary view of music and the mind entails going beyond the standard model of a unidirectional arrow from brain to mind in music perception and production. While technologies that would realize music directly from the brain in transparent, unidirectional ways have appeared in science fiction, the practice of developing and working with such tools supports a co-constitutive view. Brain-computer music interfaces are informed by a rich tradition in experimental music (e.g., see Alvin Lucier's 1965 *Piece for Solo Performer*; Leslie's chapter 13 in this volume). Recent developments continue to open up novel possibilities for musical creativity, testing theories in the music sciences such as embodiment and extended mind theory (cf. Witek's chapter 7 in this volume) while offering practical applications for music production in those whose physical limitations may hinder more conventional forms of music making. As a model for new directions in the science-music borderlands, we asked Eduardo Miranda, an expert in brain-computer music interfacing, to describe some of his work in this area.]

Introduction

Imagine if you could play a musical instrument with signals detected directly from your brain. Would it be possible to generate music that represents brain activity? What would the music of our brains sound like? These are some of the questions addressed by research into music neurotechnology,[1] a relatively new field of investigation that is emerging at the crossroads of neurobiology, engineering sciences, and music. Systems that interact directly with the human nervous system (Rosenboom, 2003), sonification methods to diagnose brain disorders (Vialatte et al., 2012), and biocomputing devices (Braund & Miranda, 2015) are emerging as plausible technologies for musical creativity. Such things were unthinkable until very recently.

Many recent advances in the neurosciences, especially in computational neuroscience, have led to a deeper understanding of the behavior of individual and large groups of biological neurons. This enables artists and musicians to apply biologically informed functional paradigms to problems of creativity, design, and control, such as building mind-controlled musical instruments.

An increasingly better understanding of the brain, combined with the emergence of sophisticated brain scanning technology, is enabling the development of brain-computer interfaces (BCIs). BCIs have tremendous potential to facilitate active music making by people with severe physical impairments, such as paralysis after a stroke or an accident damaging the spinal cord. In addition, BCIs present new ways to harness creative practices.

This chapter discusses two projects being conducted at the University of Plymouth's Interdisciplinary Centre for Computer Music Research in the UK. One concerns the development of methods to compose music inspired and informed by neurobiology. More specifically, we created *Symphony of Minds Listening*, an experimental symphonic piece in three movements based on functional magnetic resonance imaging (fMRI) brain scans. Then we introduced our work into developing brain-computer music interfacing (BCMI) technology and created a composition and performance using that technology.

Listening to Minds Listening

Symphony of Minds Listening is based on the fMRI brain scans from three persons—a ballerina, a philosopher, and a composer[2]—while they listened to the second movement of Ludwig van Beethoven's Seventh Symphony.

In a nutshell, we deconstructed the Beethoven movement to its essential elements and stored them with information representing their structural features. Then we reassembled (or remixed) these elements into a new composition, but with a twist: the fMRI information influenced the process of remixing the music. However, *Symphony of Minds Listening* was scored for the same instrumentation as Beethoven's Seventh Symphony.

The fMRI brain scanning method measures brain activity by detecting changes in blood flow. The brain images were collected using equipment and parameters that are typical in the field of cognitive neuroscience. The measurements can be presented graphically by color-coding the strength of activation across the brain. Each scanning session generated sets of fMRI data, each of which was associated with one measure of the second movement of Beethoven's Seventh Symphony.

The score of Beethoven's movement was deconstructed with custom-made artificial intelligence software that extracted statistical information about the structure of the music (Gimenes & Miranda, 2011). We used this information to reconstruct the Beethoven movement, but the reconstruction process was influenced by the fMRI data; effectively, the fMRI data altered the original music.[3] Not surprisingly, the fMRI scans differed among the three listeners. Therefore, activity from three different brains yielded three different movements for the resulting composition. Each of the movements displayed varying degrees of resemblance to the original symphony.

The Compositional Process

The compositional process involved manual and computer-automated procedures. Historically, there have been two approaches to using computer-generated materials in composition: *purist* and *utilitarian*. The purist approach to computer-generated music tends to be more concerned with the correct application of the rules programmed into the system than with the musical results per se. In this case, the output of the computer is considered the final composition. The composer would not normally modify the music at this point, as this would meddle with the integrity of the model or the system. At the other end of the spectrum is the utilitarian approach, adopted by those composers who consider output from the computer raw material for further work. These composers would normally tweak the results to fit their aesthetic preferences, to such an extent that the system's output might not be identifiable in the final composition. Obviously, the line dividing these two approaches is blurred, as practices combining aspects of both are common. Although *Symphony of Minds Listening* was composed with a balanced approach, it tends toward the utilitarian. This author advocates the use of computers as assistants to the creative process, rather than as autonomous composing machines. (For a discussion of how science and technology can inform and inspire the act of musical composition, see Miranda, 2013, 2014b; chapters 13 and 14 in this volume.)

The composition of the symphony evolved in tandem with the development of a piece of generative music software, referred to as MusEng. MusEng was programmed with artificial intelligence to learn musical information from given pieces and use this information to generate new music. The system has three distinct phases of operation: *learning*, *generative*, and *transformative*.

The learning phase takes a musical score and analyzes it to determine a number of musical features. A data set comprising these features and rules representing the likelihood of given features appearing in the data are then stored in memory. During the generative phase, these data inform the generation of new sequences, which ideally should resemble the sequences used to train the system in the first phase. Finally, at the transformative phase, the outcome from the generative phase is modified according to a number of transformation algorithms. It is during this final phase that the fMRI information is used to influence the resulting music. Note that we are not interested in a system of rules that reproduces an exact copy of the original music. Rather, we are interested in producing new music that resembles the original. Hence the transformative phase was added to further modify the results from the generative phase. The role of fMRI information is to control the extent of the transformation. Essentially, stronger activity in a given statistical component of the fMRI data results in greater transformation of the musical outcome.

For the composition of *Symphony of Minds Listening*, the first step was to deconstruct the score of Beethoven's composition into a set of basic materials. These materials were then given to MusEng for processing.

First, Beethoven's piece was divided into 13 sections, ranging from 5 to 26 measures in length. The 13 sections informed the overarching structure of each of the three movements of the new symphony. This provided a template for the new piece, which preserved the overall form of the original Beethoven movement.

Note that MusEng did not learn the whole Beethoven piece at once. Rather, it was trained on a section-by-section basis. The musical sequences for the respective new sections of the new movements were generated independently from each other. For instance, section 1 of the movement *Ballerina* has 26 measures and was composed based on materials from the first 26 measures of Beethoven's music. Section 2 has 24 measures and was composed based on materials from the next 24 measures (27–50) of Beethoven's music, and so on.

A block diagram portraying the compositional process is shown in figure 10.1. The blocks with thicker borders represent procedures that can be influenced and/or controlled by fMRI results. After the music's segmentation into 13 sections, the flow of action bifurcates into two possibilities: manual handling of the segments (left-hand side of figure 10.1) or computerized handling with MusEng (right-hand side of figure 10.1). A discussion of manual handling is beyond the scope of this chapter.

Finally, once a new segment has been generated, it is orchestrated and appended to the respective score of the new movement. Occasionally, the fMRI results also influenced instrumentation and orchestration. For instance, in *Philosopher*, the second movement, different independent components (ICs) were associated with groups of instruments in the orchestra (IC 25 = violins and violas, IC 15 = trumpets and horns, and so on); these associations changed from section to section. So, if the flute is to play in measure *x* of *Philosopher*, the IC value of the respective component in measure *x* of Beethoven's music would define how the flute player should produce the notes.

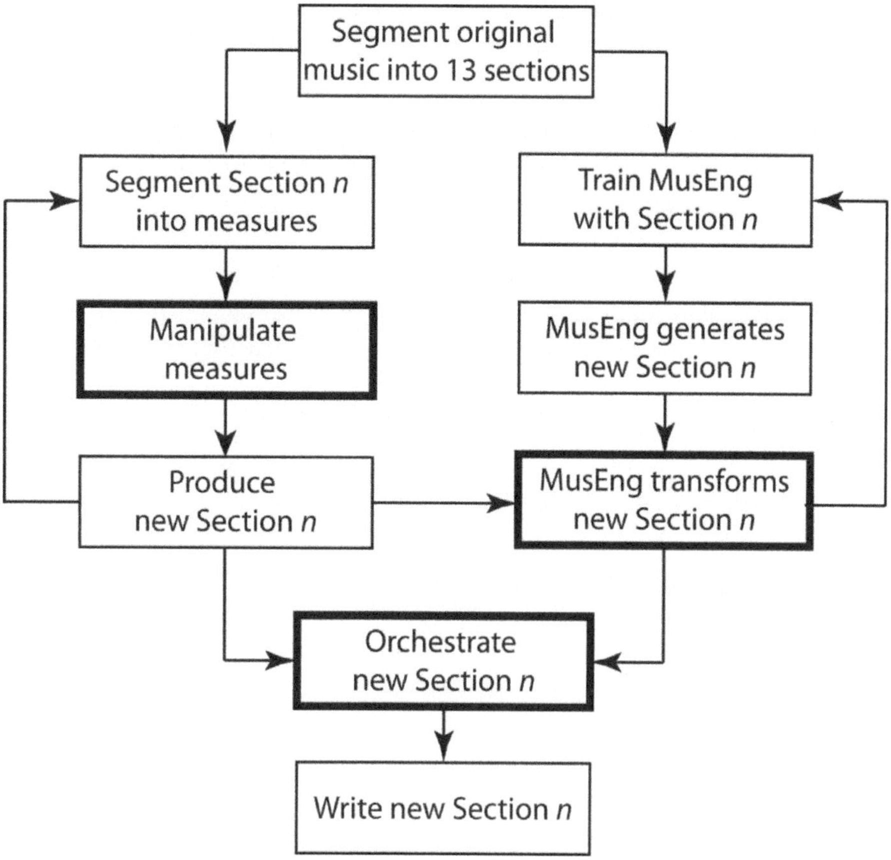

Figure 10.1
Block diagram of the overall compositional process.

We defined various tables mapping IC activity to instrumental playing techniques and other musical parameters, such as onto a continuum of musical dynamics. A detailed technical explanation of the learning and generative phases of the MusEng system is beyond the scope of this chapter (the reader is invited to consult Miranda, 2014a, for more information).

Brain-Computer Music Interfacing

BCI technology allows a person to control devices by commands expressed as brain signals, which are detected through brain monitoring technology (Dornhege et al., 2007).

We are interested in developing BCI technology for music (BCMI).[4] Our research is aimed at music therapy and people with special needs, particularly those with severe physical disabilities but with relatively preserved cognitive functions. Severe brain injury, spinal cord injury, and locked-in syndrome result in weak, minimal, or no active movement, which curbs the ability to play a musical instrument. People with these conditions are currently either excluded from music recreation and therapy or able to engage only in a less active manner through listening or receptive methods (Miranda et al., 2011).

Currently, the most viable and practical method of detecting brain signals for BCMI is the electroencephalogram (EEG), which records electrical signals through electrodes placed on the scalp. The EEG expresses the overall electrical activity of millions of neurons. It is a difficult signal to detect because it is extremely faint. Moreover, the signal is filtered by the membranes that separate the cortex from the skull, the skull itself, and the scalp. To be used in BCI, this signal needs to be amplified significantly and harnessed through signal processing techniques (Miranda, 2010; Miranda et al., 2014).

In general, power spectrum analysis is the most commonly used method of analyzing the EEG signal. In simple terms, power spectrum analysis breaks the EEG signal into different frequency bands and reveals the distribution of power between them. This is useful because it is believed that specific distributions of power in the spectrum of the EEG can encode different cognitive behaviors.

As far as BCI systems are concerned, the most important frequency activity in the EEG spectrum lies below 40 Hz. Recognized bands of EEG activity below 40 Hz, also referred to as EEG rhythms, are associated with specific states of mind. For instance, the frequencies falling between 8 and 13 Hz, referred to as alpha rhythms, are usually associated with a state of relaxed wakefulness, such as during meditation. The exact boundaries of these bands are not clearly defined, and the meanings of these associations can be contentious. In practice, however, the exact meaning of EEG rhythms is not crucial for a BCI system. What is crucial is the ability to establish whether users can voluntarily produce power within distinct frequency bands. For instance, we have used alpha rhythms in an early proof-of-concept BCMI system that enabled a person to switch between two types of generative algorithms to produce music on a musical instrument digital interface (MIDI)–controlled Disklavier piano in the style of Robert Schumann (when alpha rhythms were detected in the EEG) and Ludwig van Beethoven (when alpha rhythms were not detected) (Miranda, 2006).

Broadly speaking, there are two approaches to manipulating EEGs for BCI: *conscious effort* and *operant conditioning*. Conscious effort induces changes in the EEG when the subject engages in specific cognitive tasks designed to produce specific EEG activity (Miranda et al., 2004; Curran & Stokes, 2003). The cognitive task used most often is motor imagery because it is possible to detect changes in the EEG of a subject who is imagining moving a limb, such as a hand (Dornhege et al., 2007). Operant conditioning involves the presentation of a task in conjunction with some form of feedback, which allows the user to develop a somewhat unconscious control of the EEG (Kaplan et al., 2005).

A steady-state visual evoked potential (SSVEP) is a robust paradigm for BCI, as long as the user is not severely visually impaired. Typically, visual stimuli representing tasks to be performed are presented to a user on a computer monitor; such tasks might include spelling words from an alphabet or selecting in which direction a wheelchair moves. Each target is encoded by a flashing visual pattern reversing at a unique frequency. To select a target, the user simply directs his or her gaze at the flashing pattern corresponding to the action to be performed. As the user's spotlight of attention falls on a particular target, the frequency of the unique pattern reversal rate can be accurately detected in the EEG through spectral analysis. It is possible to classify not only a user's choice of target but also the extent to which the user is attending to the target. This allows SSVEP-based BCI systems in which each target is not a simple binary switch but represents an array of options, depending on the user's level of attention. Effectively, each target of such a system can be implemented as a switch with a potentiometer.

An Initial SSVEP BCMI System

In 2011 we completed the implementation of our first SSVEP-based BCMI system, which we tested on a patient with locked-in syndrome at the Royal Hospital for Neurodisability in London. The system comprised four targets, shown on a computer screen in front of the patient. Each target image represented a different musical instrument and a sequence of notes (figure 10.2). Each image flashed, reversing its color (in this case, red) at different frequencies: 7 Hz, 9 Hz, 11 Hz, and 15 Hz, respectively. For instance if the person gazed at the image flashing at 15 Hz, the system activated the xylophone and produced a melody using a sequence of six notes associated with this target; these notes were set beforehand, and the number of notes could be other than six. The more the person attended to this icon, the more prominent the magnitude of the brain's SSVEP response to this stimulus, and vice versa. This produced a varying control signal, which was used to make the melody. Also, it provided visual feedback to the user; the size of the icon increased or decreased as a function of this control signal.

The melody was generated as follows: The sequence of notes was stored in an array whose index varied, in this case, from one to six. The amplitude of the SSVEP signal was normalized so that it could be used as an index sliding up and down through the array. As the signal varied, the corresponding index triggered the respective musical notes stored in the array.

Figure 10.2
Each target image is associated with a musical instrument and a sequence of notes.

The system required just three electrodes on the user's scalp: a pair placed on the region of the visual cortex and a ground electrode placed on the front of the head. Filters were programmed to reduce noise interference and artifacts such as those generated by blinking eyes or moving facial muscles. SSVEP data were then filtered via band-pass filters to measure the band power across the frequencies correlating to the flashing stimuli.

The patient took approximately fifteen minutes to learn how to use the system, and she quickly mastered how to make melodies by increasing and decreasing the level of her SSVEP signal. We collected suggestions and criticisms from the hospital staff and the patient with respect to improvements and future developments (Miranda et al., 2011). An important challenge emerged from this exercise: our system enabled a one-to-one interaction with a musical system, but it was immediately apparent that it would be desirable to design a system that would promote interaction among the participants.

Activating Memory and The Paramusical Ensemble

To address the above-mentioned challenge, we adopted a slightly different research methodology. We started by imagining a musical composition and a performance scenario. Only then did we consider how that would work in practice with our BCMI technology. To tackle the problem of lack of expressivity, we decided to have the user generate a score on the fly for a human musician to sight-read, instead of relaying it to a synthesizer for playback.

To promote group interaction, we determined that the composition had to be generated collectively by a group of participants. However, the generative process had to be simple and clearly understood by the participants. Also, the controlling-brain participants had to clearly feel that they were in control of what was happening with the music. Obviously, these were not trivial tasks. In the end, we established that the act of collectively generating the music in real time could be like playing a musical game, but with no winners or losers. We thought of designing something resembling a game of dominoes—that is, musical dominoes played by sequencing blocks of precomposed musical phrases selected from a pool. Finally, we created the concept of a musical ensemble where severely physically disabled and nondisabled musicians made music together: The Paramusical Ensemble. The result was the composition *Activating Memory*, a piece for eight participants: a string quartet and a BCMI quartet.

A new version of the SSVEP-based system was built. Each member of the BCMI quartet was furnished with a unit of the new SSVEP-based BCMI system. The system enabled them to generate a musical score in real time. Each participant generated a part for the string quartet, which was displayed on a computer screen for the respective string performer to sight-read during the performance (figure 10.3).

Figure 10.3
A rehearsal of The Paramusical Ensemble, with locked-in syndrome patients performing *Activating Memory*.

The new system worked similarly to the one described in the previous section, with the fundamental difference being that the visual targets were associated with short musical phrases. Moreover, instead of flashing images on a computer monitor, we built devices with flashing LEDs and LCD screens to display what the LEDs represented.

The LED devices increased the SSVEP response to the stimuli because we were able to produce more precise flashing rates than those produced using standard computer monitors. Moreover, the LCD screens provided an efficient way to change the set of options available for selection. And it promoted the notion that one was using a custom-made musical device to interact with others, rather interacting via a computer.

Activating Memory was generated on the fly by sequencing four voices of predetermined musical sections simultaneously. For each section, the system provided four choices of musical phrases, or riffs, for each part of the string quartet, which were selected by the BCMI quartet. The selected riffs for each instrument were relayed to the computer monitors facing the string quartet for sight-reading. While the string quartet was playing the riffs for a section, the system provided the BCMI quartet with another set of choices for the next section. Once the current section had been played, the new riffs chosen for each instrument were relayed to the musicians, and so on. To allow enough time for the BCMI quartet to make choices, the musicians repeated the respective riffs a few times. The system followed an internal metronome, which guaranteed synchronization. The Paramusical Ensemble's first public performance of *Activating Memory* took place on July 17, 2015, at the Royal Hospital for Neuro-disability in Putney, London.[5]

Concluding Remarks

This chapter examined how the neurosciences can be harnessed to develop technologies and methodologies for composing music. We introduced two new pieces of music and the respective technomethodologies developed to compose them.

With *Symphony of Minds Listening*, we introduced an approach to musical composition inspired by the notion that the neural patterns and corresponding mental images and events around us are creations of the brain prompted by the information we receive through our senses.

Even though humans have identical mechanisms for processing the basics of sound, music is a construction of the brain. There is increasing evidence that this construction differs from person to person. When we listen to music, sounds are deconstructed as soon as they enter the ear. Different streams of neuronally coded data travel through distinct auditory pathways toward cortical structures, such as the auditory cortex and beyond, where the data are reconstructed and mingled with data from other senses and memories into what is perceived as music (Thaut & Hodges, 2019; Arbib, 2013; Peretz & Zatorre, 2003).

Metaphorically speaking, the compositional approach we developed to compose *Symphony of Minds Listening* did to Beethoven's score what our hearing system does when we listen to music: sounds are deconstructed as they enter the ear and relayed through various pathways toward cortical structures, where the data are reconstructed into what is perceived as music.

The BCMI research behind *Activating Memory* has come a long way since the 1960s. Today, meaning derived from EEGs is better understood and easier to detect. However, it is still difficult to retrieve useful EEG data. Signal interference from external sources, unpredictable EEG information, and other physiological input are widely reported by the BCMI research community.

More generally, this chapter presented approaches to leverage our understanding of the brain to compose music. Every now and then, composers have been inspired by science to compose: works such as Gustav Holst's *The Planets Suite* (1918) and Philip Glass's *Einstein on the Beach* (1976) come to mind. Beyond compositions inspired by science, however, we advocate music *informed* by science. The compositions presented in this chapter prompted the author to become conversant with neuroscience and medical engineering. This created opportunities to gain insight and make scientific contributions. Thanks to increased access to scientific information and discovery (e.g., open access to research journals and freely available online repositories for academic prepublications), musicians and artists in general have an unprecedented opportunity to engage with the scientific community, not only to inform their creations but also to establish partnerships for the development of interdisciplinary projects that can impact both the arts and the sciences.

Acknowledgments

The work presented in the chapter would not have been possible without the expertise and active input from a number of collaborators, including Dan Lloyd (Trinity College, Hartford, Connecticut, USA), Zoran Josipovic (New York University, USA), Wendy Magee and Julian O'Kelly (Royal Hospital for Neuro-disability, London), John Wilson and Ramaswamy Palaniappan (University of Sussex, UK), and Duncan Williams and François Guegen (University of Plymouth, UK). I am indebted to my former PhD student Joel Eaton for his unparalleled software development skills.

Notes

1. The term *music neurotechnology* first appeared in print in 2009 in an editorial in *Computer Music Journal, 33*(1), 1.

2. The composer is this author.

3. We used a musical instrument digital interface (MIDI) representation of the score to process the music.

4. The phrase *brain-computer music interfacing*, or BCMI, was coined by this author to denote BCI systems for musical applications, and it has been adopted by the research community.

5. A video documentary is available at https://vimeo.com/143363985. A recording of one of the millions of possible renderings of *Activating Memory* is available at https://soundcloud.com/ed_miranda/activating-memory.

References

Arbib, M. A. (Ed.). (2013). *Language, music, and the brain*. MIT Press.

Braund, E., & Miranda, E. (2015). Biocomputer music: Generating musical responses with Physarum polycephalum–based memristors. *Proceedings of 11th Computer Music Multidisciplinary Research (CMMR15): Music, Mind, and Embodiment*. Plymouth University, Plymouth, UK.

Curran, E. A., & Stokes, M. J. (2003). Learning to control brain activity: A review of the production and control of EEG components for driving brain-computer interface (BCI) systems. *Brain and Cognition, 51*(3), 326–336.

Dornhege, G., del Millan, J., Hinterberger, T., McFarland, D., & Muller, K.-R. (Eds.). (2007). *Toward brain-computer interfacing*. MIT Press.

Gimenes, M., & Miranda, E. R. (2011). An ontomemetic approach to musical intelligence. In E. R. Miranda (Ed.), *A-life for music: Music and computer models of living systems* (pp. 261–286). A-R Editions.

Kaplan, A., Ya Kim, J. J., Jin, K. S., Park, B. W., Byeon, J. G., & Tarasova, S. U. (2005). Unconscious operant conditioning in the paradigm of brain-computer interface based on color perception. *International Journal of Neurosciences, 115*, 781–802.

Miranda, E. R. (2006). Brain-computer music interface for composition and performance. *International Journal on Disability and Human Development, 5*(2), 119–125.

Miranda, E. R. (2010). Plymouth brain-computer music interfacing project: From EEG audio mixers to composition informed by cognitive neuroscience. *International Journal of Arts and Technology, 3*(2/3), 154–176.

Miranda, E. R. (2013). On computer-aided composition, musical creativity and brain asymmetry. In D. Collins (Ed.), *The act of musical composition*. Ashgate.

Miranda, E. R. (2014a). Creative music technology with *Symphony of Minds Listening*. In E. R. Miranda & J. Castet (Eds.), *Guide to brain-computer music interfacing* (pp. 271–295). Springer.

Miranda, E. R. (2014b). *Thinking music*. University of Plymouth Press.

Miranda, E. R., Lloyd, D., Josipovic, Z., & Williams, D. (2014). Brain-computer music interfacing: Interdisciplinary research at the crossroads of music, science and biomedical engineering. In E. R. Miranda & J. Castet (Eds.), *Guide to brain-computer music interfacing* (pp. 1–28). Springer.

Miranda, E. R., Magee, W., Wilson, J. J., Eaton, J., & Palaniappan, R. (2011). Brain-computer music interfacing (BCMI): From basic research to the real world of special needs. *Music and Medicine*, *3*(3), 134–140.

Miranda, E. R., Roberts, S., & Stokes, M. (2004). On generating EEG for controlling musical systems. *Biomedizinische Technik*, *49*(1), 75–76.

Peretz, I., & Zatorre, R. J. (Eds.). (2003). *The cognitive neuroscience of music*. Oxford University Press.

Rosenboom, D. (2003). Propositional music from extended musical interface with the human nervous system. In G. Avanzini et al. (Eds.), The neurosciences and music. *Annals of the New York Academy of Sciences*, *999*, 263–271.

Thaut, M. H., & Hodges, D. A. (Eds.). (2019). *The Oxford handbook of music and the brain*. Oxford University Press.

Vialatte, F., Dauwels, J., Musha, T., & Cichocki, A. (2012). Audio representations of multi-channel EEG: A new tool for diagnosis of brain disorders. *American Journal of Neurodegenerative Disease*, *1*(3), 292–304.

III Beyond Reductionism

Volume Editors

Perhaps no issue is a more obvious flashpoint for humanistic and scientific approaches than reductionism. The opposing tendencies to favor either irreducible complexity or isolable causal factors, to prize the particular or the generalizable, come to the fore around cultural phenomena such as music. The potential utility and pitfalls of reductionism operate at multiple levels of research design and interpretation, from the isolation of discrete, measurable musical parameters to the explanation of complex results in terms of single deterministc causes. As Iyer has observed, something as commonplace as discussing a study's results in terms of what "people" tended to do—when the study recruited as participants only undergraduates at a single institution—can, at best, constitute a faulty overgeneralization and, at worst, seem to imply that only certain kinds of responses are "human" (personal communication, May 21, 2021; see also Iyer, 2016; Sears, 1986).

Examples of this kind of interpretive slippage abound. Broesch et al. (2020) trace the history of a task commonly used to assess whether children have developed the capacity for self-recognition. A mark is surreptitiously placed on their forehead, and if they reach up to inspect it after catching a glimpse of their reflection in a mirror, it suggests that they understand they're seeing an image of themselves. Most children attain this developmental milestone at around eighteen months of age. But when scientists started using this paradigm cross-culturally, even older children did not show this response. It turned out that in these cultural contexts, the experimenter who placed the mark on the forehead was viewed as an authority the children did not want to offend by seeming to protest or even notice the mark. Without cultural knowledge to inform interpretation, these results would falsely indicate the absence of a cognitive ability. Similarly, in a study based on field interviews in the Rimrock area of Arizona, McAllester (1954) posed a question he understood to be about aesthetic response: "how do you feel when you hear a drum?" Participants (who knew drums as instruments accompanying songs in particular ceremonies and healing rites) understood this question as an inquiry about their state of health. Without substantive knowledge about culture, experimental tasks designed to measure one thing might in fact reveal the influence of another.

The same principle holds true when designing stimuli and selecting samples. Consider, as a hypothetical case, a study that asked participants from a single US university to rate how much they enjoyed excerpts that had been manipulated to sound more like either jazz or electronica. Let's say the participants consistently rated the jazz-like excerpts higher. The researchers summarize their findings as indicating a preference for jazz over electronica, and media outlets run with the claim that science has shown the superiority of jazz. Such an interpretation ignores numerous variables relevant to the participants' ratings. It could be that the modifications to the electronica stimuli were less stylistically successful than the jazz modifications—the electronica stimuli were less electronica-like and the jazz stimuli were more jazz-like. Or it could be that the student body at this university listens extensively to jazz and little to electronica, whereas in other communities (not studied by the researchers), the opposite is true. To a humanist who has devoted a career to studying music within social and cultural frameworks, the problems with the relationship between experimental design and claimed findings are blaringly obvious: not just reductive but fundamentally wrong.

We see this state of affairs less as a crisis and more as an opportunity. Collaboration at the outset of a project between people with humanistic, scientific, and musical expertise can make it more likely that studies are framed, designed, and interpreted accurately. Proposing best practices for sustained collaborations among globally distributed researchers, Savage et al. (chapter 18) outline some of the considerations necessary to realize such integrations of diverse expertise, from the epistemological and methodological (e.g., the potential need not just for language translation but also for conceptual reframing of questions to suit local communities and cultural contexts) to the ethical and infrastructural (e.g., how to appropriately credit different kinds of research from contributors working within different structures of value). Leslie's chapter 13 argues that when composers are setting up musical experiences, they are often acting as a kind of music cognition researcher, and when music cognition researchers design stimuli and procedures for experiments, they are acting as a kind of composer. Leslie, an electronic musician and music cognition researcher who develops musical brain-computer interfaces, explores how the relationship between artistic and scientific practice might illuminate alternatives to reductionism. Pamela Z (chapter 15) offers insight from her experience as an experimental musician into the open-ended possibilities of what might be considered music and the wide range of audience reaction to her work.

Conversations about the dangers of reductionism aren't just coming from humanities and the arts; scholars from the cognitive sciences have been exploring new methods to address some of these limitations. Hasson, Nastase, and Goldstein observe that traditional approaches to controlled experiments in neuroscience rely "on a core commitment to the assumption that the neural computations supporting . . . cognitive functions can be decontextualized . . . into a handful of latent features that . . . are . . . human interpretable and can be manipulated in isolation, and that the piecemeal recomposition of these features will yield a satisfying understanding of brain and behavior"—an assumption that has not proved true (2020, p. 416). They argue for overparameterized, direct-fit models as a way forward. Similarly, observing that emotion categories do not seem to have distinct, nonoverlapping physiological fingerprints, Siegel et al. (2018) propose population thinking and the systematic study of variability as an antidote to a reliance on essentialism in research design. Faber and McIntosh (chapter 12) point to the power of complex systems as another possibility. They observe that approaches from complex systems allow for nonlinearities and complex interactions that experimental methods reliant on studying features in isolation cannot hope to explain. Deutsch's chapter 14 looks back at a long-standing controversy between rational and empirical approaches to music theory—the former favoring reduction to mathematical rules, the latter the complications of perceptual experience—and discusses how the advent of computer technology beginning in the 1950s dramatically changed the possibilities for the empirical study of music perception and cognition.

One challenge for reductive designs in music cognition is ecological validity. Raising similar concerns to our hypothetical jazz example, studies that reduce music to individual manipulable parameters gain experimental control but risk findings that are untenable or inapplicable to music—the phenomenon the experiment was trying to understand. When individual features such as pitch or timbre are studied outside their ordinary context, what suggests that they will operate the same way when reassimilated into a rich, complex, real-world musical experience? Williams and Sachs, neuroscientists who work on music, report in chapter 11 on new tools for analyzing responses to full-fledged musical excerpts rather than discrete elements drawn from them. They assess how these tools can and cannot address the challenges associated with reductionism.

Together, the chapters in this section chart a path forward for integrating humanistic and scientific approaches to study music empirically in ways that are more robust and likely to yield richer insights.

References

Broesch, T., Crittenden, A. N., Beheim, B. A., Blackwell, A. D., Bunce, J. A., Colleran, H., Hagel, K., Kline, M., McElreath, R., Nelson, R. G., Pisor, A. C., Prall, S., Pretelli, I., Stieglitz, J., & Mulder, M. B.

(2020). Navigating cross-cultural research: Methodological and ethical considerations. *Proceedings of the Royal Society B, 287*, 20201245.

Hasson, U., Nastase, S., & Goldstein, A. (2020). Robust-fit to nature: An evolutionary perspective on biological (and artificial) neural networks. *Neuron, 105*(3), 416–434.

Iyer, V. (2016). Improvisation, action understanding, and music cognition with and without bodies. In G. Lewis & B. Piekut (Eds.), *The Oxford handbook of critical improvisation studies*, Volume 1 (pp. 74–90). Oxford University Press.

McAllester, D. P. (1954). *Enemy way music: A study of social and esthetic values as seen in Navaho music*. Peabody Museum.

Sears, D. O. (1986). College sophomores in the laboratory: Influences of a narrow data base on social psychology's view of human nature. *Journal of Personality and Social Psychology, 51*(3), 515–530.

Siegel, E. H., Sands, M. K., Van den Noortgate, W., Condon, P., Chang, Y., Dy, J., Quigley, K. S., & Barrett, L. F. (2018). Emotion fingerprints or emotion populations? A meta-analytic investigation of autonomic features of emotion categories. *Psychological Bulletin, 144*(4), 343–393.

11 Combating Reductionism in Music Neuroscience with Ecologically Valid Paradigms: What Can (and Cannot) Be Gained?

Jamal Williams and Matthew Sachs

Introduction

Why, across cultures and history, do we consistently find music to be an integral part of every human society? One reason must be because of its temporal and structural complexity, its ability to at times surprise us and at times reassure us, to connect us with our past as well as bring us into the present moment, to teach us about ourselves as well as others. When we listen to or perform music in our daily lives, we rarely pay attention to only one aspect of sound. Rather, music is experienced as a collective whole embedded within our own psychodynamic context. It develops and changes over time, and we develop and change over time as we listen. Yet in the field of cognitive science, music is often studied in isolated contexts, with the sounds themselves broken down into controlled or confined components (e.g., beats vs. tones). It has been argued that studying any complex processes in such a componential manner may lead to artificially constrained or inaccurate theories of the mind (Friston et al., 1996). Neural populations may very well behave differently when processing highly controlled stimuli in the lab versus multimodal stimuli in the natural world (Hasson et al., 2004). This all begs a very important question: what can studying individual components of music in the lab really tell us about how the mind works and how music interacts with it in our daily lives?

Part of the reason neuroscientific explorations have focused on individual components of music, rather than the collective whole, pertains to the technological limitations of the tools available for imaging the brain. The earliest studies of the human brain's physiological responses to music used electroencephalography (EEG), a noninvasive technique that involves placing electrodes on the scalp to record the electrical signal produced by a collection of neurons in the brain. EEG has high temporal resolution, meaning that a reliable signal can be recorded every 0.5 to 5 msec (2000–200 Hz). With EEG, scientists can directly measure neural activity that is time-locked to a single change in the environment or stimulus, termed an event-related potential (Biasiucci et al., 2019). Because a response to a single event tends to be small and noisy, EEG studies typically require repeated trials of the same stimulus. Consequently, the first EEG studies with music used highly repetitive and rhythmically standardized musical sounds with a very clear moment of change, such as monophonic musical phrases that ended with either a harmonically "correct" or "incorrect" note (Besson & Macar, 1987).

As new imaging techniques developed, research on music adapted to the new technological limitations. With functional magnetic resonance imaging (fMRI), for example, researchers can use the electromagnetic properties of molecules in the brain to measure changes in blood flow (technically, deoxygenated hemoglobin) in spatially precise cubes, or voxels, within the brain (typically 1–3 mm^3). This fine spatial resolution comes at the cost of temporal resolution, wherein a measure of blood-flow change (referred to as fMRI signal) is recorded on the order of seconds rather than milliseconds (typically 0.8–2 sec). Because of this limitation, fMRI analyses typically involve temporal averaging of the brain signal recording during a task (generally about 15–200 sec in duration) and subtracting this average from the average brain signal recorded during a control condition that does not involve the cognitive process being studied (the *subtraction method*). The univariate methods and highly constrained stimuli required for certain EEG and fMRI analyses afford a high degree of experimenter control, which is useful for testing certain hypotheses. Findings with such paradigms have therefore yielded useful and informative insights into aspects of certain cognitive processes that are likely involved in the musical experience.

That being said, more recent schools of thought argue that focusing on mean activation across time in a particular area of the brain may be misrepresenting how the brain functions in the natural world (Nastase et al., 2020; Poldrack, 2012). Mapping a single component of a stimulus onto a single area of the brain—a practice referred to as *localization of function* or, more colloquially, *blobology* (Poldrack, 2012)—has not proved fruitful, leading many to contend that cognition is likely the result of multiple brain regions functioning in tandem. Given these considerations, there has been a push in neuroscience to adapt more naturalistic paradigms and stimuli into the study of the brain and mind (Sonkusare et al., 2019; Zaki & Ochsner, 2009). While there is no single definition of what makes research studies naturalistic, they typically involve (1) stimuli that are dynamic, complex, and as unconstrained as possible, conveying situations, actions, and behaviors that mirror how we might encounter them outside the lab (e.g., film clips, spoken narratives, full musical pieces); and/or (2) unintrusive tasks that require participants to do little more than engage with the stimulus as they would in everyday life (e.g., free-viewing, free-listening paradigms).

Listening to or playing music while inside an MRI scanner or while wearing EEG electrodes on the scalp is certainly not the most natural way to experience music.[1] The process of collecting fMRI data is particularly abnormal. The participant needs to restrict movements while inside a narrow tube. Plus, the noise emanating from the scanner while collecting images (continuous beeps ranging from 85 to 105 dB, depending on the magnetic strength and sequence parameters) can be particularly problematic in auditory studies (Gaab et al., 2007). Typically, researchers have addressed these challenges by including rest conditions (scanner noise alone without a stimulus) and contrasting the signal between a music condition and the rest condition. Other techniques include using sparse temporal sampling[2] or headphones with active noise-canceling abilities (Dewey et al., 2021). Because EEG does not require the magnetic pulses that cause scanner noise and is less susceptible to movement artifacts, it can be used in situations that more closely mirror how we experience music outside the lab. That being said, the spatial advantages provided by fMRI may justify its use, despite these constraints.

In this chapter, we juxtapose the findings from research taking a reductionist approach (optimizing for more experimenter control) with those taking a more naturalistic approach (optimizing for ecological validity) to the study of music and the brain. Rather than draw a clear line between research that is naturalistic and that which is not, we take the stance that it is more instructive and more accurate to consider the methodologies, tasks, and stimuli as separate components that lie somewhere along a continuum of ecological validity and experimenter control. The technology used to measure some component of neurophysiology is a third consideration when assessing the overall naturalistic nature of study. For the sake of clarity, this chapter is organized into sections covering a single cognitive process, even though these cognitive processes have some degree of overlap in terms of neural correlates and subjective experience and likely influence one another. We conclude with a discussion of how more ecologically valid paradigms allow researchers to integrate findings across cognitive domains and then consider how future studies can avoid the common pitfalls associated with designing more ecologically valid studies to enhance our collective understanding of musical experience and its neural underpinnings.

A Note on Methodology and Statistical Approaches

As neuroscientists move away from the constrained, univariate designs traditionally used with neuroimaging, they are embracing computational analytical approaches that do not require making hard assumptions about the structure and timing of the expected signal. These methods allow increased flexibility and generalizability and are therefore more suited for assessing neural activation patterns associated with less-controlled stimuli in which the timing of modeled events or changes are rarely defined a priori or even known. Furthermore, some of these methods can determine changes within and across brain regions, as well as across time, within a single model. These models seek to relate patterns of multivariate brain activity measurements with cognitive processes, based on the assumption that aspects of a stimulus are "represented" by a collection of brain voxels (presumably, populations of neurons) whose activity can overlap in time and location (Nili et al., 2014).

No matter what model is used to analyze fMRI data (hypothesis driven or data driven), the question is always how best to determine what information is shared across people's brains. Most processing pipelines involve some sort of spatial normalizing, or warping each individual person's anatomical data into a standard space. This process, however, can wash away potentially important microanatomical and topographical differences across people. Some researchers therefore opt to use a functional localizer, an additional imaging task used to locate a spatially contiguous set of voxels that are functionally selective to a particular stimulus or cognitive demand, such as looking at faces (Kanwisher & Barton, 2011) or sentences (Scott et al., 2017). The voxels significantly activated by this localizer task are then used to test specific hypotheses about their role in a subsequent task of interest. Importantly, the voxels used can vary across participants. In this way, individual differences in anatomy can be preserved, while ensuring that the function of these voxels is shared across people (Saxe et al., 2006). That being said, depending on how it is employed, a functional localizer approach assumes that brain function can be clearly mapped onto a set of discrete spatial clusters, which goes against the evidence that multiple structures in the brain can perform the same function (i.e., degeneracy; Friston & Henson, 2010). For that reason, this approach is not commonly employed with more complex, naturalistic stimuli.

Researchers have recently started to incorporate algorithms that can project subject-specific functional data in response to a naturalistic task into a common "informational" space.[3] The naturalistic task for alignment is kept separate from the task of interest, although it should be similar enough that it can presumably trigger brain states that are likely shared across people (Haxby et al., 2020). Hyperalignment and shared response modeling (Chen et al., 2015) are two such algorithms that have been employed with relative success and have certain advantages over functional localizers, in that they involve fMRI tasks that are more engaging and suitable for specific populations (movies versus flashing images with localizers) and are designed to activate multiple brain areas that respond to many ecologically valid examples and categories (Jiahui et al., 2020).

It is not necessary for readers to fully comprehend these analytical techniques; rather, it is important to understand that while these methods provide new ways to assess the functioning of the brain, they involve a number of drawbacks that can hinder or obfuscate the discovery process if they are not employed carefully. For interested readers, a more thorough discussion of the advantages and disadvantages of various methods used to analyze naturalistic neuroimaging data can be found in Nastase et al. (2020) and Poldrack (2012).

Long-Timescale Processing of Musical Structure

Some of the earliest neuroscience studies of music were dedicated to understanding the sensory systems of the brain that allow the processing of tonality, rhythm, timbre, and harmony. Although these studies provided fundamental insights into functional specialization of the human brain, their ecological validity is questionable, given their use of highly controlled musical stimuli, such as short, isolated melodic sequences (Milner, 1962; Shankweiler, 1966; Zatorre et al., 1998) and simple rhythmic patterns (Kester et al., 1991; Penhune et al., 1998). In addition to quick sensory changes, music contains information at longer timescales, such as the reintroduction of a theme or chorus or the transition from one segment to the next. Somehow, our brains are able to process and maintain information that is changing over the course of minutes (Sridharan et al., 2007; Williams et al., 2022). This type of music processing can be assessed only with stimuli that convey these structural changes over time.

In light of this, researchers have used longer, more dynamic musical pieces to understand aspects of music perception that evolve over longer timescales. In one study, participants listened to four 9.5-minute symphonic excerpts during fMRI, as well as scrambled versions that preserved the pitch, loudness, and spectral information but disrupted the temporal structure. This study showed that activity patterns in the inferior frontal cortex and superior and medial portions of the auditory cortex were most correlated across participants in response to temporally coherent music rather than scrambled music (Abrams et al., 2013). To further assess how the brain processes the different timescales embedded in music's inherently hierarchical structure (musical notes make phrases, which make measures, which make movements), Farbood et al. (2015) played approximately four minutes of a Brahms piano concerto to expert pianists undergoing fMRI. The excerpt was presented to each participant in its intact form as well as in versions scrambled at different levels: one at the section level, the phrase level, and the measure level and a completely time-reversed version. They found that anatomically distinct areas of the cortex were involved in processing the hierarchical structure of music. The early auditory cortical areas responded to all conditions, whereas the neighboring superior temporal gyrus (STG), as well as parts of the parietal and frontal gyri, became engaged only in the more coherent, higher-level temporal conditions (phrase and measure scrambles, as well as the intact version). Although research with less ecologically valid music has also demonstrated the involvement of these brain areas in at least some aspects of music processing (Zatorre et al., 1994; Janata et al., 2002; Menon et al., 2002), studies that incorporate extended, polyphonic music have been able to dissociate the brain regions involved in processing rapid, low-level temporal information from those that process longer timescale information.

That being said, these studies still relied on a contrast condition with very unnatural-sounding music—that is, music played in reverse and temporally scrambled music. A recent investigation applied data-driven analytical techniques to show that it is possible to use intact musical recordings to investigate the brain regions that represent the high-level temporal structure of music. Williams et al. (2022) had fMRI participants listen to jazz and classical excerpts ranging from one to four minutes and then had a separate group of behavioral participants mark when they felt that a meaningful transition had occurred within each piece. The authors subsequently applied a data-driven statistical approach (hidden Markov models, or HMM[4]) to participants' fMRI data and found that in addition to the auditory cortex, the angular gyrus, medial prefrontal cortex (MPFC), and posterior medial cortex (PMC)[5] showed HMM-defined transition points that aligned with the transition points provided by the behavioral participants. These HMM-identified events were significantly longer in the MPFC than in the lower-order auditory cortex, PMC, and angular gyrus.[6] Taken together, the results suggest a hierarchical neural representation of musical event structure, with the auditory cortex, PMC, and angular gyrus processing shorter-term transitions and the MPFC processing longer-term transitions.

To summarize, when we encounter music in everyday life, we likely experience it as a collective whole, recognizing that events are unfolding over the course of minutes. Evidence from more ecologically valid music-listening paradigms indicates that higher-order areas of the brain, in addition to the sensory processing auditory cortex, together represent the hierarchical structure of music, stitching together information occurring at various timescales (e.g., from short to long). Earlier studies using simple musical sequences to investigate the neural basis of music perception have proved limited in their ability to tell us how the brain constructs these holistic musical representations.

Memory

We've all encountered a piece of music that has instantaneously transported us back to the past, stirring up vivid and evocative memories of who we once were. Despite music's powerful ability to connect us to our past, many studies investigating music's influence on memory focus on recalling simplistic things, such as images of faces (Proverbio et al., 2015) or lists of words (Wallace, 1994; Balch & Lewis, 1996; Cournoyer Lemaire, 2019). Even if some of these studies incorporate more complex and dynamic musical examples, they still pair the music with these less naturalistic stimuli (Ferreri et al., 2013, 2014). In our everyday lives, however, we typically experience music alongside continuous, multimodal, and dynamic sensory information. Therefore, the results from these studies cannot address how the brain uses music to encode and bind events from an individual's life.

In an attempt to measure this binding effect under more ecologically valid conditions, Janata (2009) selected thirty songs from the *Billboard* Top 100 Pop and R&B charts for the years the participants were between the ages of seven and nineteen years. Participants rated how autobiographically salient these excerpts were and then listened to truncated (30-second) versions of the songs during fMRI. The author found that activity in the dorsal medial prefrontal cortex (DMPFC) was positively correlated with the mnemonic salience of the different pieces, suggesting that this region is a key mediator for the retrieval of music-evoked autobiographical memories. The DMPFC was significantly involved in tracking the tonality of the pieces as well, suggesting that this region provides a mechanism by which structural features of the music are used as retrieval cues for personal memories. Follow-up neuroimaging studies with similar musical stimuli confirmed the DMPFC's critical role in connecting musical cues with autobiographical memories and in accessing the full sensory context during event retrieval (Belfi et al., 2018; Ford et al., 2016).

Although these studies incorporated popular music from participants' past, their ecological validity is still limited because the content of the music-evoked memories was unobserved. Moving forward, laboratory studies that are capable of probing music-related episodic events from everyday life will be necessary to better understand the brain changes that enable the formation of memories. Recent investigations not focused on music have tried to address this issue by having participants navigate real-world environments outside the lab using life-logging devices (e.g., video recorders, portable microphones) and subsequently cueing these recorded memories during fMRI (Nielson et al., 2015; Rissman et al., 2016; Chow & Rissman, 2017). These studies have already provided evidence that the prefrontal cortex (including the DMPFC), the hippocampus, and the parietal cortex play important roles in episodic memory retrieval for real-world events. Employing such designs and using musical moments from a person's life may be a promising avenue for better understanding how complex brain networks integrate music with our subjective experience.

Reward

Music is often cited as one of life's most enjoyable activities (Zatorre & Salimpoor, 2013), and neuroimaging tools have allowed us to uncover the brain activation patterns involved in the rewarding aspects of musical engagement. At the same time, music can be a useful tool for uncovering how our brains learn to seek and obtain stimuli that make us feel good. Brain recordings from animals learning to perform an action that results in getting a treat have led to the theory that reward processing actually contains two separable mechanisms: wanting versus liking. *Wanting*, sometimes referred to as incentive salience, refers to the motivational process or desire to obtain or pursue an attractive stimulus, whereas *liking* refers to the positive, hedonic feelings upon consummation of a rewarding stimulus.[7] When we refer to human feelings of pleasure or enjoyment, we are typically referring to the liking mechanism; in contrast, the wanting mechanism would apply to cravings or yearnings or the pursuit of pleasure (Berridge & Kringelbach, 2015). We know that both mechanisms involve several areas located deep within the brain (subcortical brain regions such as the nucleus accumbens [NAc], ventral pallidum, and parts of the insula), as well as other areas closer to the surface (cortical brain regions such as the orbitofrontal cortex [OFC] and parts of the anterior cingulate cortex [ACC]). Reward also involves a third component, *learning*, which is the process of associating a stimulus with a positive (or negative) valence over time. The learning phase appears to involve mainly dopaminergic neurons in subcortical regions, such as the ventral striatum (which includes the NAc), amygdala, ventral putamen, and pallidum (Garrison et al., 2013).

One empirically sound hypothesis for why music is such a rewarding experience is that its temporal, rhythmic, and harmonic structure motivates the listener or performer to naturally form expectations and predictions that can then be violated or confirmed (Vuust & Frith, 2008). An fMRI study that used highly controlled musical stimuli (four-part Bach chorales played on a musical instrument digital interface [MIDI] piano at 75 beats per minute [bpm]) showed that musical violations can induce activity related to reinforcement prediction errors (RPEs) in the NAc during a reinforcement-learning paradigm in which the chorales ended as written (consonantly) or with a manipulated, unexpected dissonant chord (Gold et al., 2019). Furthermore, RPE-related activity in the NAc was correlated with a self-reported measure of music reward sensitivity—that is, the degree to which one finds various forms of musical engagement enjoyable. By experimentally controlling tempo, timbre, and form, the authors of this study were able to show that listening to music can motivate reinforcement learning by setting up and subsequently subverting expectations. It is therefore plausible that reinforcement learning is one mechanism by which music is experienced pleasurably.

Several limitations of this study weaken the conclusions that can be drawn about how the brain imbues music with reward value. First, participants were not asked to report how pleasurable they found the musical stimuli. Second, because of the highly controlled nature of the stimuli (MIDI versions of manipulated Bach chorales), it is unlikely that repeatedly listening to these musical excerpts could result in feelings of enjoyment comparable to hearing one's favorite song on the radio. Addressing the neural systems involved in the conscious experience of pleasure likely requires using less controlled, more familiar music. This poses a challenge because, with more complex stimuli, it becomes difficult to isolate the neural components that are specific to reward and not conflated with other processes that are likely co-occurring, such as memory.

Another fMRI study attempted to balance these concerns by using chord progressions extracted from recorded pop songs featured on the *Billboard* charts (Cheung et al., 2019). To create the stimuli, the melody and rhythm were removed, transposed to a common key, played on a single MIDI instrument, and elongated so that each chord lasted for the duration of the temporal resolution of fMRI. With these controlled stimuli, the authors were able to disentangle the interrelated components of musical expectations: *surprise*—how predictable or unpredictable a particular chord is within the musical environment—and *uncertainty*—how predictable or unpredictable an upcoming chord is, based on the established musical environment. They found that the interaction between the amount of surprise and uncertainy of a chord predicted subjective ratings of pleasure. This interaction effect was reflected in fMRI signal changes recorded from the amygdala, hippocampus, and auditory cortex. Surprisingly, striatal activation (NAc and caudate) was involved in coding only for uncertainty, not for surprise, suggesting that this system plays a role in something akin to the wanting mechanism, by which we are motivated to gain subsequent information that resolves the musical uncertainty. The involvement of the NAc, as well as the rest of the striatum, has also been shown in response to rhythmic violations in piano chords (Matthews et al., 2019, 2020).

By breaking down musical stimuli into these basic components, researchers were able to show how two temporal aspects of musical processing relate to rewarding experience. But once again, it is unlikely that the enjoyment experienced while listening to decontextualized chord sequences equates to the enjoyment of listening to one's favorite music (Laeng et al., 2016). Moments of peak enjoyment in music, often co-occurring with chills (Mori & Iwanaga, 2017), are associated with changes in the music that unfold over periods longer than a few seconds, such as crescendos, builds, and the entrance of singers and the human voice (Bannister, 2020). It would be difficult to study this fascinating phenomenon using stimuli in which the vocals, dynamics, and instrumentation had been regularized, manipulated, or edited in such noticeable ways. Studies using more familiar musical clips have shown that the NAc (in particular, dopamine released in the NAc) is linked with the experience of peak enjoyment of music, whereas other areas of the striatum (e.g., the caudate) are likely involved in the anticipation of reward (Salimpoor et al., 2011, 2013). Furthermore, the more rewarding the listening experience, the more likely it was that activity in the NAc was correlated with activity in higher-order brain regions such as the ventromedial prefrontal cortex (VMPFC) and OFC.

More recent studies have attempted to address the higher-level experience of pleasure by playing pieces of music in full, with no cuts or edits and requiring no explicit tasks other than listening. Although it is always somewhat unnatural to listen to music in an unfamiliar setting while being monitored in some way, the hope is that such paradigms move closer to how we experience music outside the lab so that the multiple complex processes involved in musical enjoyment can be assessed. The neural findings from such studies vary somewhat from those using less ecologically valid paradigms, showing that coactivation patterns in large-scale brain networks, both within an individual (Lehne et al., 2014; Singer et al., 2016) and across individuals (Sachs et al., 2020), track changes in subjective feelings in response to the music. Specifically, such studies have shown that the default mode network (DMN; a large-scale brain network consisting of regions in the parietal, temporal, and frontal cortices that tend to function together) is involved in the subjective experience of enjoyment in response to music.

At this point, we can only speculate about the DMN's role in musical enjoyment, but existing theories argue that it is actively involved in integrating external information from the stimulus with internal information from the body. EEG findings provide evidence for this active role of the DMN in reward processing: in participants with heightened reward sensitivity to music, there was less of a match between the temporal complexity of the brain signals recorded in DMN areas (ACC and inferior parietal lobule) and the temporal complexity of the musical stimuli. When these same participants were explicitly told to attend to perceptual features of the music, however, the match between neural and musical signal was higher, suggesting that the DMN might be involved in shifting focus away from the external information coming from the stimuli and toward the internal process of generating expectations and emotions related to the piece as a whole (Carpentier et al., 2020). Such insights into the significance of this clearly intrinsic yet poorly understood brain network can result only from studies involving more complex, more ecologically valid musical stimuli.

In sum, studies that have isolated and manipulated specific components of music (e.g., harmony, rhythm) have laid the groundwork for our understanding of the process by which music can become preferred and liked and the role of subcortical regions in this learning process. Studies with polyphonic, unmanipulated music indicate the involvement of higher-order brain regions, particularly the OFC and DMN, during the subjective experience of music-evoked pleasure. The extent to which subcortical and cortical brain systems interact with each other during pleasurable experiences with music requires further exploration with techniques that allow the use of self-selected music.

Emotions and Feeling

Enjoyment and pleasure can be viewed as one of many possible affective responses to music. But music can evoke many feelings.[8] Indeed, music's ability to both convey and induce a range of emotions makes it a useful tool for studying the neural systems involved in emotional responses (Juslin, 2013). However, despite decades of neuroscience research, the definition of emotions and the appropriate way to measure them are still fiercely debated (Barrett & Satpute, 2019). While there is evidence that the emotions evoked by music across genres and cultures cluster into categories that mirror those evoked by faces and video clips (Cowen et al., 2020), there is also evidence that what music evokes can be accurately captured by dimensional models corresponding to how pleasant or unpleasant, arousing or unarousing the music feels, rather than basic emotional categories such as happy or sad (Cespedes-Guevara & Eerola, 2018).

Part of this discrepancy stems from the issue of how to establish a "ground truth" when studying emotions and feelings: that is, how do we define an agreed-upon, ideal outcome with which we can compare the actual findings. There is not one method of assigning an emotional label to a stimulus. One way of dealing with this empirical uncertainty is to focus on the individual, not on what is common across people but on what is different, and try to relate these differences to some aspect of that individual's unique experience. Another way is to obtain multiple measures that describe the stimulus (e.g., acoustic and musical analysis, subjective ratings, composers' intentions) and use models that combine these features in different ways to predict a particular response. Although a purely objective ground truth may never be realized when it comes to emotions, certain methodological approaches can afford a higher degree of certainty with regard to how an observed response relates to some aspect of a stimulus or cognitive process. For now, since there is no clear definition of *emotion*, we consider the concept broadly, including studies that operationalize emotions as discrete categories as well as continuous dimensions.

Despite these limitations, dozens of studies have used musical stimuli to assess the neural systems involved in music-*evoked* emotions—that is, how we feel when listening to a piece, as opposed to what emotion we perceive in the music. One meta-analysis reported that across these studies—some of which used emotional stimuli that were highly controlled (computer-generated music or chord sequences) and some of which used unmanipulated excerpts—music listening was consistently associated with activation patterns across the entire brain, including subcortical areas (amygdala, hippocampus, ventral striatum–NAc, dorsal striatum–caudate nucleus, inferior colliculus), auditory areas extending into the parietal operculum and secondary somatosensory cortex, and medial cortical areas (OFC, ACC, middle posterior cingulate cortex; Koelsch, 2020). Many of these regions overlap with those found to be associated with aspects of emotions outside the realm of music (Adolfi et al., 2017). Given that the studies included in the meta-analysis varied with regard to the behavioral measures used to determine the emotional response and included mostly musical examples from the European classical world, the investigation of music-evoked emotions appears to be plagued by some of the same methodological issues and ethnocentrism as other nonmusical neuroscience investigations of emotions. There is thus little indication that using musical stimuli helped resolve outstanding issues in terms of the conceptualization of emotions.

Naturalistic listening paradigms might provide new insights into the neuroscientific understanding of emotions by exploring the brain systems sensitive to emotion dynamics. Despite the fact that our feelings and moods are constantly in flux, the temporal component of emotions in the brain is relatively unexplored. Full-length pieces of music that traverse different themes, motifs, rhythms, and melodies might allow researchers to determine whether and how regions of the brain that respond to music-evoked emotions in isolation track emotional dynamics. Recent studies using naturalistic listening paradigms have been able to extend these findings to show brain regions and networks in which time-varying patterns of activation and connectivity reflect changes in emotional responses to music. Continuous ratings of felt emotions outside the scanner in response to music were associated with time-varying coactivation patterns in the amygdala, caudate, hippocampus, thalamus, insula, and cingulate (Sachs et al., 2020; Singer et al., 2016). Brain activation in many of these regions was associated with nondynamic, less naturalistic emotional stimuli as well, but by assessing temporal patterns, these results clarify the role of these regions in integrating emotionally relevant information over time.

Another way that naturalistic musical paradigms could lead to new discoveries in the neuroscience of emotions is to focus on complex emotional experiences that occur infrequently in everyday life and typically only in the realm of aesthetics (Juslin, 2013; Mori & Iwanaga, 2017). These could include chills or frisson (as previously discussed), as well as feelings of being moved (Menninghaus et al., 2015) or *kama muta* (Fiske, 2019), awe, nostalgia, entrancement or transcendence, wonder (Juslin, 2013; Mori & Iwanaga, 2017), pleasurable sadness (Sachs et al., 2015), and general aesthetic appreciation. Music is not the only art form that can induce these emotions, but it is certainly more easily accessible to most people on a daily basis than fine arts, movies, or nature.

Finally, the issue of ground truth with emotional stimuli could be minimized by incorporating non-Western musical systems with a clearer mapping between musicality and feeling. North Indian classical music, for example, uses a musical system, in which specific tonal relationships, rhythms, and tempos are capable of eliciting distinct, predictable emotions (Valla et al., 2017). The expansion of music cognition research to include musical stimuli from a broad range of cultures could be used to validate, or potentially establish, a more robust relationship between brain function and the phenomenological experience of emotions. Moving forward, neuroscientific investigation of emotions might benefit from embracing these deeply meaningful and deeply human affective experiences that tend to occur uniquely within the realm of music.

Incorporating the Social Context

Listening to or performing music is rarely a solo experience. Until recent technological innovations changed the way music can be distributed and consumed, people almost exclusively experienced music in the company of others. Throughout this chapter, we have circled around this fundamental question of whether breaking down the experience of engaging with music into testable components is a viable strategy for understanding the associated bodily and neural reactions. Perhaps the strongest criticism against this reductionist approach is that with such an empirical design, it is difficult or impossible to assess how other people might influence the cognitive and behavioral processes involved in the experience.

Listening to music can range from the less naturalistic (sine tones) to the more naturalistic (full-length recorded music), and studies designed to probe the social component of the musical experience range along this same axis. Some use very simple tasks that probe one particular element of shared music making, such as joint tapping (Cui et al., 2012), whereas others try to capture the holistic experience, such as recording brain activity while professional musicians are performing together (Müller et al., 2018). With recent technological advances (e.g., mobile systems for recording EEG, movement, and physiological responses), it has become easier to obtain rich, multisensory data from multiple music performers and listeners simultaneously (see McMaster University's LIVE Lab concert hall founded by Dr. Laurel Trainor; Chang et al., 2019). Incorporating these technologies into real-world spaces has the potential to provide an ecologically valid understanding of how our brains and bodies change as a function of engaging with music in group settings.

Because relatively few studies have used joint music paradigms, it is difficult to draw any strong conclusions about how the naturalistic nature of the task influences neural processing. Furthermore, the issue of naturalism in tasks that require group dynamics presents an additional methodological challenge associated with the cost and portability of the various neuroimaging techniques. The sheer size of MRI scanners makes it quite challenging to image the brains of two or more people engaged simultaneously in an interactive task. Typically, these so-called hyperscanning studies involve two separate scanners and virtual interactions conducted via screens and headsets (but see Renvall et al., 2020, for the future possibility of dual-coil scanning). Because of the high cost and low mobility, hyperscanning studies are more commonly performed with EEG (Lindenberger et al., 2009; Osaka et al., 2015) or functional near-infrared spectroscopy (fNIRS; Duan et al., 2015), both of which are more portable, more mobile, and less susceptible to movement artifacts than MRI. These studies have focused on the implications of correlated neuroimaging signals across musicians' brains for performance quality (Greco et al., 2018), as well as prosocial behaviors such as cooperation (Balconi & Vanutelli, 2017) and empathy (Babiloni et al., 2012). However, they reveal very little about which brain regions and networks vary as a function of the social context. Given evidence that real-time nonmusical social interactions involve coordination between subcortical regions (Krill & Platek, 2012; Špiláková et al., 2019), a more complete understanding of the social impact of the neural processing of music may require complementary neuroimaging techniques that can assess spatial components of the brain and body.

One potentially fruitful avenue for studying the brain networks modulated by social components is to use music tasks that compare solo listening experiences to listening with others. No study to date has used hyperscanning techniques to assess co-listening, even though such a task would evade the movement artifacts associated with playing music that currently make fMRI hyperscanning studies less tenable.[9] Studies without neuroimaging have shown that listening to music alone versus with another person has implications for self-reported emotions and enjoyment (Egermann et al., 2011; Liljeström et al., 2013; Sutherland et al., 2009). Therefore, research that involves music listening tasks with other people seems like a plausible means of furthering our understanding of sensory processing, reward, and emotions in a more naturalistic context. Such studies could provide an innovative perspective on the social bonding that occurs as a result of emotional sharing or emotion matching, an idea very much at the heart of the argument for music's cultural importance and ubiquity (Savage et al., 2020).

Conclusions and Future Directions

Throughout this chapter, we have discussed the advantages and disadvantages of incorporating more ecologically valid paradigms into music cognition research, highlighting the key insights garnered from taking a more holistic, rather than reductionist, approach. We have also suggested some areas for future research that can push forward our collective understanding of the functioning of the human brain and its relationship to musical experiences.

Given music's ability to touch on so many aspects of cognition, one additional direction for future research with naturalistic musical paradigms is to test existing theories of the structure and function of the brain as a *complex system* made up of networks and nodes that integrate information from functionally distinct neural populations (Honey et al., 2007). Evidence suggests that these networks are dynamic and adaptive; they form and dissolve in real time to meet specific computational demands (Pessoa, 2018). How an integrated conscious experience emerges from the interaction of these multiple brain systems remains an open question. Incorporating polyphonic music with analyses designed to assess connectivity and network changes over time may therefore lead to a clearer understanding of the dynamic quality of the brain.

Several studies have already been conducted along this line of inquiry. Carpentier et al. (2020) showed that the temporal complexity of the EEG signal across the entire brain matched the temporal complexity of the musical stimuli, particularly when participants were attending to perceptual aspects of the music. With naturalistic fMRI paradigms, several functional brain networks appear to be consistently and reliably involved when engaging with a variety of stimuli, yet the nature of the interactions between them varies as a function of the demands of the stimulus, such as movies versus music (Bottenhorn et al., 2019), joyful versus fearful music (Koelsch & Skouras, 2014), or changing rhythmic complexity in music (Toiviainen et al., 2020). Together, these findings provide evidence for segregated brain networks that become functionally integrated with one another as internal and external information changes while engaging with dynamic, complex stimuli. Future studies could incorporate analytical approaches that consider multiple events occurring on multiple timescales to better capture how brain networks form and dissolve over time during musical engagement.

Another promising method of evaluating the neurobiological basis of music perception is to employ more advanced computational models. Artificial neural networks, for example, learn in a data-driven manner how information is transmitted through a series of transformations (via hidden layers), similar to hierarchical information processing in the mammalian brain (Kriegeskorte, 2015). As discussed, music has a unique temporal structure, whereby representations become more abstract and complex as they travel along a cortical processing hierarchy (Farbood et al., 2015; Williams et al., 2022). Therefore, it is likely that using neural networks to construct rich music-based representations will help us more accurately model how our brains represent complex musical information. Some studies have already taken promising steps in this direction (Güçlü et al., 2016; Kell et al., 2018). However, it is important to keep in mind that although the inputs (e.g., musical stimuli) and outputs (e.g., musical behaviors) of these networks may resemble the types of representations and behaviors observed in humans, the series of input-output transformations in these networks (particularly in the case of deep neural networks) are typically intractable, which limits our ability to make claims about whether these transformations resemble the way musical information is processed throughout the human brain. As we become better at quantifying and mapping the relevant aspects of our stimuli and participants, we will become better at using advanced computational models to quantify and map brain functioning.

Another important area for future endeavors is to incorporate non-Western populations and stimuli into neuroscientific investigations to expand the generalizability of existing findings. The majority of research discussed in this chapter used stimuli and study populations within the Eurocentric framework, which, among other biases, limits the conclusions we can draw about the perceptual, social, and emotional impacts of music (Baker et al., 2020). Furthermore, the structure and cultural function of certain musical traditions might be uniquely suited to study certain aspects of cognition (Valla et al., 2017). The expansion of music cognition research to include stimuli from around the world will therefore not only result in more ecologically valid theories but also lead to a more inclusive path of inquiry.

Notes

1. It is important to note that the term *natural*, used in this context, is not meant to indicate that there is an unnatural way in which people listen to, perform, or experience music in everyday life. Rather, it is a comment on and response to the types of tasks and stimuli that have traditionally been used in lab-based scientific studies. It is also worth mentioning that despite the recent

push to use more ecologically valid musical stimuli in the lab, this largely involves using recorded music in studies, which is a relatively modern way of engaging with music and may not reflect how certain cultures or communities around the world commonly engage with music (Walther-Hansen, 2020). That being said, we use the term *naturalistic* throughout this chapter to align with other fields of neuroscience that continue to use it.

2. In sparse temporal sampling, sounds are presented while the MRI scanner is turned off (to prevent sound interference from the scanner), and brain volumes are acquired following the silence when the hemodynamic response function is at its peak (Hall et al., 1999). This approach has clear advantages when it comes to studying auditory stimuli like music, although there are severe drawbacks as well, such as lower statistical power with fewer data points.

3. *Information* in this context refers to the relationships among functional pattern vectors, typically quantified as the similarity (or dissimilarity or distance) between each pair of vectors.

4. HMM is a statistical method that learns the temporal structure of data and identifies latent state transitions within a brain region without requiring any timing information about the stimulus.

5. The PMC includes the precuneus, dorsal posterior cingulate, ventral posterior cingulate, and retrosplenial cortex (Bzdok et al., 2015).

6. Higher-order cortical areas refer to a set of brain regions that support more complex (i.e., high-level) cognitive functions such as planning, executive functions (Miller & Wallis, 2009), and integrating information across sensory inputs (Chen et al., 2017), as opposed to more automatic, less integrative neural processes such as sensory perception (i.e., low-level cognition).

7. Throughout this chapter, we use the term *pleasure* to refer to the "liking" mechanism.

8. For the sake of clarity, we do not distinguish between emotions and feelings, as others have suggested (see Damasio, 1999). Here, we use the term *emotion* to refer to any bodily state changes that accompany a change in the environment, as well as the subjective experience of those bodily state changes.

9. Given the logistical and economic constraints of hyperscanning, for any future endeavor, it is important to consider whether obtaining *simultaneous* measures of brain activation is absolutely necessary to answer the proposed research question. To aid in the design of future social neuroscience studies, we suggest following the framework proposed by Misaki et al. (2021) and Redcay and Schilbach (2019). They argue that fMRI hyperscanning is necessary only if the task involves a reciprocal interaction that is evolving, unpredictable, and difficult to reproduce. Sharing a new piece of music with someone may very well fall into this category if one is interested in both the giver's and the receiver's neural responses, although it may be possible to study this type of social interaction with music sequentially, scanning only one person at a time.

References

Abrams, D. A., Ryali, S., Chen, T., Chordia, P., Khouzam, A., Levitin, D. J., & Menon, V. (2013). Inter-subject synchronization of brain responses during natural music listening. *European Journal of Neuroscience, 37*(9), 1458–1469.

Adolfi, F., Couto, B., Richter, F., Decety, J., Lopez, J., Sigman, M., Manes, F., & Ibáñez, A. (2017). Convergence of interoception, emotion, and social cognition: A twofold fMRI meta-analysis and lesion approach. *Cortex, 88*, 124–142.

Babiloni, C., Buffo, P., Vecchio, F., Marzano, N., Del Percio, C., Spada, D., Rossi, S., Bruni, I., Rossini, P. M., & Perani, D. (2012). Brains "in concert": Frontal oscillatory alpha rhythms and empathy in professional musicians. *NeuroImage, 60*(1), 105–116.

Baker, D. J., Belfi, A., Creel, S., Grahn, J., Hannon, E., Loui, P., Margulis, E. H., Schachner, A., Schutz, M., Shanahan, D., & Vuvan, D. T. (2020). Embracing anti-racist practices in the music perception and cognition community. *Music Perception, 38*(2), 103–105.

Balch, W. R., & Lewis, B. S. (1996). Music-dependent memory: The roles of tempo change and mood mediation. *Journal of Experimental Psychology: Learning, Memory, and Cognition, 22*(6), 1354–1363.

Balconi, M., & Vanutelli, M. E. (2017). Cooperation and competition with hyperscanning methods: Review and future application to emotion domain. *Frontiers in Computational Neuroscience, 11*, 86.

Bannister, S. (2020). A survey into the experience of musically induced chills: Emotions, situations and music. *Psychology of Music, 48*(2), 297–314.

Barrett, L. F., & Satpute, A. B. (2019). Historical pitfalls and new directions in the neuroscience of emotion. *Neuroscience Letters, 693*, 9–18.

Belfi, A. M., Karlan, B., & Tranel, D. (2018). Damage to the medial prefrontal cortex impairs music-evoked autobiographical memories. *Psychomusicology: Music, Mind, and Brain, 28*(4), 201–208.

Berridge, K. C., & Kringelbach, M. L. (2015). Pleasure systems in the brain. *Neuron, 86*(3), 646–664.

Besson, M., & Macar, F. (1987). An event-related potential analysis of incongruity in music and other non-linguistic contexts. *Psychophysiology, 24*(1), 14–25.

Biasiucci, A., Franceschiello, B., & Murray, M. M. (2019). Electroencephalography. *Current Biology, 29*(3), R80–R85.

Bottenhorn, K. L., Flannery, J. S., Boeving, E. R., Riedel, M. C., Eickhoff, S. B., Sutherland, M. T., & Laird, A. R. (2019). Cooperating yet distinct brain networks engaged during naturalistic paradigms: A meta-analysis of functional MRI results. *Network Neuroscience, 3*(1), 27–48.

Bzdok, D., Heeger, A., Langner, R., Laird, A. R., Fox, P. T., Palomero-Gallagher, N., . . . Eickhoff, S. B. (2015). Subspecialization in the human posterior medial cortex. *NeuroImage, 106*, 55–71.

Carpentier, S. M., McCulloch, A. R., Brown, T. M., Faber, S. E. M., Ritter, P., Wang, Z., Salimpoor, V., Shen, K., & McIntosh, A. R. (2020). Complexity matching: Brain signals mirror environment information patterns during music listening and reward. *Journal of Cognitive Neuroscience, 32*(4), 734–745.

Cespedes-Guevara, J., & Eerola, T. (2018). Music communicates affects, not basic emotions—a constructionist account of attribution of emotional meanings to music. *Frontiers in Psychology, 9*, 215.

Chang A., Kragness H. E., Livingstone, S. R., Bosnyak, D. J., & Trainor, L. J. (2019). Body sway reflects joint emotional expression in music ensemble performance. *Scientific Reports, 9*(1), 1–11.

Chen, J., Leong, Y. C., Honey, C. J., Yong, C. H., Norman, K. A., & Hasson, U. (2017). Shared memories reveal shared structure in neural activity across individuals. *Nature Neuroscience, 20*(1), 115–125.

Chen, P.-H., Chen, J., Yeshurun, Y., Hasson, U., Haxby, J. V., & Ramadge, P. J. (2015). A reduced-dimension fMRI shared response model. *NIPS, 28*, 460–468.

Cheung, V. K. M., Harrison, P. M. C., Meyer, L., Pearce, M. T., Haynes, J.-D., & Koelsch, S. (2019). Uncertainty and surprise jointly predict musical pleasure and amygdala, hippocampus, and auditory cortex activity. *Current Biology, 29*(23), 4084–4092.e4.

Chow, T. E., & Rissman, J. (2017). Neurocognitive mechanisms of real-world autobiographical memory retrieval: Insights from studies using wearable camera technology. *Annals of the New York Academy of Sciences, 1396*(1), 202–221.

Cournoyer Lemaire, E. (2019). The effect of background music on episodic memory. *Psychomusicology: Music, Mind, and Brain, 29*(1), 22–34.

Cowen, A. S., Fang, X., Sauter, D., & Keltner, D. (2020). What music makes us feel: At least 13 dimensions organize subjective experiences associated with music across different cultures. *Proceedings of the National Academy of Sciences of the United States of America, 117*(4), 1924–1934.

Cui, X., Bryant, D. M., & Reiss, A. L. (2012). NIRS-based hyperscanning reveals increased interpersonal coherence in superior frontal cortex during cooperation. *NeuroImage, 59*(3), 2430–2437.

Damasio, A. R. (1999). *The feeling of what happens: Body and emotion in the making of consciousness.* Houghton Mifflin Harcourt.

Dewey, R. S., Hall, D. A., Plack, C. J., & Francis, S. T. (2021). Comparison of continuous sampling with active noise cancelation and sparse sampling for cortical and subcortical auditory functional MRI. *Magnetic Resonance in Medicine: Official Journal of the Society of Magnetic Resonance in Medicine/ Society of Magnetic Resonance in Medicine.*

Duan, L., Dai, R.-N., Xiao, X., Sun, P.-P., Li, Z., & Zhu, C.-Z. (2015). Cluster imaging of multi-brain networks (CIMBN): A general framework for hyperscanning and modeling a group of interacting brains. *Frontiers in Neuroscience, 9*, 267.

Egermann, H., Sutherland, M. E., Grewe, O., Nagel, F., Kopiez, R., & Altenmüller, E. (2011). Does music listening in a social context alter experience? A physiological and psychological perspective on emotion. *Musicae Scientiae, 15*(3), 307–323.

Farbood, M. M., Heeger, D. J., Marcus, G., Hasson, U., & Lerner, Y. (2015). The neural processing of hierarchical structure in music and speech at different timescales. *Frontiers in Neuroscience, 9*, 157.

Ferreri, L., Aucouturier, J.-J., Muthalib, M., Bigand, E., & Bugaiska, A. (2013). Music improves verbal memory encoding while decreasing prefrontal cortex activity: An fNIRS study. *Frontiers in Human Neuroscience, 7*, 779.

Ferreri, L., Bigand, E., Perrey, S., Muthalib, M., Bard, P., & Bugaiska, A. (2014). Less effort, better results: How does music act on prefrontal cortex in older adults during verbal encoding? An fNIRS study. *Frontiers in Human Neuroscience, 8*, 301.

Fiske, A. P. (2019). *Kama muta: Discovering the connecting emotion*. Routledge.

Ford, J. H., Rubin, D. C., & Giovanello, K. S. (2016). The effects of song familiarity and age on phenomenological characteristics and neural recruitment during autobiographical memory retrieval. *Psychomusicology, 26*(3), 199–210.

Friston, K. J., & Henson, R. N. (2010). Commentary on divide and conquer: A defense of functional localizers. In S. J. Hanson & M. Bunzl (Eds.), *Foundational issues in human brain mapping* (pp. 43–48). MIT Press.

Friston, K. J., Price, C. J., Fletcher, P., Moore, C., Frackowiak, R. S., & Dolan, R. J. (1996). The trouble with cognitive subtraction. *NeuroImage, 4*(2), 97–104.

Gaab, N., Gabrieli, J. D. E., & Glover, G. H. (2007). Assessing the influence of scanner background noise on auditory processing. I. An fMRI study comparing three experimental designs with varying degrees of scanner noise. *Human Brain Mapping, 28*(8), 703–720.

Garrison, J., Erdeniz, B., & Done, J. (2013). Prediction error in reinforcement learning: A meta-analysis of neuroimaging studies. *Neuroscience and Biobehavioral Reviews, 37*(7), 1297–1310.

Gold, B. P., Pearce, M. T., Mas-Herrero, E., Dagher, A., & Zatorre, R. J. (2019). Predictability and uncertainty in the pleasure of music: A reward for learning? *Journal of Neuroscience, 39*(47), 9397–9409.

Greco, A., Spada, D., Rossi, S., Perani, D., Valenza, G., & Scilingo, E. P. (2018). EEG hyperconnectivity study on saxophone quartet playing in ensemble. In *Conference Proceedings: . . . Annual International Conference of the IEEE Engineering in Medicine and Biology Society* (pp. 1015–1018). IEEE Engineering in Medicine and Biology Society.

Güçlü, U., Thielen, J., Hanke, M., & van Gerven, M. A. J. (2016). Brains on beats. *arXiv*. http://arxiv.org/abs/1606.02627.

Hall, D. A., Haggard, M. P., Akeroyd, M. A., Palmer, A. R., Summerfield, A. Q., Elliott, M. R., Gurney, E. M., & Bowtell, R. W. (1999). "Sparse" temporal sampling in auditory fMRI. *Human Brain Mapping, 7*(3), 213–223.

Hasson, U., Nir, Y., Levy, I., Fuhrmann, G., & Malach, R. (2004). Intersubject synchronization of cortical activity during natural vision. *Science, 303*(5664), 1634–1640.

Haxby, J. V., Guntupalli, J. S., Nastase, S. A., & Feilong, M. (2020). Hyperalignment: Modeling shared information encoded in idiosyncratic cortical topographies. *eLife, 9*.

Honey, C. J., Kötter, R., Breakspear, M., & Sporns, O. (2007). Network structure of cerebral cortex shapes functional connectivity on multiple time scales. *Proceedings of the National Academy of Sciences, 104*(24), 10240–10245.

Janata, P. (2009). The neural architecture of music-evoked autobiographical memories. *Cerebral Cortex, 19*(11), 2579–2594.

Janata, P., Birk, J. L., Van Horn, J. D., Leman, M., Tillmann, B., & Bharucha, J. J. (2002). The cortical topography of tonal structures underlying Western music. *Science, 298*(5601), 2167–2170.

Jiahui, G., Feilong, M., Visconti di Oleggio Castello, M., Guntupalli, J. S., Chauhan, V., Haxby, J. V., & Gobbini, M. I. (2020). Predicting individual face-selective topography using naturalistic stimuli. *NeuroImage, 216*, 116458.

Juslin, P. N. (2013). What does music express? Basic emotions and beyond. *Frontiers in Psychology, 4*, 596.

Kanwisher, N., & Barton, J. J. S. (2011). The functional architecture of the face system: Integrating evidence from fMRI and patient studies. In *Oxford handbooks online.*

Kell, A. J. E., Yamins, D. L. K., Shook, E. N., Norman-Haignere, S. V., & McDermott, J. H. (2018). A task-optimized neural network replicates human auditory behavior, predicts brain responses, and reveals a cortical processing hierarchy. *Neuron, 98*(3), 630–644.e16.

Kester, D. B., Saykin, A. J., Sperling, M. R., O'Connor, M. J., Robinson, L. J., & Gur, R. C. (1991). Acute effect of anterior temporal lobectomy on musical processing. *Neuropsychologia, 29*(7), 703–708.

Koelsch, S. (2020). A coordinate-based meta-analysis of music-evoked emotions. *NeuroImage, 223*, 117350.

Koelsch, S., & Skouras, S. (2014). Functional centrality of amygdala, striatum and hypothalamus in a "small-world" network underlying joy: An fMRI study with music. *Human Brain Mapping, 35*(7), 3485–3498.

Kriegeskorte, N. (2015). Deep neural networks: A new framework for modeling biological vision and brain information processing. *Annual Review of Vision Science, 1*, 417–446.

Krill, A. L., & Platek, S. M. (2012). Working together may be better: Activation of reward centers during a cooperative maze task. *PLOS ONE, 7*(2), e30613.

Laeng, B., Eidet, L. M., Sulutvedt, U., & Panksepp, J. (2016). Music chills: The eye pupil as a mirror to music's soul. *Consciousness and Cognition, 44*, 161–178.

Lehne, M., Rohrmeier, M., & Koelsch, S. (2014). Tension-related activity in the orbitofrontal cortex and amygdala: An fMRI study with music. *Social Cognitive and Affective Neuroscience, 9*(10), 1515–1523.

Liljeström, S., Juslin, P. N., & Västfjäll, D. (2013). Experimental evidence of the roles of music choice, social context, and listener personality in emotional reactions to music. *Psychology of Music, 41*(5), 579–599.

Lindenberger, U., Li, S.-C., Gruber, W., & Müller, V. (2009). Brains swinging in concert: Cortical phase synchronization while playing guitar. *BMC Neuroscience, 10*, 22.

Matthews, T. E., Witek, M. A. G., Heggli, O. A., Penhune, V. B., & Vuust, P. (2019). The sensation of groove is affected by the interaction of rhythmic and harmonic complexity. *PLOS ONE, 14*(1), e0204539.

Matthews, T. E., Witek, M. A. G., Lund, T., Vuust, P., & Penhune, V. B. (2020). The sensation of groove engages motor and reward networks. *NeuroImage, 214*, 116768.

Menninghaus, W., Wagner, V., Hanich, J., Wassiliwizky, E., Kuehnast, M., & Jacobsen, T. (2015). Towards a psychological construct of being moved. *PLOS ONE, 10*(6), e0128451.

Menon, V., Levitin, D. J., Smith, B. K., Lembke, A., Krasnow, B. D., Glazer, D., Glover, G. H., & McAdams, S. (2002). Neural correlates of timbre change in harmonic sounds. *NeuroImage, 17*(4), 1742–1754.

Miller, E. K., & Wallis, J. D. (2009). Executive function and higher-order cognition: Definition and neural substrates. *Encyclopedia of Neuroscience, 4*, 99–104.

Milner, B. (1962). Laterality effects in audition. *Interhemispheric Relations and Cerebral Dominance*, 177–195.

Misaki, M., Kerr, K. L., Ratliff, E. L., Cosgrove, K. T., Simmons, W. K., Morris, A. S., & Bodurka, J. (2021). Beyond synchrony: The capacity of fMRI hyperscanning for the study of human social interaction. *Social Cognitive and Affective Neuroscience, 16*(1–2), 84–92.

Mori, K., & Iwanaga, M. (2017). Two types of peak emotional responses to music: The psychophysiology of chills and tears. *Scientific Reports, 7*, 46063.

Müller, V., Sänger, J., & Lindenberger, U. (2018). Hyperbrain network properties of guitarists playing in quartet. *Annals of the New York Academy of Sciences, 1423*(1), 198–210.

Nastase, S. A., Goldstein, A., & Hasson, U. (2020). Keep it real: Rethinking the primacy of experimental control in cognitive neuroscience. *NeuroImage, 222*, 117254.

Nielson, D. M., Smith, T. A., Sreekumar, V., Dennis, S., & Sederberg, P. B. (2015). Human hippocampus represents space and time during retrieval of real-world memories. *Proceedings of the National Academy of Sciences of the United States of America, 112*(35), 11078–11083.

Nili, H., Wingfield, C., Walther, A., Su, L., Marslen-Wilson, W., & Kriegeskorte, N. (2014). A toolbox for representational similarity analysis. *PLOS Computational Biology, 10*(4), e1003553.

Osaka, N., Minamoto, T., Yaoi, K., Azuma, M., Shimada, Y. M., & Osaka, M. (2015). How two brains make one synchronized mind in the inferior frontal cortex: fNIRS-based hyperscanning during cooperative singing. *Frontiers in Psychology, 6*, 1811.

Penhune, V. B., Zattore, R. J., & Evans, A. C. (1998). Cerebellar contributions to motor timing: A PET study of auditory and visual rhythm reproduction. *Journal of Cognitive Neuroscience, 10*(6), 752–765.

Pessoa, L. (2018). Understanding emotion with brain networks. *Current Opinion in Behavioral Sciences, 19*, 19–25.

Poldrack, R. A. (2012). The future of fMRI in cognitive neuroscience. *NeuroImage, 62*(2), 1216–1220.

Proverbio, A. M., Nasi, V. L., Arcari, L. A., De Benedetto, F., Guardamagna, M., Gazzola, M., & Zani, A. (2015). Erratum: The effect of background music on episodic memory and autonomic responses: Listening to emotionally touching music enhances facial memory capacity. *Scientific Reports, 5*, 17237.

Redcay, E., & Schilbach, L. (2019). Using second-person neuroscience to elucidate the mechanisms of social interaction. *Nature Reviews Neuroscience, 20*(8), 495–505.

Renvall, V., Kauramäki, J., Malinen, S., Hari, R., & Nummenmaa, L. (2020). Imaging real-time tactile interaction with two-person dual-coil fMRI. *Frontiers in Psychiatry/Frontiers Research Foundation, 11*, 279.

Rissman, J., Chow, T. E., Reggente, N., & Wagner, A. D. (2016). Decoding fMRI signatures of real-world autobiographical memory retrieval. *Journal of Cognitive Neuroscience, 28*(4), 604–620.

Sachs, M. E., Damasio, A., & Habibi, A. (2015). The pleasures of sad music: A systematic review. *Frontiers in Human Neuroscience, 9*, 404.

Sachs, M. E., Habibi, A., Damasio, A., & Kaplan, J. T. (2020). Dynamic intersubject neural synchronization reflects affective responses to sad music. *NeuroImage, 218*, 116512.

Salimpoor, V. N., Benovoy, M., Larcher, K., Dagher, A., & Zatorre, R. J. (2011). Anatomically distinct dopamine release during anticipation and experience of peak emotion to music. *Nature Neuroscience, 14*(2), 257–262.

Salimpoor, V. N., van den Bosch, I., Kovacevic, N., McIntosh, A. R., Dagher, A., & Zatorre, R. J. (2013). Interactions between the nucleus accumbens and auditory cortices predict music reward value. *Science, 340*(6129), 216–219.

Savage, P. E., Loui, P., Tarr, B., Schachner, A., Glowacki, L., Mithen, S., & Fitch, W. T. (2020). Music as a coevolved system for social bonding. *Behavioral and Brain Sciences, 44*, e59.

Saxe, R., Brett, M., & Kanwisher, N. (2006). Divide and conquer: A defense of functional localizers. *NeuroImage, 30*(4), 1088–1096.

Scott, T. L., Gallée, J., & Fedorenko, E. (2017). A new fun and robust version of an fMRI localizer for the frontotemporal language system. *Cognitive Neuroscience, 8*(3), 167–176.

Shankweiler, D. (1966). Effects of temporal-lobe damage on perception of dichotically presented melodies. *Journal of Comparative and Physiological Psychology, 62*(1), 115–119.

Singer, N., Jacoby, N., Lin, T., Raz, G., Shpigelman, L., Gilam, G., Granot, R. Y., & Hendler, T. (2016). Common modulation of limbic network activation underlies musical emotions as they unfold. *NeuroImage, 141*, 517–529.

Sonkusare, S., Breakspear, M., & Guo, C. (2019). Naturalistic stimuli in neuroscience: Critically acclaimed. *Trends in Cognitive Sciences, 23*(8), 699–714.

Špiláková, B., Shaw, D. J., Czekóová, K., & Brázdil, M. (2019). Dissecting social interaction: Dual-fMRI reveals patterns of interpersonal brain-behavior relationships that dissociate among dimensions of social exchange. *Social Cognitive and Affective Neuroscience, 14*(2), 225–235.

Sridharan, D., Levitin, D. J., Chafe, C. H., Berger, J., & Menon, V. (2007). Neural dynamics of event segmentation in music: Converging evidence for dissociable ventral and dorsal networks. *Neuron, 55*(3), 521–532.

Sutherland, M. E., Grewe, O., Egermann, H., Nagel, F., Kopiez, R., & Altenmüller, E. (2009). The influence of social situations on music listening. *Annals of the New York Academy of Sciences, 1169*(1), 363–367.

Toiviainen, P., Burunat, I., Brattico, E., Vuust, P., & Alluri, V. (2020). The chronnectome of musical beat. *NeuroImage, 216*, 116191.

Valla, J. M., Alappatt, J. A., Mathur, A., & Singh, N. C. (2017). Music and emotion—a case for North Indian classical music. *Frontiers in Psychology, 8, 2115*.

Vuust, P., & Frith, C. D. (2008). Anticipation is the key to understanding music and the effects of music on emotion. *Behavioral and Brain Sciences, 31*(5), 599–600.

Wallace, W. T. (1994). Memory for music: Effect of melody on recall of text. *Journal of Experimental Psychology: Learning, Memory, and Cognition, 20*(6), 1471–1485.

Walther-Hansen, M. (2020). *Making sense of recordings: How cognitive processing of recorded sound works*. Oxford University Press.

Williams, J. A., Margulis, E. H., Nastase, S. A., Chen, J., Hasson, U., Norman, K. A., & Baldassano, C. (2022). High-order areas and auditory cortex both represent the high-level event structure of music. *Journal of Cognitive Neuroscience, 34*(4), 699–714.

Zaki, J., & Ochsner, K. (2009). The need for a cognitive neuroscience of naturalistic social cognition. *Annals of the New York Academy of Sciences, 1167*, 16–30.

Zatorre, R. J., Evans, A. C., & Meyer, E. (1994). Neural mechanisms underlying melodic perception and memory for pitch. *Journal of Neuroscience, 14*(4), 1908–1919.

Zatorre, R. J., Perry, D. W., Beckett, C. A., Westbury, C. F., & Evans, A. C. (1998). Functional anatomy of musical processing in listeners with absolute pitch and relative pitch. *Proceedings of the National Academy of Sciences of the United States of America, 95*(6), 3172–3177.

Zatorre, R. J., & Salimpoor, V. N. (2013). From perception to pleasure: Music and its neural substrates. *Proceedings of the National Academy of Sciences of the United States of America, 110*(Suppl 2), 10430–10437.

12 Hidden Repertoires in the Brain Accessed by Music in Aging and Neurodegeneration

Sarah Faber and Randy McIntosh

Introduction

When I (Sarah) was a music therapist, I worked in long-term care with individuals in the middle to late stages of Alzheimer's disease. Their referral documents showed a fairly standard pattern: "nonverbal, nonambulatory, does not attend recreation programs, likes music." "Likes music" was how I wound up with a large caseload of clients and a long list of questions. With few exceptions, my clients *did* like music—so much so that the word *like* barely seemed adequate to describe what happened during our sessions: a smile blooming across a normally expressionless face, singing in voices and languages their loved ones hadn't heard in months or years, improvising with new and familiar instruments. This goes beyond liking and into something much deeper.

Playing and listening to music, no matter how simple the melody, is complicated from a brain perspective (see Williams and Sachs's chapter 11 in this volume for an excellent walkthrough). First, there's the sound itself. Sound hits the ears and gets converted into brain signals, where the brain figures out whether the input is music, speech, or something else and starts parsing the content: what does the sound mean? If you've heard the song before, you make predictions about what happens next, and you'll notice if there's an error (e.g., if the song has been edited for radio, if it's a live version that deviates from the album version). You might even experience memories related to that song if it's from a significant time in your life. If you're playing the music yourself, other brain networks are involved in physically operating the instrument and planning your next steps—while listening to the sound and rapidly correcting course if something goes wrong. There are many balls to juggle, yet most of us are able to do this effortlessly. How? Well, practice helps, but it's all thanks to an intricate system of brain networks.

This system of brain networks is established early in development and grows and changes as we age. Adaptation continues into adulthood, which is seen as individual differences in cognitive aging and response to illness and injury. Successful prediction of how different individuals will respond to treatment for a stroke or progress through a neurodegenerative disease may benefit from understanding how these brain networks operate during real-world behavior. Music fits in nicely here, as it is broadly enjoyable, cognitively complex, and accessible to individuals with cognitive decline in a way that traditional text-based assessments are not. Music and music therapy have been shown to benefit individuals in long-term care (see Särkämö & Sihvonen, 2018), and by adapting tools from complex systems research, we can leverage music's complexity to capture the variability and richness of brain dynamics in aging.

Overview

In the following sections, we outline temporal and spatial concepts from complex systems research, introducing the space-time structure of the brain. We review how information is extracted from a complex system using multiscale measures and briefly discuss recent findings from studies across the human life span. We then expand this information to a manifold framework, showing how multiple configurations, or repertoires, can be active and lead to complex behavior. We also review the hidden states that are physiologically possible but accessed only with injury or illness and discuss this framework's utility in studying individual variability in aging using music, itself a complex system.

Complexity

The science of complexity covers a wide range of systems, from biology to the study of group behavior and social systems (Holland, 2014). The common characteristics of complex systems, such as nonlinearities and emergence, provide a better understanding of the similarities between systems and open the possibility of linking the complexity of one system with the complexity of another.

Perhaps one of the essential features of complex systems is their dependence on time. Complex systems are considered synonymous with dynamic systems: the system's characteristics depend on when it is observed. Previous work has demonstrated that the particular behavior of a system is dependent on its initial conditions. Even with identical core components, a slight variation in where the system begins can lead to a qualitatively different outcome as it evolves. This makes complex systems challenging to study because their behavior is hard to predict. With simple systems, in contrast, we can easily predict the outcome for a given set of inputs. We use the *simple* designation most often with linear systems, where we can estimate the output as a weighted combination of the inputs. In complex systems, because of their nonlinearities, the best we can do is define a space over which a given set of initial conditions will lead to a distribution of potential outcomes.

The time aspect of complex systems plays out at another level, wherein the system's evolution has multiple timescales that can influence one another. Slow timescales usually act as a background over which faster timescales are expressed. Let's imagine a piece of music. The harmonic structure that makes up the tonality of the piece (whether it's major, minor, modal, or tonally ambiguous) relies on how individual notes are organized in time. Although individual chords can express tonality, how we understand the broader tonality of a piece is expressed over a slow timescale that relies on consistent harmonic patterns over time. Melody can then sit on top of the slow harmonic structure, shifting between consonance and dissonance with the underlying harmonic structure at a faster timescale without changing the piece's tonality.

In the brain, this temporal interplay is seen in the variation of signals that underlie cognitive function. For most operations during cognition, fast and slow processes can be seen in the changes in high and low frequencies, respectively, in brain signals. Sometimes lower frequencies dominate, acting as "carriers" for higher frequencies. The exchange of information through oscillations at different frequencies is thought to be the basis for communication in the brain (Buzsaki & Draguhn, 2004).

It is hard to discuss time dependence in complex systems without considering space. This is especially true in the brain, where we talk about networks of regions that interact with one another during different behaviors. These networks have a particular topology that spans a differing amount of space and therefore differing times. Brain regions that are connected over a long distance take more time to interact than regions that are close. This intersection of space and time is often referred to as the *space-time structure*, a widely used term in fields from quantum physics to biology. In the brain, the space-time structure nicely captures the notion that complex systems have an architecture that constrains but does not completely determine the system's emergent behavior from a process (Deco et al., 2011). Stated differently, the space-time structure constrains the range of what could happen, but what does happen is determined by the initial conditions.

The brain's space-time structure changes with maturation and aging. The connections between cells and regions change with experience. Long-range white matter connections become more effective insulators, enabling rapid signal propagation, while local connections are formed and pruned to support adaptive behavior. This process is referred to as the progression between *segregation* and *integration*. Segregation occurs when the local configuration of neurons in an area changes to become specialized for particular types of information. A simple example is when areas connected along the auditory pathway become more specialized to segregate auditory information into parcels, such as pitch and timbre. However, the information needs to be integrated to ensure that the full percept is formed and that it can be integrated with other systems, such as the vision and motor systems, so that appropriate actions can be executed. Integration is supported through the white matter connections between regions. A complex system is one that shows an optimal balance between segregation and integration. When a system moves too far in either direction, complexity declines. Too much segregation allows incoming information to be broken down, but it cannot be integrated to drive the system forward. With too much integration, the system cannot adequately parse incoming information, and it tends to show the same response even with different information. This balance is vital for the adaptability of the brain. As information comes in, it can be appropriately parsed but also linked to potential responses. Stimulus-response mapping undergoes constant evaluation, and mappings can be updated to optimize adaptation. This is the hallmark of a complex *adaptive* system like the brain.

Information Processing in Space and Time

Because of the ever-present balance of segregation and integration, complex systems generate signals that are mixtures of processes across timescales. As a result, there tends to be more information in signals from complex systems than from simple linear systems. Indeed, complex systems can show a combination of entirely random noise mixed with activity from legitimate sources. The noise in complex systems is a benefit, as it provides the "kinetic energy" to drive the navigation of possible network configurations, but it is a curse in linear systems, as it obscures the input-output relationship (Faisal et al., 2008).

One way to assess the information content of a signal is to measure its *entropy*, which we define here as a measure of predictability. More predictable signals have low entropy, while signals that contain a mixture of scales have high entropy. However, this latter feature is a bit of a problem because signals that are only noise and signals that contain information may have similar levels of entropy. One solution takes advantage of the space-time structure of complex systems. Information may be available across multiple scales in a complex system, while noisy systems with random space-time structures have little information across scales. We used a metric called multiscale entropy (MSE; Costa et al., 2005) that measures entropy as a function of temporal scale. Entropy of the original signal is calculated; then the signal is successively down-sampled, and at each sample, entropy is calculated again. For a complex signal, entropy remains

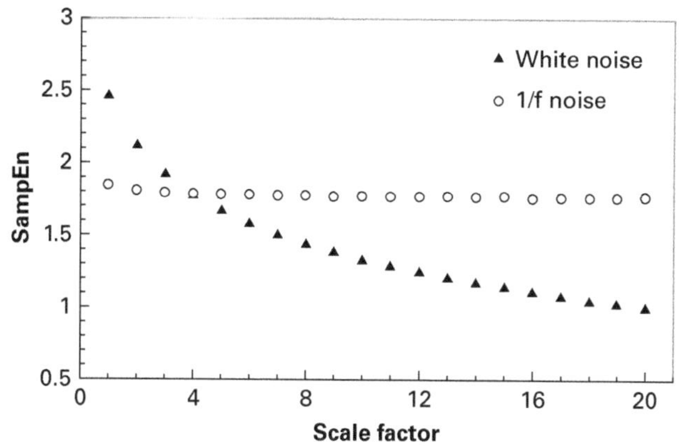

Figure 12.1
Example of MSE curves for two signals. The y-axis is entropy, and the x-axis is timescale. White noise has little information beyond the first few scales, and sample entropy (SampEn) drops. 1/f noise, which has dependencies between frequencies, has information across multiple scales, so sample entropy remains high across many timescales. Figure extracted from https://archive.physionet.org/physiotools/mse/tutorial/node2.html.

high for most down-sample levels, while entropy for a noisy signal declines as it gets progressively smoother. These observations can be depicted in an MSE curve that plots entropy by timescale (figure 12.1).

A second benefit of looking at information across scales of a complex system is that we can relate the timescale profile to determine whether there is a particular scale at which information is highest. We can then relate the shape to the space-time structure to identify the collection of networks with the highest information-processing potential. We have examined this relationship across maturation and aging to see whether the link between information and space-time changes and whether this relationship breaks down in the face of pathology such as dementia. This relationship between the space-time structure and information processing (see McIntosh, 2019) can be summarized as follows (see figure 12.2):

1. We can use healthy young adults as the reference point. Collectively, they show a gradual increase in entropy with a timescale that peaks at midrange and then declines slightly. In relation to the space-time structure, this means that information at fine scales (shorter timescales) would be lowest, with more information at medium timescales and then dropping at slower timescales.
2. Children (from infants to about ten years old) show a similar relative relationship across scales, but the overall level of entropy is lower.

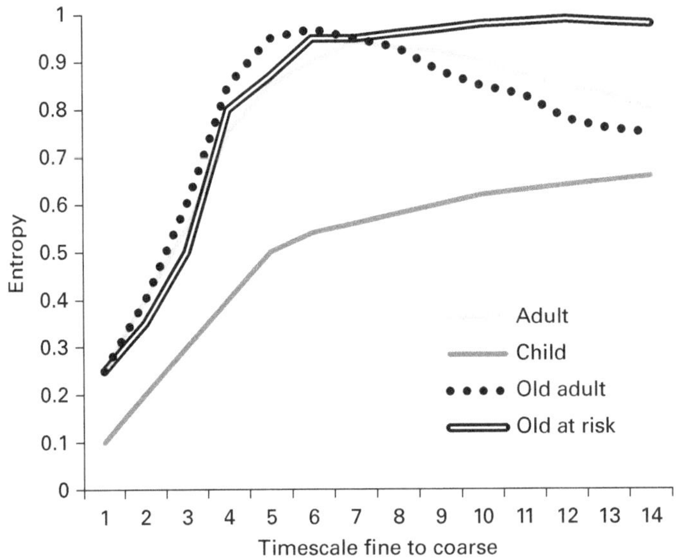

Figure 12.2
Schematic representation of the MSE curves measured from the brains of four age groups. The y-axis is entropy, and the x-axis is timescale.

3. Healthy older adults (fifty to eighty-five years) show relatively greater entropy at faster scales and less at slower scales. This pattern is associated with good cognitive status for the cohort.
4. Older adults at risk for cognitive decline show relatively less information at fine scales and more at coarse scales. This pattern can be seen before the overt expression of cognitive decline, suggesting that it may be prognostic.

We can further relate these observations to brain organization by stating the following:

1. The healthy adult brain shows optimal information processing at intermediate spatiotemporal scales.
2. Children show overall lower information processing than adults, which increases with maturation. This correlates nicely with the structural changes in development, where local processing capacity increases with neuronal pruning, complemented by increased myelin density that facilitates long-range interactions.
3. The healthy older adult brain shows optimal information integration at local spatiotemporal scales, with less integration at longer ranges.
4. Adults at risk for cognitive decline show lower information processing locally but more at longer ranges.

The details underlying these changes are not well known but are likely the result of both structural changes that occur across the life span and pathological changes that may affect local neural populations. These brain changes also occur in a rich environmental context, which opens the possibility that the particular patterns of information processing within the space-time structure reflect the intersection of biological and environmental factors.

Implications of a Complex Systems Framework for Music Neuroscience

If we accept that the brain has a rich spatiotemporal structure, we can assume that any emergent cognitive functions will show a complementary richness. This connection is captured in the notion that cognitive functions—or indeed, any function—should be considered processes rather than discrete states. The "state bias" is evident in the common practice of treating functions as static in neuroscience research. In part, this may reflect the fact that much of what happens across time in functions such as memory or attention is invisible to us, and an estimation of the function is reduced to point estimates, such as accuracy or reaction time. Much of music research follows this practice, but music's temporal structure makes it perfect for relating to the brain's concomitant temporal structure that evolves with music.

There is a challenge associated with linking the temporal structures of music and the brain. The phenomenology of music is an emergent feature of elements. Accounting for this emergence in the study of music and the brain could provide unique insights into how the emergent features of music interact with the emergent features of brain function. Past work has quantified patterns of behavioral responses to musical pitch (Krumhansl, 1990) and the neural correlates of harmonically rich music (Alluri et al., 2012), personally significant music (Salimpoor et al., 2011), improvisation (Limb & Braun, 2008), and social music making (Lindenberger et al., 2009; Müller et al., 2013; Sänger et al., 2012; Donnay et al., 2014). However, important information is lost by dividing music perception into discrete categories and a priori selecting circumscribed sets of brain regions known or theorized to respond to these categories. Such work has been fundamental to our understanding of music and the brain, but music is much more than sequences of tones, and the brain is much more than the firing of neurons.

One way to link the emergent features of music and the brain is to use metrics that are sensitive to the whole rather than the parts, such as multiscale entropy. We can characterize the complexity of music by measuring its entropy as a function of timescale. This is possible because music has a rich temporal structure. At finer timescales, information related to the acoustic or timbral properties of the sound is dominant, while properties related to structure (novelty, tonality, rhythm) are dominant at coarser scales. The advantage of using MSE to measure complexity for both music and the brain is that we can now link these complexity metrics when someone listens to music.

We obtained electroencephalograms (EEGs) in young adults while they listened to short musical excerpts under two conditions (Carpentier et al., 2020). In one condition (Perceptual), the participants were asked to track the relative pitch and tempo of the music as it unfolded. In the second condition (Emotional), the participants were asked to track their relative changes in arousal and valence to a different set of music. In essence, the Perceptual task had participants focus on the physical characteristics of the music, while the Emotional task had them focus on how the music made them feel. We then measured MSE from the participants' brains and MSE from the music they were listening to and measured the geometric distance between the shape of the two MSE curves. The idea was that the closer the participants were tracking the music, the smaller the distance between the MSE curves.

This is what happened, and a bit more. Figure 12.3 compares the changes in distance between MSE curves for music and the brain in both the Emotional and Perceptual tasks in ten-second epochs. A higher "brain score" means that the distance between curves is greater. The results show that the distance for the Emotional task is greater than that for the Perceptual task. In addition, the MSE curves for the Emotional task

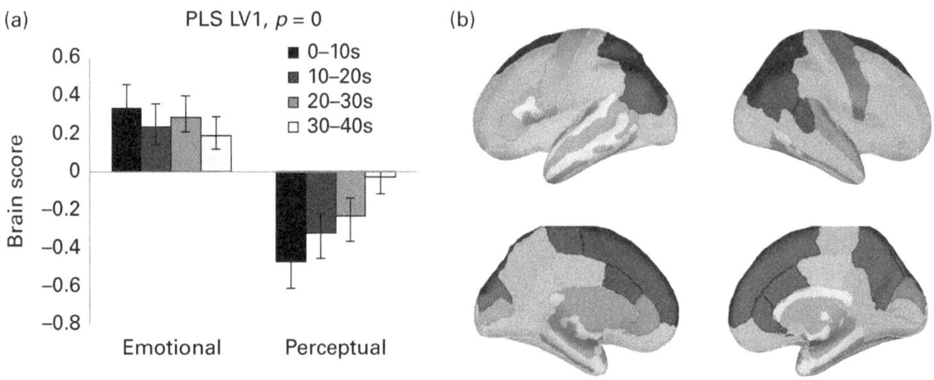

Figure 12.3

Analysis of the match between brain and music complexity. Higher scores mean more distance between brain complexity and music complexity. The bar graph in panel A shows the change in the match across ten-second epochs of music listening, where either subjective emotion (Emotional) or changes in pitch and tempo (Perceptual) are judged. Panel B shows the brain regions most strongly involved in this change, with the darkest colors indicating the most robust contribution. Many of these areas are part of the brain's default mode network, which is thought to be critical for integration with other brain networks.

are larger than those for the Perceptual task. This means that participants "added" extra information in their brains when listening to music and judging how they felt.

We also measured whether the distance of the MSE curves was related to the participants' music listening behavior in general. Interestingly, we discovered that individuals with greater distances between music and brain MSE curves found music more rewarding in general, as assessed by the Barcelona Music Reward Questionnaire (Mas-Herrero et al., 2013). Basically, the more information a participant adds to the music signal as it courses through the brain, the more personally rewarding music is to that person.

This intersection of complexity in music and the brain opens a new avenue of exploration to determine how the individual features of a person's brain operations relate to that person's current experience. We can capture spatiotemporal brain patterns more explicitly using methods such as hidden Markov modeling (HMM; Vidaurre et al., 2016). This technique works by assuming that there are several hidden states—here, defined as either brain networks or features of music (see below)—whose activities vary across time as a process evolves. For the brain, one can envision different functional networks (e.g., auditory, motor, default mode) varying in engagement when listening to music. For music, one can express states in terms of the shifting prevalence of acoustic features (e.g., loudness, roughness) or structural features (e.g., modality, novelty) derived using music information retrieval (MIR) methods such as the MIR toolbox (Lartillot & Toiviainen, 2007). In both cases, the critical output of HMM is how the states relate across time in terms of which states are more likely to follow one another.

We also applied HMM to both music and brain data (Faber & McIntosh, 2022). The brain states were divided among perpetual networks for monitoring the musical signal, goal-directed movement networks for operating the mouse tracker, and emotional or interoception networks for monitoring internal states and activation and deactivation states for the default mode network (DMN), which is ubiquitous during wakeful rest. We found that the emotional listening condition with a higher complexity score in the previous analysis now showed a repeating circuit of activity between brain states related to the deactivation of the DMN, cursor movement, decision making, and the processing of emotions tuned to structural features in the music. A similar circuit was present in the perceptual condition but tuned to acoustic features in the music, reinforcing the emergent nature of music listening. When analyzing functional connectivity states over time, we saw larger patterns of activity that weren't captured by correlating the brain data to the music or behavioral features. These results also raised some interesting questions about how the brain employs repeating patterns of networks over time to generate complex behavior. How robust are these networks? Will they adapt with age and expertise, and what happens with aging and injury or illness?

Age Effects

The HMM approach has great potential to capture the changes in spatiotemporal flows in aging and dementia. There are few examples in the literature, but the work consistently shows that both the spatial pattern of brain networks and their temporal dependencies change in aging. One of the first studies to do so was by Cabral et al. (2017), using leading eigenvector decomposition analysis (LEIDA), which is comparable to HMM. The researchers analyzed functional magnetic resonance imaging (fMRI) scans in 120 subjects aged fifty years or older obtained while they were resting in the

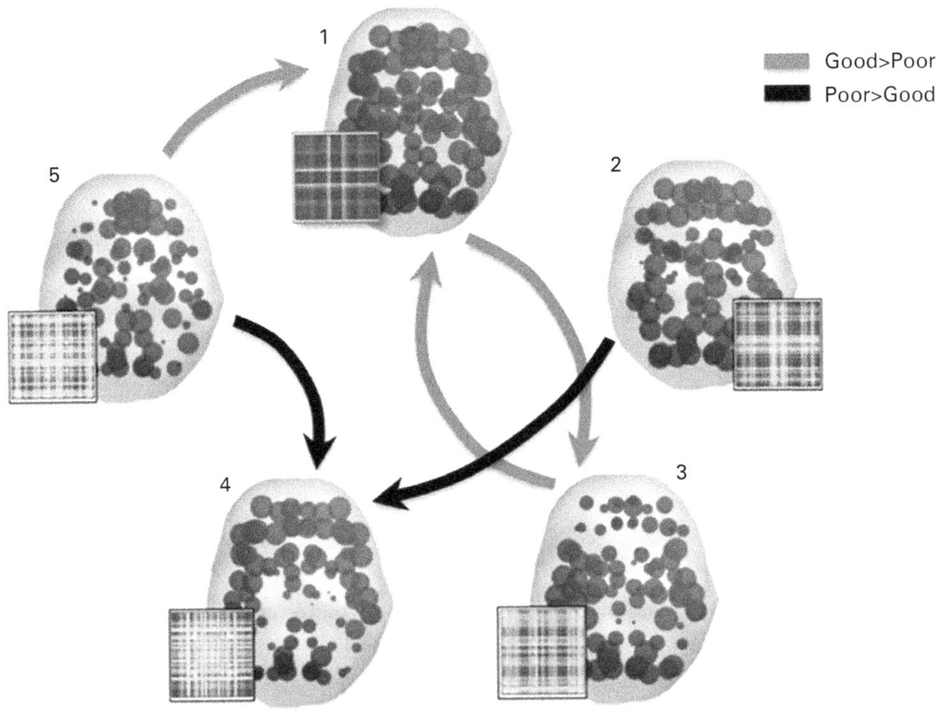

Figure 12.4
Summary of resting-state fMRI data from older adults using LEIDA. The five common states are numbered 1–5, and nodes in each state are displayed as colored balls; the color map shows the degree of correlation between nodes, with darker shades indicating a higher correlation. Arrows connecting the states indicate the paths where the transitional probability between those states differed in relation to cognitive status (gray = stronger transitional probability in persons with relatively good cognitive performance, black = stronger transitional probability in persons with relatively poor cognitive performance). (Figure excerpted from Cabral et al., 2017.)

scanner (resting state). The participants were divided into two groups based on neuropsychological memory and attention.

LEIDA identified five spatial patterns corresponding to distinct functional network states (Figure 12.4). These networks had temporal relationships that differed between groups, indicated by the transitional probabilities of moving between states. Thus, although the states may be relatively stable in aging, the way they interact across time is not. Complementary work from Battaglia et al. (2020) studied a broader age range (eighteen to seventy years). These investigators focused on the temporal switching of networks, specifically examining the speed and number of switches in relation to aging and cognitive status. One major finding was that there is less network switching across age. This does not mean it stops, but rather the switches become less frequent. A second outcome was that, among the older subjects, faster pattern switches related to better cognitive status. These two examples underscore that the key to understanding the aging brain involves more than determining whether certain brain regions are intact; the broader patterns of region interaction across the brain and over time are important as well.

The same motivation can prompt us to apply spatiotemporal analyses like LEIDA or HMM to neurodegenerative disorders in aging (Sourty et al., 2016). In that case, the emphasis would be on differentiating changes in the spatiotemporal structure related to neurodegeneration itself and to aging and analyzing the potential interaction of the two processes. Moreover, given the disparity across individuals in response to the degenerative process, a focus on individual modeling is critical (Botha et al., 2020).

To make progress, however, we need to go beyond simply characterizing differences in the brain and move toward linking the spatiotemporal flows in the brain with the concurrent flows in behavior. In our initial example, it is possible to envision that ongoing flows in the brain of a person with dementia move as that person listens to music, and the music's familiarity takes that individual to a vivid moment in his or her history. The music study in which HMM was applied to the brain and to music is a good example of how we can design studies that better link the brain and music experience.

An emerging theory of complex systems focuses on brain and behavior, emphasizing their mutual dependency. This theory, termed structured flows on manifolds (SFM), posits that behavior can be appreciated as the expression of flows bounded by manifolds (Perdikis et al., 2011a, 2011b; Pillai & Jirsa, 2017). Flows can be a series of movements, the perception of a series of sounds that combine into music, or the transition between brain states that link movement and perception. The manifold provides the constraints for which flows can be accessed—essentially, the rules. There can be variations in flows that depend on initial conditions and the inherent randomness of nonlinear systems, but the SFM approach provides a comprehensive framework to connect brain and behavior processes in terms of the available potential actions (flows). This is an important point because most brain theories are derived only from what a system does when it is observed. SFM differs, in that it focuses on what the system did at a given time (the actual flow), why that flow happened (the manifold that dictates the rules), but also what *could have* happened (the potential flows). This potentiality is essential for complex adaptive systems and can bring a new perspective on the brain's capacity to process information (McIntosh & Jirsa, 2019).

The SFM framework also provides a different perspective on brain changes in neurodegeneration. With progressive decline, the preferred navigation of flow on manifolds may be unavailable, but there may be alternative routes through different pathways. This hypothesis is exciting in terms of providing a potential therapeutic target to help maintain or recover some behaviors compromised by degeneration. If the degenerative process can be detected early, characterization of the manifold architecture for that individual can guide interventions to develop alternative flows to support equivalent behaviors. Such characterization is becoming possible with TheVirtualBrain neuroinformatics platform (Ritter et al., 2013; Sanz Leon et al., 2013), which can simulate a person's entire brain from structural and functional brain imaging data. The simulation then becomes a virtual laboratory to explore the SFM architecture for that person. If flows are compromised, manifold estimation may also yield clues on how to reintegrate brain networks, and simulations can be done to test the potential for actual intervention.

Music uniquely engenders multiple systems in the brain. Perceptual and motor networks are the usual suspects, but linguistic, social, and emotional components are indivisible from the musical experience. The social aspect links with the listener's personal history via personal memories. Understanding how we maintain our connection to music with the onset of neurodegeneration requires extensive foundational knowledge on the nature of the brain, music, and aging. How does the brain process music from a purely perceptual standpoint? How do emotion, preference, and familiarity factor in? What patterns describe an individual's brain activity in response to music, and how do we reconcile individual differences? How do these patterns change with age and neurodegeneration? Why does one individual with Alzheimer's disease still sing his or her favorite song while another does not? By applying the complex systems framework, we can start addressing some of these questions, leading to a greater understanding of the richness of individual brain health across the adult life span.

References

Alluri, V., Toiviainen, P., Jaaskelainen, I. P., Glerean, E., Sams, M., & Brattico, E. (2012). Large-scale brain networks emerge from dynamic processing of musical timbre, key and rhythm. *Neuroimage*, *59*(4), 3677–3689.

Battaglia, D., Boudou, T., Hansen, E. C. A., Lombardo, D., Chettouf, S., Daffertshofer, A., . . . Jirsa, V. (2020). Dynamic functional connectivity between order and randomness and its evolution across the human adult lifespan. *Neuroimage, 222*, 117156.

Botha, H., Graff-Radford, J., Gunter, J. L., Schwarz, C. G., Senjem, M. L., Vemuri, P., . . . Jones, D. T. (2020). Disrupted brain dynamics across the Alzheimer's disease spectrum is related to tau accumulation. *Alzheimer's & Dementia, 16*(S5), e040583.

Buzsaki, G., & Draguhn, A. (2004). Neuronal oscillations in cortical networks. *Science, 304*(5679), 1926–1929.

Cabral, J., Vidaurre, D., Marques, P., Magalhaes, R., Silva Moreira, P., Miguel Soares, J., & Kringelbach, M. L. (2017). Cognitive performance in healthy older adults relates to spontaneous switching between states of functional connectivity during rest. *Scientific Reports, 7*(1), 5135.

Carpentier, S. M., McCulloch, A. R., Brown, T. M., Faber, S. E., Ritter, P., Wang, Z., . . . McIntosh, A. R. (2020). Complexity matching: Brain signals mirror environment information patterns during music listening and reward. *Journal of Cognitive Neuroscience, 32*(4), 734–745.

Costa, M., Goldberger, A. L., & Peng, C. K. (2005). Multiscale entropy analysis of biological signals. *Physical Review E, 71*(2 Pt 1), 021906.

Deco, G., Jirsa, V. K., & McIntosh, A. R. (2011). Emerging concepts for the dynamical organization of resting-state activity in the brain. *Nature Reviews Neuroscience, 12*(1), 43–56.

Donnay, G. F., Rankin, S. K., Lopez-Gonzalez, M., Jiradejvong, P., & Limb, C. J. (2014). Neural substrates of interactive musical improvisation: An fMRI study of "trading fours" in jazz. *PLOS ONE, 9*(2), e88665.

Faber, S. E., & McIntosh, A. R. (2022). Modelling music listening in multiple modalities in space and time [Poster presentation]. Organization for Human Brain Mapping Annual Meeting, Glasgow, Scotland.

Faisal, A. A., Selen, L. P., & Wolpert, D. M. (2008). Noise in the nervous system. *Nature Reviews Neuroscience, 9*(4), 292–303.

Holland, J. H. (2014). *Complexity: A very short introduction*. Oxford University Press.

Krumhansl, C. L. (1990). *Cognitive foundations of musical pitch*. Oxford University Press.

Lartillot, O., & Toiviainen, P. (2007). A Matlab toolbox for musical feature extraction from audio [Paper presentation]. International Conference on Digital Audio Effects, Bordeaux, France.

Limb, C. J., & Braun, A. R. (2008). Neural substrates of spontaneous musical performance: An fMRI study of jazz improvisation. *PLOS ONE, 3*(2), e1679.

Lindenberger, U., Li, S. C., Gruber, W., & Müller, V. (2009). Brains swinging in concert: Cortical phase synchronization while playing guitar. *BMC Neuroscience, 10*(1), 1–12.

Mas-Herrero, E., Marco-Pallarés, J., Lorenzo-Seva, U., Zatorre, R., & Rodriguez-Fornells, A. (2013). Individual differences in music reward experiences. *Music Perception, 31*, 118–138.

McIntosh, A. R. (2019). Neurocognitive aging and brain signal complexity. In *Oxford research encyclopedia of psychology*. Oxford University Press.

McIntosh, A. R., & Jirsa, V. K. (2019). The hidden repertoire of brain dynamics and dysfunction. *Network Neuroscience*, 1–34.

Müller, V., Sänger, J., & Lindenberger, U. (2013). Intra- and inter-brain synchronization during musical improvisation on the guitar. *PLOS ONE, 8*(9), e73852.

Perdikis, D., Huys, R., & Jirsa, V. (2011a). Complex processes from dynamical architectures with time-scale hierarchy. *PLOS ONE, 6*(2), e16589.

Perdikis, D., Huys, R., & Jirsa, V. K. (2011b). Time scale hierarchies in the functional organization of complex behaviors. *PLOS Computational Biology, 7*(9), e1002198.

Pillai, A. S., & Jirsa, V. K. (2017). Symmetry breaking in space-time hierarchies shapes brain dynamics and behavior. *Neuron, 94*(5), 1010–1026.

Ritter, P., Schirner, M., McIntosh, A. R., & Jirsa, V. K. (2013). TheVirtualBrain integrates computational modeling and multimodal neuroimaging. *Brain Connectivity, 3*(2), 121–145.

Salimpoor, V. N., Benovoy, M., Larcher, K., Dagher, A., & Zatorre, R. J. (2011). Anatomically distinct dopamine release during anticipation and experience of peak emotion to music. *Nature Neuroscience, 14*(2), 257–262.

Sänger, J., Müller, V., & Lindenberger, U. (2012). Intra- and interbrain synchronization and network properties when playing guitar in duets. *Frontiers in Human Neuroscience, 6*, 312.

Sanz Leon, P., Knock, S. A., Woodman, M. M., Domide, L., Mersmann, J., McIntosh, A. R., & Jirsa, V. (2013). TheVirtualBrain: A simulator of primate brain network dynamics. *Frontiers in Neuroinformatics, 7*, 10.

Särkämö, T., & Sihvonen, A. J. (2018). Golden oldies and silver brains: Deficits, preservation, learning, and rehabilitation effects of music in ageing-related neurological disorders. *Cortex, 109*, 104–123.

Sourty, M., Thoraval, L., Roquet, D., Armspach, J., Foucher, J., & Blanc, F. (2016). Identifying dynamic functional connectivity changes in dementia with Lewy bodies based on product hidden Markov models. *Frontiers in Computational Neuroscience, 10*, 60.

Vidaurre, D., Quinn, A. J., Baker, A. P., Dupret, D., Tejero-Cantero, A., & Woolrich, M. W. (2016). Spectrally resolved fast transient brain states in electrophysiological data. *Neuroimage, 126*, 81–95.

13 Composing at the Border of Experimental Music and Music Experiment

Grace Leslie

Music cognition and experimental music are practices that emerge from separate disciplines yet share several overarching goals. Their creators use the same algorithms and hardware to generate stimuli or compositions. These compositions often use the simplest building blocks of sound: sine waves, clicks, and pulses. Such stimuli probe singular facets of human auditory perception and music experience and force the listener to encounter the fallible machinery that makes the act of listening possible. Using shared tools, music cognition researchers and composers have explored the same terrain for decades. Karlheinz Stockhausen, a pioneer of electronic music in postwar Germany, wrote of the need to blur the boundaries between the two disciplines to bring electronic music practice into a new era: "The musician, therefore, for whom the question of research in sound had for the first time become acute, has been obliged to undertake a considerable amount of this research himself. He has had to expand his métier, and to study acoustics, in order to better the acquaintance with his material" (Eimert & Stockhausen, 1959, p. 60).

Electronic and computer music techniques enable the exploration of perception through sculpting of sound in its time and frequency domains to match our engineering preferences, bypassing the limitations of acoustic instruments and concert halls. Our music and sound perception, whether through evolution or experience, expects sounds to have certain spectral shapes due to the physical characteristics of resonant bodies (Theunissen & Elie, 2014)—but we can circumvent these expectations by engineering acoustic patterns directly.

I compose this circumventory music as an auditory-music experiment, inspired by two parallel research paths. The field of music perception and cognition is tasked with mapping and understanding the roles that various perceptual mechanisms and anatomies play in perceiving, understanding, remembering, and responding to music. It's a composer's thought experiment to wonder how one might create music for the express purpose of driving the perceptual mechanism and physiology in a particular way, let alone how to use this engineered music as a means to benefit the listener.

Here, I present a part of my musical work that exists at this borderland between experimental music and music experiment. I first acknowledge the scientist-composers whose work established the conceptual framework on which I build. These are examples in which control and exploration of the perceptual mechanism are the goals: a composer creates a piece with the goal of probing the inner workings of the mind, and a scientist creates a stimulus to map the brain's response to sound.

Music cognition researchers often electronically generate or manipulate auditory stimuli to elicit particular perceptual experiences, with the goal of explaining one or more specific aspects of music cognition; however, in focusing on these methods, I do not mean to imply that they are employed only in this field. As explored by Williams and Sachs (see chapter 11 in this volume), other strains of music cognition research use naturalistic paradigms that attempt to re-create everyday music listening experiences for each participant, aligning with the earlier Gestalt positioning of complex experience as more than the sum of its simplest components.

Max Mathews's (1963) 1957 MUSIC I program was the first to perform musical tones and compositions with a digital computer. Roger Shepard (1964) was among the first to use computer-generated musical tones to demonstrate aspects of music perception and cognition, in his case, the circular nature of relative pitch discrimination. Diana Deutsch (see chapter 14 in this volume) has composed many musical illusions and paradoxes that reveal individual differences in the perception of musical patterns. In her octave illusion, the right and left channels of a stereo composition each contain a sequence of two alternating tones spaced one octave apart (Deutsch, 1981). The two sequences are arranged such that the higher tone is played in one channel when the lower tone is played in the other channel. Although the stimulus is a two-tone chord, with the ear of presentation switching with each repetition, the perception of this stimulus varies among listeners. Some always hear the high tone in the same ear, often corresponding to their handedness, while other listeners exhibit even more complex illusions. Deutsch conjectured that this ambiguity arises from two competing mechanisms attempting to form a combined percept—one responsible for pitch perception, and the other responsible for location perception. In the presence of these two contradictory streams of information, we end up suppressing the pitch perception of one stream in order to weave together a more probable explanation. The scale illusion (Deutsch, 1974) operates on a similar conjunction of pitch and location. Because it is much more likely for a single musical source to play with stepwise motion in a fixed location, we perceive this arrangement of pitch and location rather than the correct but improbable interpretation. Jean-Claude Risset (1978, 1985) created similar illusions and incorporated them into his electroacoustic compositions.

In Iannis Xenakis's electronic music compositions, huge masses of sound are created by specifying the statistical trends of randomly generated sound features over the course of a piece, rather than individually generating and controlling musical voices (Xenakis, 1992). *Concrete Ph* is an exemplar of this approach: a tape composition intended to evoke a large "cloud" of sound (Bridoux-Michel & Xenakis, 2005). *Concrete Ph* was written as an interlude during the premiere of Varése's *Poeme electronique* at the 1958 World's Fair in Brussels (Treib & Felciano, 1996). Xenakis's approach to *Concrete Ph* involved splicing, looping, mixing, and speed-changing one-second fragments of recordings of burning wood (Roads, 2004). These small pieces of sound were layered to such a degree that a singular cloud-like sound mass resulted from the mix. Xenakis wrote of his intuitive approach: "Start with a sound made up of many particles, then see how you can make it change imperceptibly, growing and developing, until an entirely new sound results" (Brody, 1970). However, the final composition was written for eleven channels and diffused over the loudspeaker system at the fair's Philips Pavilion, creating a blurry distinction between individual sound "grains" and the composite "cloud," thus becoming an informal experiment in spatial audio perception. With greater distance between channels, the human auditory system is better able to distinguish between multiple sound sources (Blauert, 1997). Thus, we travel along a "phase state" curve between the sound cloud and the many separate voices (Schick, 2007). It can only be imagined that as spectators walked around in the space, their changing locations allowed them to travel along this "phase state" curve, experiencing the fragile boundary between sound fragments and clouds and thus experimenting with the limits of their own auditory perception. Xenakis used this technique in many of his percussion compositions, such as *Persephassa* and *Rebonds*; however, the flexibility of the diffused electroacoustic medium allowed him the fine-grained control to experiment with the smooth transition between these varying states.

Fred Lerdahl famously straddled the composition-research divide with his (and Jackendoff's) generative theory of tonal music. His practice as a composer led to key research insights to shift music-generating algorithms toward outputs relevant to the internal mental representations created by the listener. For instance, he proposed that composers of contemporary serial music who adopt artificial grammars do so in violation of the intuitive constraints that govern the music listening process (Lerdahl, 1992).

These researchers and composers wrote compositions to probe the inner workings of the musical and auditory mind, and my recent compositions attempt something similar: they are engineered to play the brain and the body as an instrument. These tracks are algorithmically generated and designed to be played continuously at low volume throughout the listener's day, creating an ambient presence in the surroundings and subtly shifting the body by inviting a parallel internal entrainment through the auditory perceptual mechanism and autonomic nervous system.

These works contain computer-generated tones employed as tuning forks to find the sympathetic resonances in the body's rhythmic functions: brain dynamics, respiration, heartbeat. Neuroscientists and doctors have developed stimulation protocols using electricity and magnetism to noninvasively produce rhythmic brain patterns to target particular wavelengths, frequencies, and anatomy. This line of research echoes medical practices from centuries ago built on the belief that the nervous system can resonate sympathetically with external vibratory stimuli (see Raz's chapter 5 in this volume). More recently, this concept has been extended to sensory neurostimulation: the use of sensory stimulation—light and sound—to create sympathetic rhythms in the brain and nervous system (Adaikkan & Tsai, 2019).

My performance practice is the result of years of "n = 1" experiments in which I discovered which kinds of flute playing and breathing affected my physiology to the extent that I could incorporate my own body signals into live electronic music compositions. I present these concepts along with a record of scientific experiments my colleagues and I have undertaken during my time at MIT, Dartmouth, and Georgia Tech. Taken together, these results are a first validation of this brain-body music concept.

Breathing Music

Breathing is a unique autonomic function, in that it is simultaneously consciously and unconsciously controlled. In this way, it can act as a gateway for the conscious manipulation of the autonomic state. Music has been written with the express purpose of training listeners to regulate their breathing patterns (Bernardi et al., 2006; Siwiak et al., 2009), inducing relaxation (Yu et al., 2018) and meditative states (Vidyarthi & Riecke, 2014), reducing muscle tension (Robb, 2000), and influencing both electrodermal activity and heart rate variability (Bhandari et al., 2015). This intentional control of breathing may regulate deeper autonomic states, and although breathing remains under autonomic control, we can override this control through our own efforts or environmental influences (Moraveji et al., 2011).

In my own performances, I find that the way I breathe has a profound effect on the composition, triggering a cascade of changes in other physiological signals (e.g., heart rate and electrodermal activity) that I record using sensors. Breathing itself is translated into sound when I play long tones into the flute, one tone per breath, or when

I amplify the sound from a stethoscope placed on my chest. The underlying tempo of this pattern is reinforced by looping algorithms that synchronize with this rate.

In 2017, at the MIT Media Lab under the auspices of Rosalind Picard's Affective Computing Group, my colleague Asma Ghandeharioun and I created an experiment to test whether an ambient music track designed to mimic the time course of relaxed breathing could invite listeners to change their own breathing to match the track (Leslie et al., 2019). I composed an algorithmically generated musical composition that breathed softly along with the listener, either according to a predetermined "ideal" breathing rate or in response to data streamed online from a respiration sensor worn by the listener. During the experiment, we did not inform listeners that the mass of tones surrounding them were mimicking their own fluctuations of breath or were designed to influence their breathing. The listeners were given an unrelated reaction-time task and told that the music was being provided to prevent boredom. We recorded the respiration trace, electrocardiogram (ECG), electroencephalogram (EEG, to measure brain waves), and electrodermal activity (EDA) of listeners as they experienced this subtle breathing presence in the room. We found that listeners' breathing rates slowed to match the musical ambience, and their other autonomic functions followed suit, indicating a general relaxation response. The EEG data showed an impact on the contingent negative variation (CNV) measured between the warning and imperative stimuli during the reaction-time task they completed, indicating that slower breathing influenced the level of cortical arousal in the brain elicited by this task. Interestingly, the music was most effective at slowing breathing when the tempo was fixed at a rate slightly slower than each listener's natural breathing rate, suggesting an added benefit to personalizing the music based on the listener's body signal information. Given that we did not instruct the listeners to match their breathing to the music, we concluded that some part of the sympathetic slowing was due to the unconscious influence of the music. As a performer, I continue to use this technique with the intention of creating synchrony, broadly defined, with an audience.

Brain Music

I have performed for several years using my EEG signals by applying a sonification algorithm that imprints their spectral quality onto a bank of stored flute and voice sounds, which I mix with my own live flute improvisation (Leslie, 2021). This performance practice emerged from an exploration of various breathing control methods intended to both create changes in the sonified sound and initiate a cascade of body and brain events. The nature of the musical composition itself played a role in my ability to sustain attention for the time required to create an expressive arc with this method: my long practice sessions formed a singular open composition. The slow and diatonic qualities of the music fed back into my mental state, making it possible to sustain the performance. After I began to perform this piece publicly, audience members reported that they found it very relaxing, indicating that some sympathetic process was at work. I began to imagine this performance practice as a sharing of a particular brain and body state that the sonified sound helped communicate to my audience. I took this idea of sympathetic response one step further by producing a series of music tracks designed to stimulate the brain within the EEG spectrum to produce beneficial effects.

There are precedents for this idea in medicine and medical research, specifically for the treatment of epilepsy and other neurological disorders. Epilepsy is characterized by an intense regularity emerging from networks of neurons firing synchronously in the brain (Duncan et al., 2006). Over time, this intermittent synchrony results in channels where these streams of voltage run through the brain, damaging memory and cognition. Doctors have discovered that inserting a competing, regular electrical pulse into this network can interrupt this synchrony and prevent a seizure (Schulze-Bonhage, 2017). Vagus nerve stimulation protocols (Ben-Menachem, 2002) are preprogrammed for the patient, creating an open-loop system (González et al., 2019). In contrast, responsive neurostimulation is engineered as a closed-loop system, in that it continuously monitors brain activity for signs of epileptiform activity and automatically introduces targeted brain stimulation at the site of seizure generation in the brain (Morrell, 2011; Sun et al., 2008). Thalamic stimulation also has anticonvulsive effects (Fisher et al., 2010). While the mechanism of action remains to be explained (Vonck et al., 2001), these therapeutic protocols are thought to promote plasticity and desynchronize cortical rhythms contributing to epileptiform activity (Sohal & Sun, 2011). Multiple stimulation protocols using various waveforms have been compared, including DC stimulation, single pulses, phase resetting, and low- or high-frequency periodic stimulation (Durand & Bikson, 2001; Schiller & Bankirer, 2007).

Scientists are also developing sensory neurostimulation protocols using flickering light and pulsatile audio stimuli to evoke oscillations at specific frequency bands. Gamma-band oscillations induced with flickering lights reduced signs of Alzheimer's pathology in mouse models (Martorell et al., 2019), and similar work in the audio domain is under way. While a postdoctoral fellow at Dartmouth in 2017, I proposed a musical sensory neurostimulation protocol inspired by this work. I produced musical samples containing low-frequency acoustic energy that overlapped the spectrum of gamma-band neural activity and played these to postoperative brain surgery patients at Dartmouth-Hitchcock Medical Center. Dr. Barbara Jobst, Robert Quon, and others in the Epilepsy and Cognition Lab have conducted further experiments using these musical neurostimulation stimuli (Quon et al., 2021). By analyzing the electrocorticography (ECoG) data of these patients as they listened to the stimuli, we found that these tones reduced interictal epileptiform activity in patients who normally demonstrated high levels of this indicator. I have incorporated similar low-frequency acoustic tones in electronic compositions such as *Sonic Tonic* (2014) and *Fais de moi un instrument* (2021). Although the technique is far from being adopted by the medical community, the results of the experiment may ground the neurostimulation metaphor developed first in my performance practice and then in my electronic compositions.

Heart Music

Of all the body signals I've incorporated into my brain-body music performance practice, directly amplifying the heart using a microphone placed inside a stethoscope tube has had the most direct effect on my listeners. During one listening session in Tokyo in 2015, I asked the audience to write down what thoughts the performance evoked for them. One audience member recalled being in his mother's womb; another described seeing a spiritual teacher returned to life. This performance technique mimics the body's own interoceptive awareness. By playing my own heartbeat onstage, I create an experience that gives listeners the sense that they are inhabiting my body, as they hear my heartbeats as their own.

For his PhD dissertation at Georgia Tech, Mike Winters (2020) studied how listening to heartbeats affected empathic states. He played synthesized heartbeat sounds to subjects as he recorded their ECGs and EEGs. He also tested their empathic reactions to this listening exercise. He found that listeners' heartbeats slowed to match the heartbeats of the imaginary person they were hearing, and this process of listening to heartbeats caused them to shift their empathic reactions to expressive faces (Winters et al., 2021). Most interestingly, his analysis of the heartbeat evoked potential (HEP) revealed a difference between most empathic and less empathic trials: the intensity of the HEP curve decreased with greater empathy, suggesting fewer neural resources supporting interoception, or the awareness of one's own heartbeat (Winters, 2020). Winters's research suggests several avenues by which heartbeats can be used to invite and sustain empathic reactions between individuals.

This result also leads to a question: can the unconscious transfer of body awareness from oneself to another through the imitation of bodily functions at least partially explain how music arouses emotional responses? Juslin (2019) lists thirteen different mechanisms by which emotion may be elicited in music listeners. The effect examined here could be categorized as both "rhythmic entrainment" and "contagion." However, results suggest that interoception may be an underlying brain-body mechanism that supports both modes of emotional arousal.

In all three examples—breathing, brain, and heart music—both formal experiments and informal experimental music reached similar conclusions, supporting the idea that sympathetic shifts in physiological oscillations may play a role in the listener's musical experience. Both approaches have their limitations. In the case of our scientific experiments, the stimuli are too simple to be considered "music" by most listeners. In fact, in the epilepsy experiments, several patients found the low-frequency acoustic stimuli unpleasant to listen to. Conversely, my informal compositional experiments lack the experimental controls necessary to extend these results to a broader theory or conjecture about music perception. These performances contain stimuli that are too naturalistic and complicated to properly tease out the effects of individual acoustic features. Important work is being done to develop formal yet naturalistic experiments to study the neural and body dynamics supporting live music performance, such as the recent experiments undertaken at the LiveLAB at McMaster University (Dotov et al., 2021).

In this chapter I've attempted to map the borderland between music experiments and experimental music. There are important historical precedents in both fields, and current work owes much to the scientists and composers who designed these initial experiments. It is also important to acknowledge the institutional support that made the work of these composer-researchers possible. In addition to the previously mentioned universities and research centers that supported the works described in this chapter, others such as the Institut de Recherche et Coordination Acoustique/Musique in Paris, the Center for Interdisciplinary Research in Music Media and Technology in Montreal, and the Max Planck Institute for Empirical Aesthetics continue to support scientists and artists working in these borderlands.

References

Adaikkan, C., & Tsai, L.-H. (2019). Gamma entrainment: Impact on neurocircuits, glia, and therapeutic opportunities. *Trends in Neurosciences, 43*(1), 24–41.

Ben-Menachem, E. (2002). Vagus-nerve stimulation for the treatment of epilepsy. *Lancet Neurology, 1*(8), 477–482.

Bernardi, L., Porta, C., & Sleight, P. (2006). Cardiovascular, cerebrovascular, and respiratory changes induced by different types of music in musicians and non-musicians: The importance of silence. *Heart, 92*(4), 445–452.

Bhandari, R., Parnandi, A., Shipp, E., Ahmed, B., & Gutierrez-Osuna, R. (2015). *Music-based respiratory biofeedback in visually-demanding tasks* [Paper presentation]. Proceedings of the International Conference on New Interfaces for Musical Expression, Baton Rouge, LA, United States.

Blauert, J. (1997). *Spatial hearing: The psychophysics of human sound localization*. MIT Press.

Bridoux-Michel, S., & Xenakis, I. (2005). *Autour de Concret Ph* [Paper presentation]. International Symposium, University of Athens.

Brody, J. (1970). *Iannis Xenakis electro-acoustic music* [Liner notes]. Nonesuch Records H-71246.

Deutsch, D. (1974). An illusion with musical scales. *Journal of the Acoustical Society of America, 56*(S1), S25.

Deutsch, D. (1981). The octave illusion and auditory perceptual integration. In J. V. Tobias & E. D. Schubert (Eds.), *Hearing Research and Theory, 1*(99), vol. 1 (pp. 99–142). Academic Press.

Dotov, D., Bosnyak, D., & Trainor, L. J. (2021). Collective music listening: Movement energy is enhanced by groove and visual social cues. *Quarterly Journal of Experimental Psychology, 74*(6), 1037–1053. https://doi.org/10.1177/1747021821991793.

Duncan, J. S., Sander, J. W., Sisodiya, S. M., & Walker, M. C. (2006). Adult epilepsy. *Lancet, 367*(9516), 1087–1100.

Durand, D. M., & Bikson, M. (2001). Suppression and control of epileptiform activity by electrical stimulation: a review. *Proceedings of the IEEE, 89*(7), 1065–1082. https://doi.org/10.1109/5.939821.

Eimert, H., & Stockhausen, K. (1959). *Die Reihe: A periodical devoted to developments in contemporary music*. T. Presser Company.

Fisher, R., Salanova, V., Witt, T., Worth, R., Henry, T., Gross, R., . . . The SANTE Study Group. (2010). Electrical stimulation of the anterior nucleus of thalamus for treatment of refractory epilepsy. *Epilepsia, 51*(5), 899–908. https://doi.org/10.1111/j.1528–1167.2010.02536.x.

González, H. F. J., Yengo-Kahn, A., & Englot, D. J. (2019). Vagus nerve stimulation for the treatment of epilepsy. *Neurosurgery Clinics, 30*(2), 219–230. https://doi.org/10.1016/j.nec.2018.12.005.

Juslin, P. N. (2019). *Musical emotions explained: Unlocking the secrets of musical affect*. Oxford University Press.

Lerdahl, F. (1992). Cognitive constraints on compositional systems. *Contemporary Music Review, 6*(2), 97–121.

Leslie, G. (2021). Inner rhythms: Vessels as a sustained brain-body performance practice. *Leonardo, 54*(3), 325–328. https://doi.org/10.1162/leon_a_01963.

Leslie, G., Ghandeharioun, A., Zhou, D., & Picard, R. W. (2019). *Engineering music to slow breathing and invite relaxed physiology* [Paper presentation]. 8th International Conference on Affective Computing and Intelligent Interaction (ACII).

Martorell, A. J., Paulson, A. L., Suk, H. J., Abdurrob, F., Drummond, G. T., Guan, W., . . . Tsai, L. H. (2019). Multi-sensory gamma stimulation ameliorates Alzheimer's-associated pathology and improves cognition. *Cell, 177*(2), 256–271 e222. https://doi.org/10.1016/j.cell.2019.02.014.

Mathews, M. V. (1963). The digital computer as a musical instrument. *Science, 142*(3592), 553–557.

Moraveji, N., Olson, B., Nguyen, T., Saadat, M., Khalighi, Y., Pea, R., & Heer, J. (2011). *Peripheral paced respiration: Influencing user physiology during information work* [Paper presentation]. 24th Annual ACM Symposium on User Interface Software and Technology.

Morrell, M. J. (2011). Responsive cortical stimulation for the treatment of medically intractable partial epilepsy. *Neurology, 77*(13), 1295–1304.

Quon, R. J., Leslie, G. A., Camp, E. J., Meisenhelter, S., Steimel, S. A., Song, Y., . . . Jobst, B. C. (2021). 40-Hz auditory stimulation for intracranial interictal activity: A pilot study. *Acta Neurologica Scandinavica, 144*(2), 192–201.

Risset, J.-C. (1978). *Paradoxes de hauteur.* Centre Georges Pompidou.

Risset, J.-C. (1985). Computer music experiments 1964–. *Computer Music Journal, 9*(1), 11–18. http://www.jstor.org/stable/4617918.

Roads, C. (2004). *Microsound.* MIT Press.

Robb, S. L. (2000). Music assisted progressive muscle relaxation, progressive muscle relaxation, music listening, and silence: A comparison of relaxation techniques. *Journal of Music Therapy, 37*(1), 2–21.

Schick, S. (2007). *Xenakis edition 7—complete percussion works* [Liner notes]. Mode Records.

Schiller, Y., & Bankirer, Y. (2007). Cellular mechanisms underlying antiepileptic effects of low- and high-frequency electrical stimulation in acute epilepsy in neocortical brain slices in vitro. *Journal of Neurophysiology, 97*(3), 1887–1902. https://doi.org/10.1152/jn.00514.2006.

Schulze-Bonhage, A. (2017). Brain stimulation as a neuromodulatory epilepsy therapy. *Seizure, 44*, 169–175. https://doi.org/10.1016/j.seizure.2016.10.026.

Shepard, R. N. (1964). Circularity in judgments of relative pitch. *Journal of the Acoustical Society of America, 36*(12), 2346–2353.

Siwiak, D., Berger, J., & Yang, Y. (2009). *Catch your breath—musical biofeedback for breathing regulation* [Paper presentation]. Audio Engineering Society Convention 127.

Sohal, V. S., & Sun, F. T. (2011). Responsive neurostimulation suppresses synchronized cortical rhythms in patients with epilepsy. *Neurosurgery Clinics, 22*(4), 481–488. https://doi.org/10.1016/j.nec.2011.07.007.

Sun, F. T., Morrell, M. J., & Wharen, R. E. (2008). Responsive cortical stimulation for the treatment of epilepsy. *Neurotherapeutics, 5*(1), 68–74. https://doi.org/10.1016/j.nurt.2007.10.069.

Theunissen, F. E., & Elie, J. E. (2014). Neural processing of natural sounds. *Nature Reviews Neuroscience, 15*(6), 355–366. https://doi.org/10.1038/nrn3731.

Treib, M., & Felciano, R. (1996). *Space calculated in seconds: The Philips Pavilion, Le Corbusier, Edgard Varèse* (Vol. 41). Princeton University Press.

Vidyarthi, J., & Riecke, B. E. (2014). Interactively mediating experiences of mindfulness meditation. *International Journal of Human-Computer Studies, 72*(8–9), 674–688.

Vonck, K., Van Laere, K., Dedeurwaerdere, S., Caemaert, J., De Reuck, J., & Boon, P. (2001). The mechanism of action of vagus nerve stimulation for refractory epilepsy: The current status. *Journal of Clinical Neurophysiology, 18*(5), 394–401.

Winters, R. M. (2020). *Empathic effects of auditory heartbeats: A neurophysiological investigation* [Unpublished doctoral dissertation]. Georgia Institute of Technology.

Winters, R. M., Walker, B., & Leslie, G. (2021). *Can you hear my heartbeat? Hearing an expressive biosignal elicits empathy* [Paper presentation]. CHI Conference on Human Factors in Computing Systems, Yokohama, Japan.

Xenakis, I. (1992). *Formalized music: Thought and mathematics in composition*. Pendragon Press.

Yu, B., Funk, M., Hu, J., & Feijs, L. (2018). Unwind: A musical biofeedback for relaxation assistance. *Behaviour & Information Technology, 37*(8), 800–814.

Interludes

14 Music Theory and Experimental Science

Diana Deutsch

[*Editors' note*: Diana Deutsch (figure 14.1) is widely regarded as a pioneer of the interdisciplinary study of music and science and for her creative approaches to perception and cognition. In addition to her classic discoveries of auditory perceptual illusions (including the speech-to-song illusion, the tritone paradox, and the octave illusion), Deutsch is known for her work on auditory working memory, hemispheric asymmetry, attention and perceptual organization, absolute pitch, and musical illusions that connect music and language. She edited three volumes of *The Psychology of Music* and was the inaugural president of the Society for Music Perception and Cognition. Looking back on four decades of developments in the science of music, Deutsch offers her perspective of the historical forces in empirical and rational thinking that shaped the landscape of the science-music borderlands.]

The relationship between music theory and experimental science has a very long history, marked by controversy between those who strongly espoused the empirical method, on the one hand, and those who argued for a rationalistic approach to music theory, on the other.

Pythagoras (c. 570–497 BC) is credited with some of the first studies in experimental science. For instance, he showed that the pitch of a vibrating string varies inversely with its length—those with lengths standing in a ratio of 1:2 sounded an octave, 2:3 sounded a perfect fifth, and 3:4 sounded a perfect fourth. Figure 14.2 shows one imagining of Pythagoras's experimentation with bells, glasses, strings, and pipes.

However, Pythagoras and his followers later lost faith in the experimental method and instead endeavored to explain all music by the contemplation of numerical relationships. Boethius (c. AD 477–524), the leading music theorist of the Middle Ages and a strong follower of Pythagoras, wrote:

> For what need is there of speaking further concerning the error of the senses when this same faculty of sensing is neither equal in all men, nor at all times equal within the same man? Therefore anyone vainly puts his trust in a changing judgement since he aspires to seek the truth. (1967)

Figure 14.1
Diana Deutsch (front row, second from right) with her graduating class in experimental psychology at the University of Oxford, 1959.

The mathematical approach of the Pythagoreans led to numerous attempts to construct musical systems by mathematical deduction from a minimal number of musical facts. However, this approach stems from a false analogy with geometry. Euclidean geometry starts with a few axioms that are considered self-evident and arrives by deduction at theorems that are not self-evident. But for music, this analogy is based on a logical error: We cannot proceed from one musical fact to another musical fact—rather, musical facts can be used as a basis for hypotheses about other musical facts, which need to be verified empirically.

Although later Greek theorists adhered to the numerical approach, there were some exceptions. In particular, Aristoxenus (c. 375–225 BC) argued in his musical treatise *Harmonic Elements* that musical phenomena are basically perceptual and cognitive in nature. He wrote:

Figure 14.2
Woodcut by Gafurius entitled *Theorica Musicae,* 1492. This depicts Pythagoras experimenting with the pitch of vibrating strings.

> It is plain that the apprehension of a melody consists in noting with both the ear and intellect every distinction as it arises in the successive sounds—successive, for melody, like all branches of music, consists in a successive production. For the apprehension of music depends on these two faculties, sense-perception and memory; for we must perceive the sound that is present and remember that which is past. In no other way can we follow the phenomenon of music. (Aristoxenus, 1902)

However, the music theorists of the Middle Ages generally embraced the numerological approach. It was not until the Renaissance that the empirical study of music perception was undertaken seriously. Particularly notable among the musical empiricists of the sixteenth century was the renowned music theorist, lutenist, and composer Vicenzo Galilei (1520–1591). His approach, particularly his strong espousal of the empirical method in the study of music—including tuning systems, consonance and dissonance, and the physics of vibrating strings and air columns—had a profound influence on his son, Galileo Galilei (1564–1642), who assisted him in his research on issues concerning music perception and later elaborated on his father's theoretical work (Galileo, 1638/2002).

Others in the early stages of the Scientific Revolution with a strong interest in music perception included Johannes Kepler (1571–1630), Rene Descartes (1596–1650), and Marin Mersenne (1588–1648). In particular, Mersenne carried out brilliant studies on the physics and perception of musical sound, using virtually no equipment. In one of his remarkable experiments, he determined empirically the number of vibrations per second made by a string as it sounded in unison with a musical note. He achieved this by working with very long strings that vibrated slowly enough to "allow leisure to count the vibrations" and then scaling down the lengths of the strings. Mersenne (1636) wrote:

> There is no difficulty in finding the number of returns for each string proposed, for if one extends to 10 or 12 times the length of a monochord . . . one can easily count its returns, the more so if it makes a very small number, for example, 2 or 3 in each second. But one needs two or three [observers] to note exactly the number of these returns, one to count the returns while the other counts the seconds, whence if one divides the number of seconds into the number of returns, one will know how many it makes in each second. And if one extends a string from a spinet or a lute to 100 or 120 feet, as I have done, one will find that each return of that string occurs a second, and that half the same string makes two vibrations in a second, that a quarter of it makes 4, the eighth part 8, the sixteenth 16, the thirty-second 32, and so on.

Using this scale model technique, Mersenne was able to determine by extrapolation the lowest vibration frequency of a string that could be heard by the human ear—a remarkably accurate deduction. Other contributions by Mersenne included his identification by ear of the first four upper partials of a harmonic series, the phenomenon of beats, the orderings of consonances both by "sweetness" and by "agreeableness," and studies of tuning and temperament. He also carried out a voluminous correspondence with foremost scholars of the time and facilitated the exchange of information among them.

Kepler was immersed in music and wrote about tuning systems, musical structure, and musical consonance. Importantly, he derived the harmonious relations between the motions of the planets from music theory. The cosmic harmony he obtained—the third law of planetary motion—reflected the relative minimum and maximum angular velocities of the planets as they would appear from the sun.

Kepler's transcriptions of the music of six planets based on his astronomical observations are shown in figure 14.3. This celestial choir consisted of a tenor (Mars), two basses (Saturn and Jupiter), a soprano (Mercury), and two altos (Venus and the Earth).

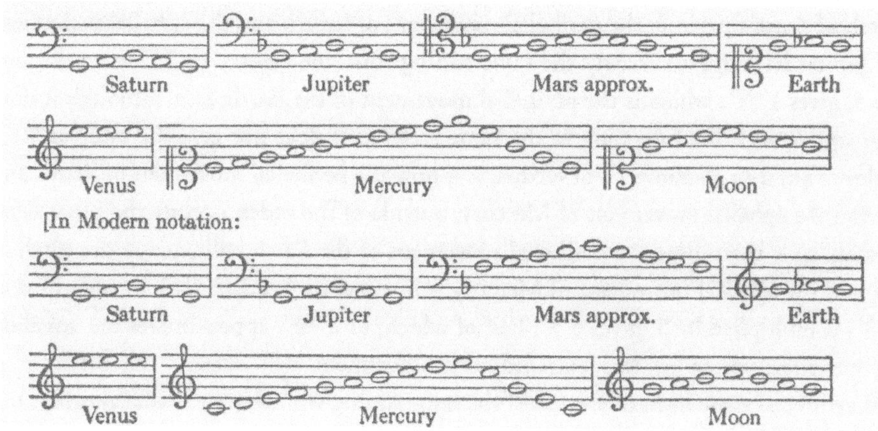

Figure 14.3
Kepler's notation of the music of six planets, based on his astronomical observations. (From Kepler, 1619/2002.)

As he wrote: "the movements of the heavens are nothing except a certain everlasting polyphony" (Kepler, 1619/2002).

Although the experimental study of music played a central role among Renaissance scientists, the strong connection between music perception and experimental science waned during the eighteenth century, when the study of the physical properties of sound became acknowledged as a scientific discipline. Among those who contributed to this development was Ernst Chladni (1756–1827), who is described as the "father of acoustics." Among his many achievements, Chladni invented a procedure for visualizing patterns of vibration on a metal plate. By scattering sand on the plate and then bowing on the edge of the plate with a violin bow, he discovered striking shapes that depended on the dimensions of the plate. This method was later applied to the bodies of violins and guitars and was used by instrument makers to tune their resonances. Later work on the production and transmission of sound by Michael Faraday (1791–1867) and Charles Wheatstone (1802–1875) continued to lay the foundations of modern acoustics.

In contrast, music theorists returned to the study of numerology in an attempt to understand musical structure. The composer and theorist Jean-Philippe Rameau (1683–1764) was particularly influential in this development. He posited the overtone series as the fundamental principle underlying tonal music and proceeded to derive an entire musical system by mathematical deduction from this one musical fact. He wrote: "Music is a science which ought to have certain rules; these rules ought to be derived from a self-evident principle; and this principle can scarcely be known to us without the help of mathematics" (Rameau, 1722/1950).

Rameau's systemization forms the basis of traditional Western music theory as it is taught today. It involves operations such as the root generation and the root progression of chords. However, Rameau turned to physics to justify his ideas, replacing the harmony of the spheres with the overtone series as an explanatory device.

Later, the development of chromaticism in the nineteenth and twentieth centuries, such as in the music of Wagner, Debussy, and Mahler, required a new conceptualization of harmony, since it became clear that the notion of a tonic did not have sufficient explanatory value for these new compositions. In consequence, the twelve-tone system developed by Arnold Schoenberg (1874–1951) became an influential framework for music theory and composition in the early and mid-twentieth century. In this framework, a tone row, defined as a specific linear ordering of the twelve tones of the chromatic scale, retains its perceptual identity when it is transposed to a different pitch range (transposition), when all ascending intervals become descending ones and vice versa (inversion), when it is presented in reverse order (retrogression), and when it undergoes both these transformations (retrograde-inversion). Schoenberg (1951) also argued that the identity of a tone row is preserved when its component tones are placed in different octaves. His illustration of this theory of equivalence relations between pitch structures is shown in figure 14.4.

Twelve-tone theory became the basis for much of the system building in the mid-twentieth century, with Milton Babbitt (1916–2011) being its foremost proponent. However, twelve-tone theorists did not submit their theoretical claims to perceptual validation; instead, they assumed—in accordance with Pythagorean tradition—that the mathematical relationships set forth by twelve-tone theory served as a sufficient explanatory device for music.

In contrast to developments in music theory, visual artists beyond the Renaissance continued to draw on experimental science to develop their ideas. This dissimilarity in the study of visual and auditory perception was probably due largely to differences in the available technology. From earliest times, vision researchers had at their disposal tools that allowed exact measurements—including rulers, dividers, and compasses—which enabled them to produce and study elaborate and carefully specified visual arrays. However, equivalent tools were not available for the study of sound patterns.

In addition, visual arrays can be perused in detail, so that a drawing or painting can be scrutinized at leisure. This led several artists during the Renaissance and later to argue that visual perception is not simply a matter of recording the information

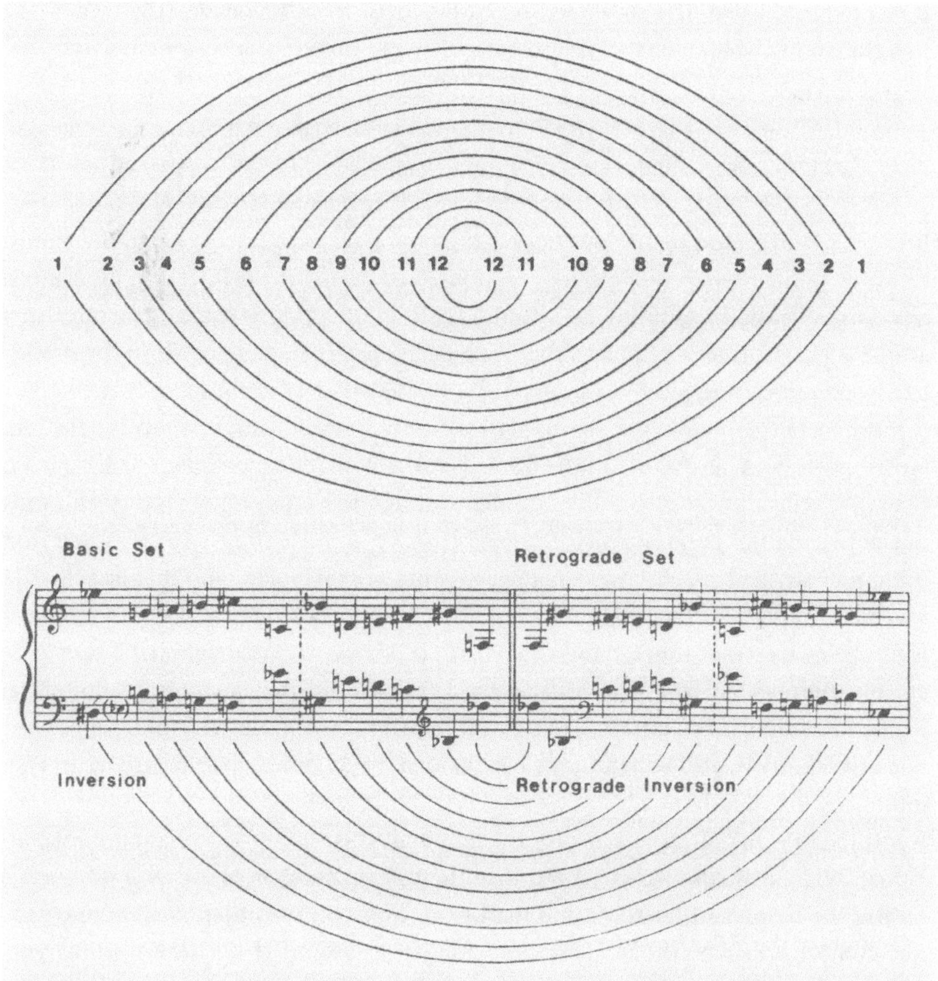

Figure 14.4

Illustration of Schoenberg's theory of equivalence relations between pitch structures, taken from his *Wind Quartet*, Op. 26. (From Schoenberg, 1951.)

presented to the eye. The seventeenth-century artist Nicolas Poussin (1594–1665) put this eloquently when he wrote that there are two procedures for viewing objects:

> One is simply seeing, and the other ponders them attentively. Simple seeing is nothing other than the natural reception in the eye of the form and resemblance of the seen object. But to see an object with deliberation . . . we search with a particular procedure for a way to understand that same object properly. (quoted in Kemp, 1990, p. 127)

Because sound is fleeting, it does not enable prolonged scrutiny. So while the empirical study of high-level visual perception flourished during the eighteenth, nineteenth, and early twentieth centuries, the empirical study of complex sound structures such as music had to await techniques that were developed by psychologists in the mid- to late twentieth century.

Another development that increased the rift between scientists, on the one hand, and artists, poets, and musicians, on the other, was the Romantic movement. It emerged in the late eighteenth century and continued through the nineteenth century and most of the twentieth. The Romanticists argued that art, at its highest level, was free from rules, that artistic creativity was innate and could not be taught, and that imagination was more important than reason. As the artist and poet William Blake wrote: "*Genius begins where rules end*" (quoted in Kemp, 1990, p. 252). Associated with this belief is the claim—still made by some today—that the left brain is verbal, analytical, and logical, while the right brain is holistic, intuitive, and imaginative. Likewise, some people, particularly scientists, are thought to be left-brained, while others, particularly artists and musicians, are right-brained.

Although it has always been easy to show that this simplistic explanation of individual differences is mistaken, it fed into the unease between artists and humanists, on the one hand, and scientists and mathematicians, on the other. This situation was forcefully criticized by the novelist and scientist C. P. Snow (1905–1980), who argued that the divide between the two cultures had widened disastrously. He wrote:

> At the heart of thought and creation we are letting some of our best chances go by default. The clashing point of two subjects, two disciplines, two cultures—of two galaxies, so far as that goes—ought to produce creative chances. In the history of mental activity that has been where some of the break-throughs came. The chances are there now. But they are there, as it were, in a vacuum, because those in the two cultures can't talk to each other. (Snow, 1959, p. 16)

Technological advances, beginning with the work of Hermann von Helmholtz (1821–1894) and others in the mid-nineteenth century, enabled scientists to study low-level sound perception with precision (Helmholtz, 1885/1954). It became possible, for example, to measure auditory thresholds and to devise scales of pitch and loudness. However, it was still extremely difficult to produce well-specified sequences of tones or to generate tones with well-defined time-varying spectra. So while psychoacousticians carried out important basic experiments on sound perception, musicians regarded the perception of simple auditory stimuli as of little interest.

However, beginning in the 1950s, striking developments in computer technology completely changed the picture. In 1957 Max Mathews, then at Bell Labs, created the software MUSIC I, which enabled a computer to play a seventeen-second musical composition—an astonishing feat at the time. In an article published in *Science* he wrote: "There are no theoretical limitations to the performance of the computer as a source of musical sounds, in contrast to the performance of ordinary instruments" (Mathews, 1963).

Mathews's software enabled Roger Shepard (1964) to produce eternally ascending and descending scales and Jean-Claude Risset (1969) to create versions using gliding tones and to synthesize sounds with various instrument timbres, such as the trumpet and the flute. John Chowning discovered FM synthesis in 1973. This led to the design of computer synthesizers, which enabled sophisticated experiments on sound perception.

My initial experiments on music perception and cognition in the late 1960s were also made possible by the development of computers. I controlled function generators to produce long sequences of sine wave tones that were precisely specified in terms of frequency, amplitude, and duration. I first controlled a function generator to produce single sequences of tones, which enabled me to study memory for musical tones in a sequential setting (Deutsch, 1970) and to produce the "mysterious melody" illusion (Deutsch, 1972). Soon after I controlled two function generators to produce two sequences of tones, one delivered to the right ear and one to the left, which led me to discover the octave illusion (Deutsch, 1974), the scale illusion (Deutsch, 1975), and other variants (Deutsch, 1987).

Once the generation of music by computer became widespread, the empirical study of music perception and cognition developed rapidly—first by psychologists and increasingly by new generations of composers and music theorists, who recognized its importance to understanding musical phenomena. Much of this research is covered in other chapters of this book. Here, I review my discoveries of musical illusions and discuss their implications for how we perceive and understand music in general. (For detailed discussions of these illusions, see Deutsch, 2019.)

Illusions are often regarded as entertaining anomalies that shed little light on the normal process of perception. In fact, however, they reveal important characteristics of perceptual processes that might otherwise go unrecognized. Although this has been acknowledged for centuries among visual artists and scientists, only in the last few decades has this issue been investigated in music.

A major feature of such illusions is that there are prominent differences in the way listeners perceive them, indicating striking discrepancies in the perception of musical patterns. For some of these illusions—the octave illusion, the scale illusion, and the glissando illusion—disagreements tend to arise between right-handed and left-handed individuals, indicating that they reflect differences in brain organization. Other illusions reflect environmental influences. For example, perception of the tritone paradox varies with the language or dialect to which the listener has been exposed, particularly in childhood (Deutsch, 1992). Perception of the phantom words illusion (Deutsch, 2003) also varies with the listener's language, experiences, and emotional state.

Many of these musical illusions illustrate the power of unconscious inference in perception. Some illusions occur when two sound sequences are presented, one coming from the left and the other from the right. When listening to the scale illusion or one of its variants, we reject the correct but improbable interpretation that two different sources are producing sounds that jump around in pitch. Instead, we reorganize the tones in space so that we hear two smooth melodies, each in its own pitch range—with one melody appearing to come from one source and the other melody from a different source. The phantom words illusion shows a strong influence of unconscious inference on the perception of speech; when listening to these ambiguous words and phrases, we draw on our knowledge and experience of speech and language.

The tritone paradox provides another example of unconscious inference, since we hear these ambiguous tritones as either ascending or descending, depending on our dominant language or dialect, particularly in childhood, and thus on our knowledge and expectations concerning sounds (Deutsch, 1986, 1992). Yet another example is the mysterious melody illusion. A well-known melody is not recognized when it is played so that the pitch classes are preserved but the tones are placed randomly in different octaves. Yet when the listeners are told the identity of the melody, they create a perceptual image of how it should sound and, by referring to this image, they can perceive the melody correctly (Deutsch, 1972).

The speech-to-song illusion (Deutsch, 2003; Deutsch et al., 2011) sheds light on the relationship between speech and music—an issue that has been debated for centuries. Here, a spoken phrase is heard as sung, without altering its physical parameters in any way, simply by repeating it several times in succession. This illusion sheds light on the relationship between speech and music—an issue of considerable interest to music theorists, psychologists, and neuroscientists.

My work on musical illusions is an example of the blending of music theory and psychology, but this is only a small part of the music research and theorizing that have occurred over the last few decades. In a book chapter I wrote more than thirty years ago on the relationship between psychology and music, I predicted that psychology would soon have a firmly established place in music theory (Deutsch, 1984). Important developments that followed included the establishment of the biennial International Conference on Music Perception and Cognition, which held its first meeting in Kyoto in 1989; the North American Society for Music Perception and Cognition, which first met in Los Angeles in 1992; the European Society for the Cognitive Sciences of Music, which first convened in Trieste in 1991; and several other societies worldwide. I can now state with confidence that the empirical study of perception and cognition has an established place in music theory and that the work of music theorists and composers has an established place in psychology and neuroscience.

References

Aristoxenus. (1902). *The harmonics of Aristoxenus*. (H. S. Macran, Trans.). Clarendon Press.

Boethius. (1967). *Boethius' the principles of music*. (C. M. Bower, Trans.). University of Michigan Press.

Deutsch, D. (1970). Tones and numbers: Specificity of interference in immediate memory. *Science, 168*, 1604–1605.

Deutsch, D. (1972). Octave generalization and tune recognition. *Perception & Psychophysics, 11*, 411–412.

Deutsch, D. (1974). An auditory illusion. *Nature, 251*, 307–309.

Deutsch, D. (1975). Two-channel listening to musical scales. *Journal of the Acoustical Society of America, 57*, 1156–1160.

Deutsch, D. (1984). Psychology and music. In M. H. Bornstein (Ed.), *Psychology and its allied disciplines* (Vol. 1, pp. 155–194). Erlbaum.

Deutsch, D. (1986). A musical paradox. *Music Perception, 3*, 275–280.

Deutsch, D. (1987). Illusions for stereo headphones. *Audio Magazine*, 36–48.

Deutsch, D. (1992). Paradoxes of musical pitch. *Scientific American, 267*, 88–95.

Deutsch, D. (2003). *Phantom words, and other curiosities* [Compact disc and booklet]. Philomel Records.

Deutsch, D. (2019). *Musical illusions and phantom words: How music and speech unlock mysteries of the brain*. Oxford University Press.

Deutsch, D., Henthorn, T., & Lapidis, R. (2011). Illusory transformation from speech to song. *Journal of the Acoustical Society of America, 129*, 2245–2252.

Galileo Galilei. (2002). *Dialogues concerning two new sciences* (H. Crew & A. de Silvia, Trans.). In S. Hawking (Ed.), *On the shoulders of giants: The great works of physics and astronomy*. Running Press. (Original work published 1638.)

Helmholtz, H. von. (1954). *On the sensations of tone as a physiological basis for the theory of music*. Dover. (Original work published 1885.)

Kemp, M. (1990). *The science of art: Optical themes in Western art from Brunelleschi to Seurat*. Yale University Press.

Kepler, J. (2002). *Harmonice mundi* [The harmony of the world]. In S. Hawking (Ed.), *On the shoulders of giants: The great works of physics and astronomy*. Running Press. (Original work published 1619.)

Mathews, M. V. (1963). The digital computer as a musical instrument. *Science, 142*(3592), 553–557.

Mersenne, M. (1636). *Harmonie Universelle, Contenant la theorie et la pratique de la musique*. Sebastien Cramoisy.

Rameau, J.-P. (1950). Traite de l'harmonie reduite a ses principes naturels. In O. Strunk (Ed.), *Source readings in music history*. Norton. (Original work published 1722.)

Risset, J. C. (1969). Pitch control and pitch paradoxes demonstrated with computer-synthesized sounds. *Journal of the Acoustical Society of America, 46*, 88A.

Schoenberg, A. (1951). *Style and idea*. Williams and Norgate.

Shepard, R. N. (1964). Circularity in judgments of relative pitch. *Journal of the Acoustical Society of America, 36*, 2345–2353.

Snow, C. P. (1959). *The two cultures*. Cambridge University Press.

15 Conversation with Pamela Z

Pamela Z, Psyche Loui, and Deirdre Loughridge

[*Editors' note*: Composer, vocalist, and multimedia artist Pamela Z has won numerous awards, including the Rome Prize and a Guggenheim Fellowship. She is known as a "wild virtuoso" (*New York Times*, January 7, 2021) for her experimental extended vocal techniques and the use of text, sampled sounds, found objects, and innovative gesture-to-sound programming. We spoke with Pamela Z to learn about her perspective on embodied cognition, the intersection of music and language, and the role of experimentation in creativity—topics that recur throughout this volume.]

What is your background as a musician, and how did you integrate technology into your practice? In my first public performance, I was five years old, and it was the elementary school talent show. I've been singing all my life and I used to play with found objects. I used to try to make sound-making objects out of the little pods that fall out of trees and, when they dry, they turn into maracas. I used to string rubber bands on the knobs of my dresser drawers to make a stringed instrument when I was a kid. Pretty early on, while I was still in elementary school, I started playing with machines.

When I was really young, my father bought us—they were a newfangled thing—cassette tape decks. They were the old-fashioned ones that had a long shape with buttons along the front edge that you could press down and then a little spring lid popped up to put the cassette in. I was trying to learn how to multitrack by singing into one and then playing it back and recording it to the other one and back and forth and so on.

By the time I was in high school, I was sort of a singer-songwriter and I was writing songs and playing guitar. At the same time, I was singing in the concert choir and the show choir and performing in school plays and musicals. When I went to college, I started as a music major. I was a voice major and I was playing in clubs by night as a singer-songwriter and also in a band, and by day I was singing opera arias and art songs. It wasn't until after I got out of school that I started working at the local public radio station.

I had a show called "The Tuesday Afternoon Sound Alternative," and I used to do a free-form program that combined all these different musics that I liked, which were completely broad and crazy. So in one show I would segue from Varèse to The Ramones to The Roches to Laurie Anderson. It was through that radio program that I became aware of a lot of experimental music. I got more interested in trying to do something more avant-garde, and I became disillusioned with the music I was playing because, to me, it seemed really conventional.

The cover tunes I was doing were starting to be more like David Byrne and Tom Waits. I was including more interesting things, but I still felt like the music on my turntable at that time was completely unrelated to what I was actually playing. I was listening to text-sound compositions and people doing experimental music, like Pauline Oliveros and Brian Eno and his collaborations with people like David Byrne.

I bought a four-track cassette recorder and was creating pieces that were different from what I was playing in clubs. They were things that used layers of my voice, and then I discovered the digital delay. That was the day I found my voice as an artist, when I started playing with layering, looping my voice in real time, and discovering all kinds of things about repetition, timbre, texture, and the harmonic, melodic, and rhythmic structures of speech sounds. All that became much more evident because I had this live sampling device.

I moved to San Francisco in 1984 and kind of remade myself. At that point I had been living in Boulder for ten years and was making a solid living as a musician, but I was playing in clubs as a singer-songwriter, doing folk rock and cover tunes that were emulating those other people. All of a sudden I was looping my voice and wasn't interested in playing the guitar anymore. I found this amazing contemporary music and performance art community, and that was the beginning of the way I work now as an artist.

How did your conceptualization of music and instruments change after you began to experiment with musical elements? At that time, pop music to me was really a commercial, bubblegum kind of music. I never thought of what I did as being pop, but it was definitely more in the popular realm in terms of the audience, the accessibility, and this idea of writing, learning, and singing songs that everybody could learn to sing along. I felt like my listening to music and sound opened up. I realized that there's a broad spectrum of music, and I love all of it. I stopped investing myself into that particular niche because this other area was so much more moving for me to do, to experiment with sound, to think about music, to know that it doesn't rely just on melody, harmony, and rhythm but instead that music is more about timbre, texture, and processes changing over time. Music practices or ideas became more interesting to me, things that were more experimental. Then I also became interested in combining acoustic sound with sound that was either electronically manipulated or produced.

Are there specific kinds of responses you aim to provoke in listeners through your performances? I don't really think about music in terms of a way of expressing emotions. My goal is for the work to be good, for it to be affecting in some way, but I don't have a strong need for it to be affecting in any particular way. There are some pieces I do that make people laugh. There are some pieces I do that make people tear up. It's not because I set out to make them tear up, and when they laugh, it's not because I set out to make them laugh.

Sometimes I don't know until I perform the work in front of an audience what the reaction is going to be. I just hope they will be moved by it in some way, or they could also just be moved intellectually. Maybe the work just makes them think about things they hadn't thought about before. I don't start my composing with an agenda of how I want them to feel when they hear the work. I think the work will discover that.

I definitely want there to be a connection. I want them to feel something or notice something or be riveted by it for whatever reason. I think it varies not just from artist to artist but also from piece to piece within an artist in what way listeners will be riveted or in what way they're going to react. The key is that I hope it's not so definable that you could write it down as a mission statement.

In your experience, how do different tools allow different interactions, especially in terms of embodiment? I like to think of the tools I use as being no different from more conventional tools. With the early musical instruments that had a plectrum for plucking a string, the only way you could change the dynamic was to be plucking, just one string or two. When they invented the pianoforte, suddenly the nuance of how you hit that key changed everything.

I know there are a lot of people who, when they play repeated notes, never use the same finger to play each iteration of that note. I think part of the reason is to give them more speed, but more than that, I think it's to give each iteration of that repeating note a slightly different feeling.

There's a large humanness to that repetition and less machineness. Somebody who studied all their life has gained some deep knowledge of their instrument. They have a relationship with it and can interact and tease out all kinds of different qualities of sound.

With my instruments, there's a certain amount of that. Like with the gesture controllers, for instance, there's a way of using them so that when you do one thing, it always has a certain result. Then there are things where you learn to be more nuanced in your approach and it gives you more nuanced results.

I think one of the biggest differences is that if your instrument is constantly changing because technology is constantly changing, it keeps you on your toes in a different way. Let's say you're a cellist and every six months a luthier comes to your house and reworks your bridge so that your strings are a slightly different distance apart. Now you've got to start all over again and gain some facility with an instrument you're not used to. Most pianists have to play a different instrument in every concert, so they have to get used to differences in slightly different instruments.

That's the big difference between more electronic-based instruments versus acoustic instruments. With traditional instruments, you can gain a level of virtuosity without these constant interruptions because the instrument itself is changing radically. Those musicians get to spend a whole lifetime establishing an intimate connection with and physical understanding of the same instrument. When you're working with electronics, half the time you're just futzing around: "I had to install the upgrade to this plug-in, and now it's not playing nicely with this other thing," or "Now I have to rewrite this because it doesn't work anymore, and now I won't be able to do this one function that was an important part of that piece when I performed it."

What is the role of found objects and their affordances in your work? If your mind has the patience to notice sounds and organize them in the way you're hearing them, they become the music. The found object and found sound ideas are an extension of that kind of thinking, but I also think it has to do with broadening the definition of what you can call music. I'm sort of broadening the definition of what you call an instrument—going from the idea that an instrument is something an instrument maker designed for its qualities and how well it can produce sound to the idea that an instrument is anything you use to make sound. If your medium is music or sound, then your instruments are the things you can make sound come from.

There's a fine line that can be crossed when using found objects. It can be corny when somebody is just trying to replace very specific things with other things that make similar sounds. When my sister and I sang in the talent show, we would use those long brown pods that fell from trees. After a few days, they would dry out, and then you could shake them and make a sound. Maracas were designed to be big rattles that produce a really nice, clean, shaky sound. These pods probably had more of a muted shaky sound. We were using them where you would use maracas, but that wasn't necessarily a very sophisticated use of a found object.

As an adult, instead of trying to find an object that will make a sound just like an instrument that already exists, I look for an object that might result in a new sound I haven't heard before. My thoughts about using found sound objects are connected with my enjoyment of using found text instead of a composed libretto. All those things are about discovery. It's part of those happy accidents that happen, sounds you might not have come up with if you were limited to canonical practices and prescribed objects.

A lot of found objects are so good that they end up becoming part of the arsenal of more conventional music. Thunder sheets are an example. Now you find many orchestral works in which percussionists are expected to use thunder sheets because there isn't any conventional instrument that can produce that sound.

If you were a cognitive scientist, what would you study? I'm very admiring of scientists. I love mining what they do and using those artifacts as material in my own work. I'm very interested in the different resonant qualities of different spaces and that kind of thing. I think that, sadly, I have to admit, it's a circular thing because it all folds back into what I do. I wish I knew more about resonant frequencies because I wish I knew more about what I'm doing when I'm mixing a piece and putting in EQ filters. I wish I had a better handle on that stuff.

Or maybe being a linguist would be interesting to me because I love language, and I love it on all these different levels. I love it on the level of understanding the literal meaning of languages, and I wish I could become fluent in other languages. I've tried to study a lot of languages, and my problem is that I studied too many so I never became fluent in any of them. I'll never stop studying languages.

Maybe being a linguistic researcher would interest me the most. I like language not just because of its grammar or structure but also because of the sound languages have, the rhythms that exist in different languages, and even the different regional accents within one language.

I'm interested in what's going on in the brain when it's processing language. I'm interested in this strange way that, in my work, I like to use language that's fragmented and repeats a lot. I'm fully aware that you can't strip language of its meaning, so if the people listening to your piece are native speakers of whatever language you're using, you can cut it into a lot of fragments but you can never wash it clean of its meaning. What can happen, though, is that through repetition, it can lose one meaning and start to take on another meaning, or the meanings of two different words start to morph with each other, and the sound of words repeated over and over begins to change. There's a whole host of people creating works by transcribing the pitch and rhythmic material from speech, and I never tire of that.

Another thing I would study is memory. I'm always fascinated when I hear about really bizarre incidents involving amnesia, like when people forget an entire swath of their knowledge yet remember everything else. How does that happen? You hear stories about people who forgot how to speak their main language but can still speak another language. Or musicians who forgot how to play music and had to start over, like a child learning an instrument all over again.

Is musicality unique to humans, or is it widespread among living things? The first two things that come to mind are birds and whales. The question is, is what we interpret as being music also music to them? I don't know if there's any way to know that, but I would think that people in your field would probably be able to figure out a way to determine how a bird sees songs. Are they songs? One thing that's really interesting about birds' songs is that they are taught. The adult birds teach the song to the children, and they always sing the same song. They may have variations or longer or shorter versions of it, but that is the family song and they pass it on to the young. That's interesting to me. We call that a song, but is it a song? I don't know, but it's wacky how much more complex bird "songs" are than they seem to be. They're so high in pitch and so fast that we hear it as just a chirp. But if you take that sound and stretch it out, you find that it's a complicated melody with thirty notes in it. Every time they tweet it, it's the same melody, and it's even in the same key. The pitch is very high—birds have this tiny little syrinx instead of a larynx. Frequency is determined not only by the speed of the vibrations but also by the *length* of the vocal cords. We're not equipped to hear or perform bird songs unless we time-expand them and pitch-stretch them so that they go down into an octave we can perform and they're long enough so that we can hear the intervals and rhythms involved. Then we can learn those songs, and they seem pretty musical to me. It would be weird, though, if every human family, tribe, or group had just one song and everybody had to learn it and always sang it from beginning to end in the same key. Would we call it music? Would we call them songs? All I can say is, it seems musical to me.

Whale songs are the opposite, lasting about eight hours. You could do the opposite thing and squish them. Then you could hear the contours of the song, which is a very specific one. The whales sing it again and again to one another. It is transmitted through vibrations in the water, and the other whales can hear it.

The songs of birds and whales seem closer to music to me. Other creatures make noises, but a dog barking somehow doesn't seem to fall into the category of music. I mean, in a Cagean way, you could say all sounds can be music, and many people have taken samples of dog barks and cat meows and composed music using them. On their own, without somebody organizing them in a particular way, they seem more like talking, crying, shouting, or whining than singing. Emotionally, though, a dog's whine does seem to express the same thing our whine does! Dogs don't whine when they're happy, and when a person they like leaves, they stand at the door and cry.

Birds, on the other hand, do seem to be singing. When I was growing up, my mother had canaries, and the males put on an incredible concert. These canaries would do an entire concert with all these trills, and it seemed varied and complex. As a kid, I never thought to record it and slow it down. One bird named Twerpy was the best singer of them all. After Twerpy died, my mother got another canary, but it didn't sing. I don't think it ever occurred to us that the bird was a female, because the females don't sing; they only chirp a little bit. Only the males sing these big elaborate songs. My mother went out and bought a vinyl record of the Hartz Mountain canaries and played that record for the bird, trying to get her to learn! Twerpy never needed the help of a choir.

Acknowledgments

Thanks to Yaen Chen for transcribing the interview.

IV Beyond Musicians and Nonmusicians

Volume Editors

Comparisons between musicians and nonmusicians are ubiquitous in the scientific literature on music. On the one hand, comparisons of their cognitive functions validate that persistent, lifelong engagement with music can shape the brain and mind. For example, classic findings in music psychology have revealed that musically trained individuals, such as those who started formal instrumental training at a young age or those who have had six or more years of musical training (Zhang et al., 2018), show superior encoding of speech sounds (Wong et al., 2007). Findings like these have spurred theoretical developments related to the similarities and differences between music and speech (Patel, 2011) and prompted the more general claim that "music is a resource that tones the brain for auditory fitness" (Kraus & Chandrasekaran, 2010, p. 599). On the other hand, the bifurcation of people into the categories "musicians" and "nonmusicians" relies on assumptions about the experiences and abilities that constitute proof of musicality, which may reflect researchers' cultural biases and be poorly suited to the competencies or phenomena they hope to illuminate. As Ilari and Habibi note in chapter 17, "a combination of training, occupation, self-identification, and music skills seems to be the defining feature of the musician in human behavior studies. . . . But, as many would argue, the term *nonmusician* is an oxymoron. . . . Aside from misrepresenting human potential, the term *nonmusician* devalues skills that are inherent to being musical, such as the ability to listen to and be moved by music, and it is also detrimental to identity construction and musical participation."

The reliance on a dichotomy between musicians and nonmusicians and the misunderstandings that accrue as a consequence exemplify the need for an interdisciplinary dialogue. Thus, this section of the book can be read as an extended case study of the issues and challenges raised elsewhere in the volume. By scrutinizing the musician-nonmusician dichotomy from multiple perspectives, the chapters in this section identify the specific issues at stake and chart a concrete path forward, bringing the book's more theoretical claims into vivid relief.

The prevalence of experimental designs that compare musicians and nonmusicians extends a tradition in psychological science that operationalizes large and elusive constructs by dividing them into levels of independent variables to enable rigorous study. In the case of musicality, this has resulted in discrete levels of variables that can be linked to measurable outcomes, such as cognitive and brain measures. Psychometricians continue to grapple with the challenge of devising statistically valid measures of individual differences in areas such as intelligence testing (e.g., the journal *Intelligence*), and others have moved to characterize individual differences in constructs that may be even more elusive, such as creativity (e.g., *Creativity Research Journal*). The concept of measurable differences in musicality falls somewhere along the continuum between clearly defined and elusive: it dates back to earlier notions of genius and/or talent and, troublingly, back to phrenologists (Fowler & Fowler, 1850) who applied the notion of "scientific proofs of musical genius" to constructs such as singing ability (Manning, 2015). Contemporary research continues to examine the source of giftedness, as exemplified by historiometric approaches (Simonton, 2005; Gardner, 1997) and empirical studies on topics such as childhood precociousness (Winner, 2000). One branch of music neuroscience is devoted to unusual types of musical abilities, such as absolute pitch, synesthesia, exceptional memory (Ericsson & Chase, 1982), musical savantism (Miller, 1987), and musicophilia (Sacks, 2010). Other branches of music neuroscience focus on deficits in musical ability, such as the deficits in pitch perception found in congenital amusia and the lack of appreciation for music found in musical anhedonia.

Modern measures of talent or "musical sophistication," such as the Ollen Musical Sophistication Scale (Ollen, 2006) or the Goldsmiths Musical Sophistication Index (Müllensiefen et al., 2014), are widely used in music psychology and neuroscience to capture demographic characteristics related to musical experience that might affect a study's results. Given the time constraints involved in many experiments, contemporary efforts in music psychology actually strive to reduce these measures further, in one case down to a single item for capturing individual differences in musicality (Zhang & Schubert, 2019).

This continuous refinement of tools to measure individual differences in musical behaviors, as part of a more general effort to quantify individual differences in ability, often proceeds without recognition of its problematic history and potential. For example, the history of intelligence testing reveals that this enterprise has played a major role in sustaining systemic racism. Croizet argues that "standardized testing, from its inception, has constituted an institutionalized arrangement aimed at expropriating resources from dominated groups to maintain dominant groups' privileges." (2011, p. 770). In addition to reckoning with the legacies of harm done by researchers whose work has been influential and even revered in their fields, a historical perspective allows a critical examination of the extent to which tools once used to foster inequity can be reconceptualized and repurposed for other ends.

Most operationalizations of the categories musician and nonmusician rely on heavily Westernized notions of musical training, defining *musician* as someone who has had a certain number of years of formal training. This underscores the need for culturally informed as well as cross-cultural research, where the goal is to understand the full spectrum of human experience as evolution, variation, and adaptation in their cultural context, rather than prioritizing statistical power in large and homogeneous sample sizes (Clancy & Davis, 2019).

This section begins with a historical perspective (Cowan's chapter 16), examines the current conundrum within music science (Ilari and Habibi's chapter 17) and best practices in research (Savage et al., chapter 18), and ends with a call for the exploration of new and fertile research areas (Feld et al., chapter 19). Cowan examines the historical context of the seemingly democratizing claim that "the musical mind is the normal mind," advanced by music psychology's founding figure Carl Seashore, together with "scientific" measurements of musicality and their utility for the social engineering envisioned by the American eugenics movement in which Seashore took part. Ilari and Habibi review the literature on musicians and nonmusicians, articulate its problems, and suggest avenues for future research. Savage and colleagues recommend best practices for cross-cultural research in music studies moving forward, with an emphasis on sustainable practices for collaborations. Finally, in an interview with the editors, Feld calls for a reintegration of cognitive approaches with grounded investigations in ethnography and speculates about how to motivate research in music studies that cuts across both disciplinary and cultural boundaries to enrich our understanding of musical skills and experiences.

References

Clancy, K. B., & Davis, J. L. (2019). Soylent is people, and WEIRD is white: Biological anthropology, whiteness, and the limits of the WEIRD. *Annual Review of Anthropology, 48*, 169–186.

Croizet, J. C. (2011). The racism of intelligence: How mental testing practices have constituted an institutionalized form of group domination. In L. D. Bobo, L. Crooms-Robinson, L. Darling-Hammond, M. C. Dawson, H. L. Gates Jr., G. Jaynes, & C. Steele (Eds.), *The Oxford handbook of African American citizenship, 1865–present* (pp. 770–816). Oxford University Press.

Ericsson, K. A., & Chase, W. G. (1982). Exceptional memory: Extraordinary feats of memory can be matched or surpassed by people with average memories that have been improved by training. *American Scientist, 70*(6), 607–615.

Fowler, O. S., & Fowler, L. N. (1850). *Phrenology proved, illustrated, and applied: Accompanied by a chart . . . together with a view of the moral and theological bearing of the science*. Fowlers and Wells.

Gardner, H. (1997). *Extraordinary minds: Portraits of 4 exceptional individuals and an examination of our own extraordinariness*. Basic Books.

Kraus, N., & Chandrasekaran, B. (2010). Music training for the development of auditory skills. *Nature Reviews Neuroscience, 11*(8), 599–605.

Manning, C. F. (2015). Phrenologizing opera singers: The scientific "proofs of musical genius." *19th-Century Music, 39*(2), 125–141.

Miller, L. K. (1987). Determinants of melody span in a developmentally disabled musical savant. *Psychology of Music, 15*(1), 76–89.

Müllensiefen, D., Gingras, B., Musil, J., & Stewart, L. (2014). The musicality of non-musicians: An index for assessing musical sophistication in the general population. *PLOS ONE, 9*(2), e89642.

Ollen, J. E. (2006). *A criterion-related validity test of selected indicators of musical sophistication using expert ratings* [Doctoral dissertation]. Ohio State University. https://etd.ohiolink.edu/apexprod/rws_olink/r/1501/10?clear=10&p10_accession_num=osu1161705351.

Patel, A. (2011). Why would musical training benefit the neural encoding of speech? The OPERA hypothesis. *Frontiers in Psychology, 2*(142).

Sacks, O. (2010). *Musicophilia: Tales of music and the brain*. Vintage.

Simonton, D. K. (2005). Giftedness and genetics: The emergenic-epigenetic model and its implications. *Journal for the Education of the Gifted, 28*(3–4), 270–286.

Winner, E. (2000). Giftedness: Current theory and research. *Current Directions in Psychological Science, 9*(5), 153–156.

Wong, P. C. M., Skoe, E., Russo, N. M., Dees, T., & Kraus, N. (2007). Musical experience shapes human brainstem encoding of linguistic pitch patterns. *Nature Neuroscience, 10*(4), 420–422.

Zhang, J. D., & Schubert, E. (2019). A single item measure for identifying musician and nonmusician categories based on measures of musical sophistication. *Music Perception, 36*(5), 457–467.

Zhang, J. D., Susino, M., McPherson, G. E., & Schubert, E. (2018). The definition of a musician in music psychology: A literature review and the six-year rule. *Psychology of Music, 48*(3), 389–409.

16 "The Musical Mind Is the Normal Mind": Remaking Musicianship for Eugenics

Alexander W. Cowan

What makes a musician? As the chapters in this volume explore, it can have everything to do with training, heredity, class, caste, race, gender, development, ability, and innumerable other factors, up to and including species. And as these same chapters suggest, the simple division between musician and nonmusician—a commonplace in the modern West—breaks along the same fault lines. Where it remains a factor in scientific research (see Ilari and Habibi's chapter 17 in this volume), it usually exists as a self-identified category, and usually for the sake of expedience—an artifact of the social division of (musical) labor that precedes any physiological or neurological observation. What makes a musician is, in the final analysis, something best answered by musicians themselves.

This simple division between musician and nonmusician has, however, been questioned before, from both within and without the sciences of music. This chapter examines one such instance. In the early 1920s, leading music psychologist Carl E. Seashore called for the sweeping removal of all existing work on the biological grounding of musical talent, for some of the same reasons listed above: the biases that creep in when defining musicianship along social lines. His critique was, by and large, successful. Seashore initiated a self-proclaimed paradigm shift in the psychology of music, the effects of which are still felt in research and education today (see Koza, 2021). The result was the establishment of a prevailing view in which musical talent was thought to be an amalgam of discrete psychophysical traits that were measurable, innate, and heritable. But as has now been firmly established (Cowan, 2016; Devaney, 2019; Koza, 2021), Seashore's work in music was inseparable from his involvement in a more pernicious human science: eugenics. The relationship between Seashore's revision of what it means to be a musician and the broader eugenic revision of what it means to be human is a question beyond the scope of this chapter. However, the specific argument Seashore brings to bear against a strict binary interpretation of musicianship—which he renders in somewhat crude terminology as a problematic focus on the musical "genius" and the "defective"—bears a resemblance to contemporary criticisms of this division, making it worthwhile to consider what role Seashore's eugenics played in its genesis.

Seashore's argument proceeds in three stages, some of which sound familiar based on modern arguments against the division between musician and nonmusician. First, its premise: A hard-and-fast biological division between musician and nonmusician is not, and never has been, supported by experimental evidence. Its persistence in the science of music is an anachronism and, worse than that, actively harms the recruitment and training of musicians by spreading misinformation about who would and would not benefit from such training, especially groups hitherto underrepresented among prominent musicians.

Second, a proposal: Replace this biological foundation of musicality with another—the insistence that the musician is a person like any other, that "the musical mind is the normal mind," and, although there might be genetic variations that make musical success more or less likely, we are all measurable by the same yardstick.

Third, some recommendations: As evidence mounts for the accuracy of these measurements and for the precise effects of these genetic variables, we should think about how they can be used to our advantage, to distribute more efficiently the limited resources of musical training to those best prepared, biologically and temperamentally, to make the most of them. Might it even be possible, given advances in the understanding of genetics, to selectively breed for such traits?

Here we reach the argument's sleight of hand: there is no valid reason why the rethinking of musicality from an exceptional state to a combination of individual variables should necessarily end in eugenic selection. It is therefore significant that whenever Seashore made the first two parts of his argument, he almost always made the third. The seemingly democratic disavowal of musicality as limited only to the genius was apparently inseparable from his eugenic belief in musical capacities as heritable and improvable through selective breeding. Why tear down one regime of innate musicality and raise another in its place?

I approach this apparent contradiction through an examination of Seashore's paradigm-shifting dictum "the musical mind is the normal mind," which occurs, in some variation, in three texts clustered around the turn of the 1920s, a pivotal phase in his career: the monograph *The Psychology of Musical Talent* (1919); an article titled "The Inheritance of Musical Talent," published in *Musical Quarterly* in 1920; and a speech based on the same article delivered to the Second International Congress of Eugenics in 1921, titled—with telling additions—"Individual and Racial Inheritance of Musical Talent." Taking these texts as its corpus, this chapter retraces the path of Seashore's argument through its three stages, attempting to account for its development.

I examine the existing literature that Seashore viewed as an inadequate foundation; the intellectual and social preconditions for his new vision of musical capacity; and the musician's relation to the social totality that must, he concluded, end in eugenic intervention. In each instance, the relationship between the changing scientific theories of inheritance and musicality and the changing social definitions of musicianship and (musical) labor emerges as a path by which the apparent contradictions of Seashore's argument can be resolved.

Discarding Prior Foundations

Seashore made the first complete statement of the argument outlined above in his 1920 *Musical Quarterly* article, under the heading "The Normal Mind versus the Genius and the Defective." After the clear assertion that "the normal mind is musical," Seashore sounded a note of caution against centering the "most tangible types of case; that is, on the one hand the genius and on the other the defective." "This distinction," we are advised, "is not as simple as it might seem" (Seashore, 1920, p. 588). Just as musical achievement is, in his analysis, made up of multiple different interacting capacities and learned behaviors, purported musical inability is too complex a phenomenon to be considered a single type. On account of this misplaced focus, he concluded, "we can get little or no help from works now extant on the inheritance of musical talent" (p. 593).

The critique was made in broad strokes, and in every instance, Seashore provided no specific examples of works he deemed inadequate. However, he named two categories that provide clues to his intended targets: musical biography, and biometric studies of inherited traits that take certain musical capacities into account.

He made his objections to musical biography clear in his 1919 monograph: "The comparatively large, though scattered, literature on the inheritance of musical genius is of little value because it does not deal with tangible fact. It merely essays to determine whether or not the ancestors of a given musician were or were not musical, on the whole" (Seashore, 1919, p. 69). Seashore was not the first to level this critique. Eugenics founder Francis Galton (1822–1911), in his first foray into the study of inheritance, identified this as a weakness of his own biographical method. After asserting that "the fact of the inheritance of musical taste is notorious and undeniable," Galton conceded that he found it "impossible" to achieve a workable list of "first-class" musicians from which to begin his study, for "most biographers are unusually adulatory of their heroes, and unjust to those with whom they compare them" (1869, p. 230).

Seashore pressed the critique of subjective biography further in his 1920 article, drawing on the half-century of advances in genetics since Galton's initial assessment in the 1860s. Because women were rarely permitted to achieve eminence and fame in most musical professions, data for half of every family tree would be missing, he noted, rendering them useless from the perspective of modern Mendelian genetics: "The male musical genius has often come from a mother whose extraordinary talent has passed undiscovered until it has appeared in the career of a son . . . [Acquiring information on the ability of women] has not been done in musical biography because biography deals primarily with achievement" (Seashore, 1920, p. 593). Gregor Mendel's (1822–1884) recently rediscovered work concerned inheritance in plants, but it was quickly applied to the physical attributes of animals and humans, in the latter case, most notably by Charles B. Davenport (1866–1944), a zoologist who would emerge in the same decade as America's leading eugenic scientist and become a close collaborator of Seashore's (see Davenport, 1911; on the Mendelian turn, see Bowler, 1989). Seashore noted that, although there was yet little precedent for applying Mendel's laws to the inheritance of mental traits in humans, accepting "as a general working basis, the Mendelian hypothesis" was "the only logical and economic way to proceed" (1920, p. 592). The criticism of masculinist history echoed arguments taking place in the field of genealogy, which was rapidly professionalizing and simultaneously undergoing its own Mendelian turn (Teicher, 2020, pp. 48–52). As Mendel's work proved that hereditary characteristics were an equal product of both parents, patrilineal ancestral charts became just as useless as the male-dominated realm of biography for the systematic study of heredity. Social eminence, dictated as it was by gender, could no longer substitute for empirical fact.

Biometric studies, while avoiding the errors of subjective assessment, still proved inadequate, for "none of them deal with specific capacities" (Seashore, 1920, p. 593). As in the case of biography, though, Seashore offered no specific examples, and it is possible to construct a counterarchive that suggests certain capacities were, in fact, considered in some of the literature. Gestures toward the study of heredity's role in pitch perception and the upper bounds of hearing can be found in the later work of Galton (1883) and that of his students (notably, James McKeen Cattell and Karl Pearson). There were also biometric studies of the absence of musical feeling, such as Carl Stumpf's roughly contemporaneous studies of "amusia," although these were aimed at investigating the deep past of music's evolution rather than the immediate mechanism of its inheritance (see Kursell, 2018). Nevertheless, it was not such a well-developed body of literature that Seashore could find satisfactory answers in it. A 1921 literature review bolsters Seashore's assessment, making no reference to any biometric studies other than a gesture toward Seashore himself (Weidemann, 1921).

The continual reference to the "genius" and the "defective" as archetypes suggests another broader field of discourse that Seashore had to sidestep: the long-standing cultural fascination with figures of purportedly superhuman—or subhuman—achievement. Musical genius, as a trope, long precedes the era of empirical psychology. Some versions of the musician endowed with divine inspiration or (un)natural creative power can be found as early as the sixteenth century, but the trope is typically associated with the aesthetics of Romanticism and, in music, with the figure of Beethoven in particular—at once a commanding individual subject and a vessel for some greater force (see, e.g., Lowinsky, 1964; Burnham, 1995).

Drawing a contrast with these earlier versions of divine genius, American studies scholar Gustavus Stadler has argued that the trope underwent a transformation in the middle of the nineteenth century, moving away from spiritual possession and toward "an increasingly detailed, psychologized, and sexualized notion of the individual genius: the genius as pathological subject." (2006, p. xv). This was a more conventionally threatening figure, whose formerly supernatural alterity was mapped in public consciousness onto the very human alterities of race, gender, and ability. Composers were the most common subjects of musical biography, but this new version of genius was open to performers too, and their race and gender more often set them apart from the idealized white male subject. For example, Stadler (2006) examines opera singer Jenny Lind as an avatar of a sexualized Nordic cosmopolitanism. And recent work by Lindsay Wright has shown how African American pianist Thomas "Blind Tom" Wiggins was framed using the racialized trope of the savant, one capable of extraordinary feats of technical display but little original creativity, in addition to being considered intellectually and physically disabled (see Wright, 2018; on music and the "idiot savant" trope more generally, see Straus, 2011). In a moment of increasingly sharp division of labor and increasingly strong delineation between cultural production and consumption, Stadler argues, "the geniuses' labor is to fortify middle-class men and women by taking upon their own minds and bodies the troubling, potentially shattering phenomena associated with modernity" (2006, p. xv). In his 1920 paper, Seashore retained the idea of genius as an aesthetic category, seeing it as almost self-evident. But as scientific grounding, the states of exception characterized by the "genius" and the "defective" now proved inadequate. The "pathological" subject, in other words, was out of time, and something new was needed to replace it.

A New "Normal"

It was from outside the sphere of musical heredity that Seashore began to develop his new method. Psychologists following Hermann von Helmholtz and Wilhelm Wundt had, for decades, investigated psychophysical attributes related to music, such as the perception of pitch differences or the upper and lower bounds of hearing. (On the role of auditory and musical stimuli in the development of modern psychology, see Erlmann, 2014; Klempe, 2011; Hui et al., 2020). Seashore built his reputation on the application of these measurement techniques to the assessment of musical talent, culminating in a set of Measures of Musical Talent, standardized tests that cemented his status as the foremost psychologist of music in the United States. The Measures relied on reconfiguring talent as a *capacity* rather than an *ability*; they were based on the supposedly innate psychophysical limits of musical achievement, rather than the quality of the achievement itself (see, e.g., Seashore, 1919, pp. 14–15). This strategy was thought to bypass the problems, on the subject's end, of varying levels of musical training—nurture as opposed to nature—and, on the observer's end, of impartial assessment. From the beginning, Seashore considered these capacities not only innate but hereditary. "Musical talent," he wrote, "like all other talent, is a gift of nature—inherited, not acquired; in so far as a musician has natural ability in music, he [*sic*] has been born with it" (Seashore, 1915, p. 129).

Seashore argued that the capacity for musical achievement was an aggregate of multiple discrete perceptual capacities, loosely mapped onto what he considered the core qualities of musical sound: pitch, time, and intensity. If a musician could not hear fine gradations in pitch, for example, the ability to play or sing in tune would inevitably be hampered, or so the logic goes. Musical talent, formerly thought of as an outward-facing quality discernible in performance or composition, was flipped; it was now rendered as an *internal* quality discernible not by the ear of the audience but through that of the performers themselves. Crucially, these purportedly essential traits were accessible to measurement, and the form and structure of the tests would change as Seashore developed different measuring technologies (Cowan, forthcoming). Nevertheless, they were consistent enough, and in 1919 recordings of the tests were made available commercially, for use in schools, universities, and the home.

Seashore's Measures—and their practical application for finding the "gold in the dross," as he put it (1915, p. 148)—rested on a number of preconditions, measurability being only one. The language of "traits" suggests another: a further indebtedness to Mendel, the significance of which, I argue, has been understated in assessments of Seashore's work. Mendelian genetics offered not just a mechanism of inheritance but also a total vision of the human organism—body and mind—as being composed of a number of independent variables that could be isolated and manipulated (Bowler, 1989; Teicher, 2020, p. 6). Seashore's framing of musical capacity around isolable traits believed to be linked to fundamental aspects of perception suggests that, even if he considered Mendel's theory untested with regard to mental traits, it at least offered an internally consistent way to derive measurable musical attributes from the basic sensorium. Put differently, although the heredity of musical traits required further testing, the very idea of a musical trait suggests a presupposition of Mendelian logic.

In Seashore's 1919 monograph, the phrase "the musical mind is first of all a normal mind" occurred at the end of the first section of the first chapter, "The Point of View," which outlined his principles for the study of musical talent. He continued, "Indeed, the normal mind is musical to the extent that it is normal" (Seashore, 1919, p. 6). The role of standardized testing in producing a very particular vision of "normal" has been well documented (for a version of this critique related to hearing tests in a Foucauldian frame, see Hui et al., 2020, p. 6; on psychology in general, see Rose, 1996). On the surface, though, Seashore presented this maxim first as a matter of expediency, to dispense with potentially complicating nonmusical aspects of the mind: "We must take it for granted that the musical mind is an aspect of a normal personality with endowments for a general mental life, and we must also take the general psychology of such mental life for granted," permitting a concern only with "aspects of talent which are peculiarly necessary for music" (Seashore, 1919, p. 6). This intervention is not presented as revolutionary, as such, but as a guardrail to limit the scope of inquiry. However, its position as a de facto conclusion to the book's opening section, and its implied rebuttal of the idea that musical genius is associated with mental abnormality (requiring a more holistic view of the mind, as a result), suggests that Seashore believed it had at least some significance.

The 1920 article "The Inheritance of Musical Talent" recapitulated many key components of Seashore's music psychology, including the division of musical talent into a hierarchy of component talents best approached as capacities, rather than abilities. But the turning point came soon after. "The normal mind is the average mind," Seashore wrote, "but such average does not represent a single dead level for all the various human capacities" (1920, p. 588). In what would become a refrain in his work, he eventually concluded, "This is only saying in other words, 'We normal people are so different'" (1920, p. 588). Absent was the argument's eugenic coda, although Seashore's other writings dispel any illusion that "normality" signifies equality. Seashore was clear on this point in his monograph: "Musical talent [is] an inborn gift. Musical talent is a gift bestowed very unequally upon individuals. . . . This fact presents an opportunity and places a great responsibility for the systemic inventory of the presence or absence of musical talent" (1919, p. 6). His signal contribution to music studies, the Measures of Musical Talent, were also published in 1919, providing the tool with which this systemic inventory could be conducted. The monograph thus had a dual function: to establish the psychological grounding for his theory of musicality, and to offer suggestions for the theory's practical application, embodied by the Measures.

Seashore was cautious on the question of inheritance, suggesting that it might be "reasonable to suppose" that certain musical attributes could be "inherited according to the Mendelian laws" (1919, p. 69), but he stops short of calling for intervention in this inheritance by eugenic selection. It is enough to simply uncover latent talent, without seeking to improve it. A year later, however, these reservations began to fade.

"Conscious Selection"

In 1921, hundreds of leaders in the global eugenics movement gathered at the American Museum of Natural History in New York City for a week of scientific presentations on the state of eugenic research (Davenport & Laughlin, 1923). Seashore was among the speakers.

Like many of the papers presented at the Second International Congress of Eugenics, Seashore's walked a narrow line between research report and manifesto. Some phrasing was recycled: the "genius" and the "defective" were interesting types, Seashore admitted, but again, "the musical mind is the normal mind" (Davenport & Laughlin, 1923, p. 232). Again, he claimed, "we must discard the literature on musical inheritance now extant, because it is not based on scientific conceptions of the musical mind." Seashore's argument advanced tentatively toward its end:

> The eugenist [*sic*] might rightly expect me to recite established facts on the inheritance of musical talent and present arguments showing that they should be applied. But the time is not yet ripe for either. The object of this paper is merely to present a point of view, showing that such facts can be gathered; and this is done in the anticipation that, once established, the desirability of their application will be taken for granted by those who are interested in this phase of eugenics. (Davenport & Laughlin, 1923, p. 232)

Seashore stopped short of making a definite pronouncement about the eugenic possibilities of his Measures of Musical Talent. It was enough to state that the importance of such research and the necessity of action were self-evident. He concluded: "My proposition is that if certain musical talents are heritable, as we believe them to be, it is quite within the power of future generations to enhance the quality and degree of a musical talent by conscious selection" (Davenport & Laughlin, 1923, p. 238). In offering this vision to this particular audience, Seashore made his strongest public statement in favor of an explicitly musical eugenic program: nothing less than the selective breeding of musicians.

Seashore's intervention into music psychology often changed emphases but kept a consistent form: when the musical mind is understood properly—and thus, when we begin music science anew—its development can be controlled. An obvious but necessary observation is that the eugenic conclusion was stated most fully in the version of the argument directed at an audience of eugenicists. It also bears noting that Seashore did not begin corresponding with Davenport directly until 1919, and his involvement with and understanding of eugenics were in the process of deepening during these pivotal years of his intellectual project. Nevertheless, it is clear that even in this short time, eugenics had become increasingly important in Seashore's thought: a goal that might, for now, be out of reach, but one that was always the end point of the study of musical talent.

Just as the first part of Seashore's argument presupposes scientific Mendelism, the last, I argue, presupposes what historian of science Amir Teicher recently termed "social Mendelism": a value system based on the "atomization of humans into fixed traits, and the need to prevent at any cost malignant elements from recoupling" (Teicher, 2020, p. 18). The first decades of the twentieth century saw waves of migration to the northern United States—overseas from Europe and internally from the South—which threatened existing hierarchies of race and labor (Roediger, 2005). At the same time, consolidation of industry and rapid technological development were remaking the nation's social fabric. Counter to these ruptures, psychology and biology were emerging as the sciences of a new form of social engineering: from the micromanaged movements of the Taylorist workplace to the racial hygiene of the eugenic state. It is against this backdrop that Seashore's disciplinary revolution was staged.

Eugenics as "Social Psychology"

Over the course of Seashore's argument, the musician and nonmusician were replaced by the "normal" mind-body-human, who is unevenly endowed with certain heritable perceptual capacities. As Seashore acknowledged, though, what makes a musician is as much a social question as one of psychophysical equipment. To accurately retrace the path of his intervention, then, requires an additional step: to examine his writing on the social purpose of psychological research.

Seashore's 1923 textbook *Introduction to Psychology* presents his social thought at its most expansive and systematic, tying together the psychology of the individual (the understanding of capability and talent, including musical talent), economic efficiency, and eugenics, which were present only inconsistently in his earlier publications. The textbook opens with an overview of the branches of psychology, two of which stand out: social psychology and individual psychology. Social psychology and its subfield applied social psychology are defined as "treat[ing] the social aspects of mental life," and these are the branches that Seashore thought had a role in eugenics (Seashore,

1923, pp. 4, 6). But individual psychology was the jewel in the crown. "[A]ll technical studies in psychology are merely a preliminary to [individual psychology]," he wrote, "for we are working under the slogan, 'Know thyself'" (p. 384). Individual psychology "consists of scientific analyses, tests, ratings and measurements in the identification of a given individual's psychophysic equipment, both as to original nature or inherited capacity, and nurtured or acquired abilities, for the purpose of scientific understanding, description, and motivation of the individual" (p. 384). The measurement of musical talent was merely one exceptionally well-developed method in this field, and eugenic enhancement of the population was only one possible outcome, but the dyad of measurement and action, especially as it pertained to aptitude and occupational fitness, appears throughout as the pinnacle of psychological endeavor.

The social question comes into focus at the section's conclusion. Seashore praised the fact that tests of occupational fitness—psychology as "good business"—were now commonplace, preventing economic waste, and he wrote with pride of the army intelligence tests, referring to them as a "project of human engineering" that "marked an epoch in history" (p. 394). He went on, in a preemptive rebuttal to political criticism of psychological intervention in the workplace: "Labor organizations have objected to the application of individual psychology on the ground that it would discourage uniformity of treatment. But the time has passed when one man can be regarded as good as another" (p. 407). Seashore's open disdain for labor unions was not representative of the general attitudes of psychologists, which tended to range from neutral to sympathetic (Gordon & Burt, 1981). However, in this dismissal he accurately represents the charges unions leveled against applied psychology in this period: that psychology and the broader program of scientific management together represented a new technology of extractive human engineering (Gordon & Burt, 1981, pp. 141–143). Individual psychology—as the basis of social psychology—provided a scientific grounding by which collective action could be dismissed, by centering the individual and thus eliminating the very possibility of a collective (Rose, 1996, pp. 105–107).

The connection to eugenics emerges clearly in this vision of the individual as the sum of various isolable capacities. As suggested earlier, this is a hallmark of Mendelian thinking, and in its economic form, it bears traces of the Fordist and Taylorist revolutions in production: the breaking down of complex manufacturing processes into discrete tasks and the honing of those tasks to optimal efficiency. (On Ford, Taylor, and efficiency, see Haber, 1964; Alexander, 2008; on its relation to eugenics, see Currell & Cogdell, 2006.) The language of efficiency shapes Seashore's discussion of individual psychology, which is presented as "a scientific approach to the problem of the conservation of human energies" (1923, p. 406). (On the history of industrial psychology in this period, see Van De Water, 1997.) Eugenics offered a "rational" solution to what political theorist Cedric Robinson framed as the irrational character of racial divisions of labor: the supplementation of base prejudices with a new universal metric (2000, p. 9). Vague language of racial inferiority could be replaced by talk of specific traits, of occupational and social fitness. Recalling Seashore's comment that "we normal people are so different," an expanded sphere of "normal" humanity permits ever-finer differentiation within it and opens up the entire population to the logic of eugenic improvement.

Examining Seashore's vision of what social psychology can accomplish, and the way he places his musical work within it, offers a way out of the problem with which this chapter began: the seeming contradiction between leveling the musical playing field and erecting a new hierarchy in its place. The concepts of the genius and the defective—standing for the idea of musicality as aberrant—had to be discarded because they represented a vision of human endowment incompatible with an increasingly rational world. The shift from the laissez-faire social Darwinism of the Gilded Age to the managed world of "social Mendelism" in the early twentieth century left no room for exceptions.

The elision of the scientific and the political that freely took place in this discourse is the final precondition for Seashore's paradigm shift: an intellectual clime that permitted, and encouraged, an effortless move from the statement that something can be measured to the suggestion that those measurements should be put to work. And it is from this elision that we may draw conclusions about our present exploration of the science-music borderlands. Certain species of universalism have now, rightly, been called into question—as engaged by many essays in this volume (e.g., the chapters by Mundy; Kragness, Hannon, and Cirelli; and Ilari and Habibi)—yet one can envision an epistemology of difference in place of Seashore's universalism that is equally flawed. Difference shunned in one era can be embraced in another, with the racialized division of labor looming behind both unchallenged (see, e.g., McWhorter, 2017). Put differently: an effort to narrow the scope of a claim—a frequent method of rapprochement between scientists and humanists and a common point of critique when engaging popular-press treatment of scientific work—can still operate only within the claim's original logic. If arguments over precision often function as a proxy for politics, perhaps Seashore's example is instructive, in that he was quite open about the political motivations and implications of his research. Recent work illustrates that reclaiming this strategy is not always successful, and rightfully so—for example, attempts to develop a model of educational genetics purportedly from the Left have ended up receiving no small praise from the Right (deBoer, 2020; Harden, 2021). There are institutional challenges, too, when attempting to develop a more explicitly politicized research agenda.

These include government censorship and the professional precariousness of the neoliberal university—a burden that, until recently, was associated primarily with the humanities but is now increasingly shared across the disciplines, in another example of the necessity of common cause.

However, such a strategy may offer a way out of many of the problems addressed in this volume: from avoiding the remnants of Seashore's eugenic legacy and other troubling inheritances in the history of music science to increasing the opportunities for intellectual collaboration between subfields of musical investigation and for professional cooperation in the face of threats to the university itself. The seeming contradiction with which this chapter began, between disciplinary progress and social conservatism, is thus resolved quite easily: the rhetoric of progress implies only a teleology, never a destination. And when a discipline lags so far behind the intellectual and political demands of modern life, it can take what looks like a revolution just to catch up to the status quo.

References

Alexander, J. K. (2008). *The mantra of efficiency: From waterwheel to social control*. Johns Hopkins University Press.

Bowler, P. J. (1989). *The Mendelian revolution: The emergence of hereditarian concepts in modern science and society*. Johns Hopkins University Press.

Burnham, S. (1995). *Beethoven hero*. Princeton University Press.

Cowan, A. W. (2016). *Music psychology and the American eugenics movement in the early twentieth century*. [Unpublished MMus dissertation]. King's College.

Cowan, A. W. (forthcoming). The phonograph as instrument of race betterment: Selling psychology for American eugenics. In E. MacGregor, E. I. Dolan, & A. Schwartz (Eds.), *Sonic circulations 1900–1950* [working title].

Currell, S., & Cogdell, C. (2006). *Popular eugenics: National efficiency and American mass culture in the 1930s*. Ohio University Press.

Davenport, C. B. (1911). *Heredity in relation to eugenics*. H. Holt.

Davenport, C. B., & Laughlin, H. H. (1923). *Eugenics, genetics and the family: Scientific papers of the Second International Congress of Eugenics held at American Museum of Natural History, New York, September 22–28, 1921* (Vol. 1). Williams & Wilkins.

deBoer, F. (2020). *The cult of smart: How our broken education system perpetuates social injustice*. All Points Books.

Devaney, J. (2019). Eugenics and musical talent: Exploring Carl Seashore's work on talent testing and performance. *American Music Review, 48*(2), 6.

Erlmann, V. (2014). *Reason and resonance: A history of modern aurality*. Zone Books.

Galton, F. (1869). *Hereditary genius: An inquiry into its laws and consequences*. Macmillan.

Galton, F. (1883). *Inquiries into the human faculty and its development*. Macmillan.

Gordon, M. E., & Burt, R. E. (1981). A history of industrial psychology's relationship with American unions: Lessons from the past and directions for the future. *Applied Psychology*, *30*(2), 137–156.

Haber, S. (1964). *Efficiency and uplift: Scientific management in the Progressive Era, 1890–1920*. University of Chicago Press.

Harden, K. P. (2021). *The genetic lottery: Why DNA matters for social equality*. Princeton University Press.

Hui, A., Mills, M., & Tkaczyk, V. (2020). *Testing hearing: The making of modern aurality*. Oxford University Press.

Klempe, S. H. (2011). The role of tone sensation and musical stimuli in early experimental psychology. *Journal of the History of the Behavioral Sciences*, *47*(2), 187–199.

Koza, J. E. (2021). *"Destined to fail": Carl Seashore's world of eugenics, psychology, education, and music*. University of Michigan Press.

Kursell, J. (2018). Carl Stumpf and the beginnings of research in musicality. In H. Honing (Ed.), *The origins of musicality* (pp. 321–346). MIT Press.

Lowinsky, E. E. (1964). Musical genius—evolution and origins of a concept. *Musical Quarterly*, *50*(3), 321–340.

McWhorter, L. (2017). From scientific racism to neoliberal biopolitics: Using Foucault's toolkit. In N. Zack (Ed.), *The Oxford handbook of philosophy and race* (pp. 282–293). Oxford University Press.

Robinson, C. J. (2000). *Black Marxism: The making of the black radical tradition*. University of North Carolina Press.

Roediger, D. (2005). *Working toward whiteness: How America's immigrants became white; the strange journey from Ellis Island to the suburbs*. Basic Books.

Rose, N. S. (1996). *Inventing our selves: Psychology, power, and personhood*. Cambridge University Press.

Seashore, C. E. (1915). The measurement of musical talent. *Musical Quarterly*, *1*(1), 129–148.

Seashore, C. E. (1919). *The psychology of musical talent*. Silver, Burdett.

Seashore, C. E. (1920). The inheritance of musical talent. *Musical Quarterly*, *6*(4), 586–598.

Seashore, C. E. (1923). *Introduction to psychology*. Macmillan.

Stadler, G. (2006). *Troubling minds: The cultural politics of genius in the United States, 1840–1890*. University of Minnesota Press.

Straus, J. N. (2011). *Extraordinary measures: Disability in music*. Oxford University Press.

Teicher, A. (2020). *Social Mendelism: Genetics and the politics of race in Germany, 1900–1948*. Cambridge University Press.

Van De Water, T. J. (1997). Psychology's entrepreneurs and the marketing of industrial psychology. *Journal of Applied Psychology, 82*(4), 486–499.

Weidemann, P. W. (1921). *The inheritance of musical talent* [Unpublished master's thesis]. University of California.

Wright, L. (2018). *Discourses of musical talent in American culture* [Unpublished doctoral dissertation]. University of Chicago.

17 The Musician-Nonmusician Conundrum and Developmental Music Research

Beatriz Ilari and Assal Habibi

Introduction

- A family sings *tanti auguri a te* to celebrate Grandma's birthday.
- A pianist repeats a virtuosic passage from a sonata, metronome clicking in the back. With each repetition, she picks up the tempo. After several minutes, the passage sounds nearly perfect. Her playing seems effortless, yet she does not seem satisfied. She takes a short break and then starts over.
- A DJ creates and shares beats with fellow musicians using an online platform.
- A girl sings while playing with LEGO in the corner of her kindergarten classroom. Her melodious voice goes up and down as she works on her construction project. At first it is possible to recognize fragments of a well-known pop tune, but the melody morphs into something completely different and unrecognizable.
- An elderly man teaches a young boy to play the tabla through the process of apprenticeship.

Musical engagement occurs across the life span and in multiple ways. The emergence of the term *musicking* (Small, 1998) propelled new thinking by suggesting that music is an action rather than a "thing" to be learned. Musicking refers to multiple forms of musical engagement, such as playing an instrument, composing, dancing, and listening. Small also stressed the relational nature of musicking and its meaning-making character. Though not without critics (Hesmondalgh, 2013), a main contribution of Small's work involves the recognition of different forms of human musical engagement, as well as their ubiquity. Similar to our understanding of musicking as plural, context bound, and polysemic, the term *musician* is context bound and takes on multiple meanings. While some would consider all the music makers described in the opening vignettes as musicians, others would vehemently oppose this idea, arguing that only a few could be labeled as such. The same happens in music research, where multiple—and at times contrasting—definitions of *musician* coexist. Music psychology research, in particular, tends to dichotomize and label study participants as musicians or nonmusicians. This binary is defined in arbitrary ways, such as the number of years of formal music instruction (Daly & Hall, 2018; Zhang et al., 2020); is inconsistent across studies (Cogo-Moreira & Lamont, 2018); and, in its essence, does not capture the complexity of the musical experiences of individuals (Zentner & Gingras, 2019). These terms *musician* and *nonmusician* are also grounded in Western and Westernized (i.e., conservatory, specialist school) notions of musicianship and musical expertise and, most often, on presentational forms of music making (Turino, 2008), with a focus on "skilled producers of music" (Rickard & Chin, 2017).

In this chapter, we discuss the musician-nonmusician binary that is still prevalent in music research involving humans, with a focus on developmental studies. We are two scholars—a music educator and a neuroscientist—who have collaborated for over a decade on developmental music studies (e.g., Habibi et al., 2018; Ilari et al., 2016). Conversations about the nature of musicality, musicianship, and musical development are part of our collaborative work, as our original fields of study are based on different epistemologies. For purposes of this chapter, we define musical development as changes that occur to music perception, cognition, and action over the course of one's lifetime and as a result of musical participation and affordances in varied contexts and cultures and over the course of time (Hargreaves & Lamont, 2019). Although they are sometimes conflated, we understand *musicality* (the quality of being musical) as being distinct from *musicianship* (the skills that allow one to sing, play, compose, and improvise "well").

We begin the chapter with definitions of *musician* that are commonly found in research on human behavior in four fields: music psychology, music education, ethnomusicology, and neuroscience. Next, we discuss how potential, skill development, and self-identification—three converging categories that appear in definitions in the aforementioned fields—relate to child development. We conclude the chapter with suggestions for future research.

The Definitional Conundrum

A pervasive idea in many Western and Westernized cultures is that a musician is an able-bodied adult (often male) with plenty of technical and interpretative skills and a number of years of formal music instruction (Daly & Hall, 2018; Rickard & Chin, 2017). Definitions of the musician tend to be narrow in scope and are almost always associated with singing, playing an instrument, or composing (Rickard & Chin, 2017). This is understandable, given the prevalence of Western "art" music in educational institutions and their emphases on individual traits and presentational modes of music making

(Turino, 2008). Musicians in this tradition are known to spend a considerable amount of time perfecting their craft. While beliefs regarding musical potential and talent vary (McPherson & Hallam, 2008; Blacking, 1973), most people seem to endorse the idea that formal music learning ought to begin in childhood because musical skills take time to develop and require deliberate practice and focused attention (Ericsson et al., 1993). Age at the onset of training, length of training, type of music education (e.g., private music lessons, music theory), having a degree in music, and self-identification have been used to categorize musicians from nonmusicians. Whether by training or occupation, professionalization is another criterion used by scholars to define musicians. Yet many professional musicians, especially in the popular music scene, have no formal training, despite their extraordinary musical skills. There is another category of musician who does not make a living from music yet displays a high level of musicianship: the musical dilettante or amateur. Some amateur musicians have extraordinary musical skills and may be known as professional-amateurs, or pro-am (Leadbeater & Miller, 2004). Thus, a combination of training, occupation, self-identification, and music skills seems to be the defining feature of the musician in human behavior studies.

If defining a musician is difficult, defining a nonmusician is an even more challenging task. Based on the criteria used to define a musician, a nonmusician can be defined as an individual who lacks formal training, lacks musical skills, or does not self-identify as a musician. But, as many would argue, the term *nonmusician* is an oxymoron. No one is really devoid of musical skills (Rickard & Chin, 2017; Zentner & Gingras, 2019), as all humans are endowed with musical potential (Blacking, 1973; Malloch & Trevarthen, 2009; Sloboda, 2005). Aside from misrepresenting human potential, the term *nonmusician* devalues skills that are inherent to being musical, such as the ability to listen to and be moved by music, and it is also detrimental to identity construction and musical participation (see Hargreaves & Lamont, 2019; Henley, 2017). From an epistemological standpoint, the terms *musician* and *nonmusician* also raise questions about the nature and origins of musicianship and musicality. What is human musicality? Is it innate or acquired? What enables and what constrains the realization of musical potential and the development of musicianship in childhood? These questions have been answered in various ways across disciplines, as discussed next.

Disciplinary Perspectives

Education

Education is a field that focuses on nurturing the skills of learners from different age groups and ability levels. This probably explains why most music educators are reluctant to speak of nonmusicians; they tend to value potential over talent. But this is not to say that they do not discuss talent or innate abilities. Such conversations are part of the everyday life of music education programs and institutions worldwide.

School-based programs are at the core of a large number of education studies. In such research, participation in music programs is often used as the criterion to distinguish "music students or music majors" from "non-music students or non-majors" (Elpus & Abril, 2016). Musical potential, in turn, is defined in multiple ways, depending on student age, type of program, and repertoire of practice (see Gordon, 1971; Gudmundsdottir, 2020). Music programs with a primary focus on Western music practices (e.g., orchestras, bands, individual performance) have a tendency to legitimize views of the musician outlined earlier, with a focus on specialized, individual performance. Children and youths who are precocious and develop the adult-like musical skills valued by such institutions are often given the title "musician," a practice that has existed for many years (e.g., Gjerdingen, 2020).

The field of music education has expanded considerably in the past few decades to encompass individual and collective experiences. There is now recognition that musical potential can be nurtured and developed in multiple spaces and through informal approaches (see, e.g., Green, 2017; Folkestad, 2005). Informal music learning is predicated on the idea of individuals getting together to learn music on their own by listening to recordings and playing "by ear" with peer groups (Green, 2017). Rules and repertoires of practice are defined by individuals in these peer groups, like in rock bands. Unlike in formal learning, where definitions of the musician are determined by specific proficiencies determined a priori by experts (e.g., acquiring specific repertoires, passing an audition or test), in informal learning, the social group and its shared musical-aesthetic values define the musician (Bennett, 2017). Along similar lines, many musicians are autodidacts—they develop musical skills and become musicians "on their own" (Watson, 2012). Thus, music training alone can take multiple forms, each making different demands on the individual and giving rise to different "types" of musicians.

Ethnomusicology

Ethnomusicology celebrates the plurality and richness of musical cultures. Through fieldwork and thick description (Geertz, 1973), ethnomusicologists have uncovered multiple forms of musicality, contributing to our understanding of musicianship as culturally situated. Ethnomusicological work also stresses how the characteristics of musicians vary, depending on culturally constructed notions of talent and potential, lineage, age, and particular musical skills that specific groups prioritize. Some non-Western cultures actually lack an equivalent term for musician, as all community members are known to participate in music in some way (Rickard & Chin, 2017). The Venda of South Africa (Blacking, 1973), the BaYaka of Central Africa, and the Anang Ibibo of Nigeria are examples of cultural groups in which musical participation is part of the fabric of everyday life among children and adults alike (Rickard & Chin, 2017).

Ethnomusicological work has also revealed the existence of individuals who are highly skilled in music yet do not self-identify as a musician on the basis of the parameters outlined at the beginning of this chapter. For the Griots in Mali, for example, musicianship is defined primarily by lineage (Durán, 2017). The muezzins, who are responsible for the Muslim "call to prayer," are sometimes called "criers" or "reciters" and are not necessarily considered musicians. Similarly, many children from around the world do not self-identify as musicians, despite having solid musical skills obtained in the context of everyday musicking through processes of enskillment (Durán, 2017), intent community participation, or guided repetition (Rogoff et al., 2007). These various modes of music and learning transmission indicate that cultural groups hold different values and prioritize different musical features and skills, rendering distinctive (and at times fluid) definitions of the musician. Additionally, in cultures where participatory musicking is the norm and not the exception (see Turino, 2008), the proportion of self-proclaimed musicians can be high. These communities are often collectivist (Hofstede, 2011), with the needs and goals of the group taking precedence over those of individuals. One example is the city of Salvador in northeastern Brazil, where many residents self-identify as musicians (Brasil, 2020). In this community, what differentiates musicians from nonmusicians is professionalization, or earning a living through music. Clearly, culture plays a central role in defining who is called a musician, as well as the precision (or imprecision) of that definition.

Psychology

Psychologists study the human mind and human behavior. Two contrasting views of musicians and nonmusicians are commonly seen in psychological work. Evolutionary psychologists tend to view musicality as a biological trait (e.g., Honing, 2018), whereas cognitive psychologists often define the musician (and consequently the nonmusician) based on musical skills, potential (or predisposition), or identity, with musical skills receiving the most attention (Zhang et al., 2020). This is not surprising, given cognitive psychologists' long interest in the acquisition and development of musical skills (Drake et al., 2000; Hargreaves, 1986; Hargreaves & Lamont, 2019; Zentner & Gingras, 2019). A solid body of knowledge on pitch and rhythmic perception and discrimination, fine motor skills, memory, auditory-memory integration, composition, creativity, instrumental and vocal performance, sight-singing, and improvisation has helped scholars formulate theories and models of musical expertise (e.g., Lehmann et al., 2018). Because musical skills take time to develop and demand deliberate practice, the number of years of musical training has often been used to define one's level of musicianship. Individuals who begin training early often exhibit the greatest influence on measures of auditory and motor function. Thus, age at the onset of training is often considered a determining factor in musicianship (Penhune, 2011).

Along with musical skills, musical potential, achievement, and ability are considered markers of musicianship, although each is defined in different ways. Zentner and Gingras (2019) consider aptitude (or potential) and achievement to be two components of a general musical ability. The use of standardized musical tests to study musical ability dates back to the beginning of the twentieth century (see Zentner & Gingras, 2019). The Goldsmiths Musical Sophistication Index (MSI; Müllensiefen et al., 2014), the Exposure to Music in Childhood Inventory (EMCI; Cogo-Moreira & Lamont, 2018), and the Montreal Battery for Evaluation of Musical Abilities (MBEMA; Peretz et al., 2013) have been used to examine musical ability in individuals from different age and cultural groups. Although not designed specifically for children, the MSI is likely the only measure that combines a self-reporting instrument with a listening task. The choice of "musical sophistication" as the test's focus also represents a shift from the musician-nonmusician binary. The EMCI represents one of the few attempts to develop a standardized measure of musical skills in Brazil. The MBEMA includes tests of memory, pitch, and rhythmic discrimination based on nonverbal stimuli, making it applicable to the study of musical skills in children from diverse cultural and linguistic backgrounds and with varying levels of proficiency. Although these three measures are known to be reliable, they are somewhat limited, in that they are still bound by Westernized conceptions of music and musicianship. For instance, the MSI includes statements such as "I am not able to sing in harmony when somebody is singing a familiar tune" or "I don't spend much of my disposable income on music." The EMCI includes both listening tasks and a self-reporting component. Some questions are associated with context and with time-specific urban experiences that imply access to technology (e.g., "Do you watch *The Voice Brazil*?" and "Do you download music from the internet?"). Furthermore, the EMCI does not include questions on community-based musicking. Although it is difficult to create a measure of musical engagement that is universal and all encompassing (Zentner & Gingras, 2019), the MSI, EMCI, and MBEMA represent positive steps in that direction.

Self-identification as a musician is another approach psychologists have used to distinguish musicians from nonmusicians (Rickard & Chin, 2017). Self-identification is obviously linked to issues of musical potential, skill development, and cultural norms.

Given the Western and Westernized biases of psychological research (Heinrich, 2020), it comes as no surprise that self-identification studies in the field tend to be connected with presentational forms of musicking (see Hill, 2018; Turino, 2008).

Neurosciences

The neurosciences study primarily the brain and the nervous system. The brain has the ability to adapt its structure and function in response to learning or to new experiences through the dynamic reorganization of synaptic connections, pruning, and myelinations broadly defined as neuroplasticity (Tardif et al., 2016). Although the human brain is shaped significantly during critical periods of early development, the rate of maturation varies across different regions of the brain; the visual system matures to adult levels within the first year of life, while the auditory and motor systems continue to develop through early adulthood (Moore & Linthicum, 2007). Evidence shows that experience-dependent neuroplasticity continues across the life span and that the adult brain can be significantly malleable in terms of learning new skills. Singing in a band, playing a solo musical instrument, and drumming in a circle with others are all complex tasks that simultaneously engage the motor, sensory, cognitive, and affective systems of the brain. Specifically, they all require a high degree of sensorimotor integration, where one has to connect different sounds with specific motor output and make necessary adjustments through top-down feedback from the executive and affective systems (Zatorre et al., 2007; Brown et al., 2015). Over the last three decades, accumulated scientific evidence has demonstrated that musical training has pronounced effects on the function and structure of the human brain (see Gaser & Schlaug, 2003). However, most studies of musically induced neuroplasticity use years of training and/or age at the onset of training to separate musicians from nonmusicians.

There is evidence that early experience differentially influences skill acquisition and brain structure in several domains, including language development. It has been suggested that a particularly sensitive period, when learning music can have lasting and strong effects, occurs at seven to eight years of age (Penhune, 2011). Based on this sensitive period, musicians are typically characterized as individuals who started learning music early and have had ten years or more of formal musical training. However, there is a wide discrepancy in the number of hours of training, what constitutes "formal" training, and the current status of musicianship and practice. For example, a forty-year-old adult who began playing the cello at age seven and continued playing for fifteen years but stopped after graduating from college may theoretically meet the criteria to be considered a musician, yet he may no longer self-identify as a musician after not playing music for more than a decade. To increase the possibility of observing the greatest impact of music-related neuroplasticity in studies with relatively small sample sizes, it is common to identify musicians as skilled individuals with long-term training that began at an early age. This binary approach, however, runs the risk of missing everyone in between and including individuals who no longer engage with music.

Longitudinal studies of children that seek evidence of specific changes in behavior and brain structure related to training can provide a clearer picture of the immediate and potentially lasting effects of music learning. We used a five-year longitudinal study to evaluate the influence of music learning on child development. Two years of music training in a group setting led to increased functional development of the auditory pathways and neuroplastic structural changes in the auditory cortex (Habibi et al., 2018). There were, however, individual differences in music-related neuroplastic changes within the group of children learning music, and these were most likely influenced by genetic predispositions and environmental factors. The longitudinal design allowed a more nuanced evaluation of music training as a continuous experience instead of a binary categorization of musicians versus nonmusicians. It also allowed us to examine how genetic factors, the environment, and music learning interact to shape musical skills over time for each individual.

Points of Convergence

While each discipline uses different criteria to define the musician, there are some points of convergence. These include the notion that musical engagement influences and changes the individual in various ways, resulting in different levels of musicianship. Musical potential appears to be important to all four fields, and so is the sense that musicianship develops over time, which leads naturally into issues of time and deliberate practice (Ericsson et al., 1993). This is true even in cultures where musicians are defined by lineage. Musicianship can be attained through formal and informal learning (including autodidactism) in both presentational and participatory ways (Turino, 2008). Furthermore, the issue of self-identification seems to be present in all fields to a greater or lesser extent. But how do these points of convergence relate to developmental research?

Developmental Perspectives: The Child Musician

Earlier in the chapter, we defined musical development as changes that occur to music perception, cognition, and action over one's lifetime due to musical participation and affordances in varied contexts and cultures and over the course of time (Hargreaves & Lamont, 2019). This definition of musical development is consistent with the suggestion that development is a multifaceted and biopsychosocial phenomenon that takes into account the bidirectional influential relations between individuals and the contexts in which they are situated (Lerner & Benson, 2013; Lerner et al., 2018). Individual attributes, proximal processes (e.g., sustained forms of reciprocal interactions with cultural objects and symbols and the immediate environment), context, and time all influence children's musical development (see also Kragness, Hannon, and Cirelli's chapter 8 in this volume).

Although musical development occurs throughout the life span (Hargreaves & Lamont, 2019), there is an uneven distribution of studies across age groups. In this chapter we focus on the period between infancy and adolescence, which coincides with our research interests and expertise. In the following sections we discuss potential, skill development and training, and self-identification—three factors described in the four fields discussed earlier—as they pertain to child development.

Potential

Musical potential can be defined as "a latent, but as yet unrealized capacity to do something musical" (Kemp & Mills, 2002, p. 3). Childhood is a time of discovery and development of the brain and body along with various skills, including musical ones. Put simply, all children are endowed with musical potential. Consistent with the systems view of development (Lerner et al., 2018), the realization of musical potential in childhood depends on a combination of brain, genes, environmental issues and affordances, personality, motivation, and disposition.

Manifestations of musical achievement are often taken as signs of musical potential (Kemp & Mills, 2002). One of the most obvious examples is when children exhibit precocious skills that are valued by a particular cultural group, such as playing an instrument, singing, or inventing music (with or without the use of notation). Evidence of musical potential in childhood has been gathered through studies of the musical skills of infants and children (see the next section) and retrospective data on the musical experiences of eminent musicians (e.g., Sloboda & Howe, 1991). As noted earlier, children's musical potential has also been identified through aptitude tests (Zentner & Gingras, 2019). Carl Seashore's Measures of Musical Talents, Edwin Gordon's Audio and Music Aptitude Profile, and Bentley's Measures of Musical Abilities are some well-known musical aptitude tests for children. As informative as they are, these tests are somewhat biased toward children with musical training, and the results can be misinterpreted when used outside of their intended context, such as in schools (Kemp & Mills, 2002).

Linked to the notion of childhood as a rich period of human and musical development is the belief that the realization of musical potential is likely more efficient in early childhood. This is consistent with the concept of sensitive periods of learning in childhood or "a window in development when specific training or experience produces long-term changes in behavior and the brain, above and beyond those associated with that same experience at a different time during development" (Bailey & Penhune, 2013; Cho, 2019). The identification of sensitive periods in the development of musical skills (Bailey & Penhune, 2013) and faster maturation of the brain's auditory regions and enhanced con nectivity between auditory and motor regions in children who start formal musical training before a specific age (see Habibi et al., 2018; Bengtsson et al., 2005) supports the association between musical potential and early-onset training. Beyond brain and behavioral studies, observational research with young children supports the existence of early musical potential in proto-conversations between neonates and their caregivers (Malloch & Trevarthen, 2009) and preschoolers' musical conducts (Delalande & Cornera, 2010).

But it is important to remember that the perception, identification, and celebration of musical potential are also culturally situated. As discussed earlier, musical potential may be linked to lineage (see Durán, 2017) or based on defining characteristics of the music practiced by each individual culture. As an example, while tambor de crioula players of Maranhão identify drumming, rhythmic improvisation, and endurance as defining features of musical potential and expertise, the ribeirinhos of the Brazilian Amazon believe that singing with a melodious and enchanting voice is the determining factor (Ilari, 2006). These contrasting views are consistent with the notion that musical potential—in childhood and beyond—comes in multiple shapes and forms (Kemp & Mills, 2002).

Skill Development

Children develop musically in leaps and bounds. The fragile newborn enters the world with sophisticated music perceptual abilities (Trehub, 2003), such as the capacity to detect beat violations (Winkler et al., 2009), familiar voices (DeCasper & Fifer, 1980), and melodies (Hepper, 1991). A few weeks later, babies show a remarkable ability to discriminate and categorize pitches and rhythmic structures (Trehub, 2003). Although babies typically do not entrain to the musical beat, rudiments of rhythmic entrainment are already present at the beginning of life (Ilari, 2015). Babies and young children are fast music learners who develop musical skills concomitantly with skills in other areas (e.g., auditory, motor). As children develop and grow, they gradually become more proficient in performing and responding to music in their surroundings (see Kirschner & Ilari, 2014). Importantly, formal music lessons in early childhood may be more the exception than the rule (Young, 2018).

For those who learn music formally, musical development is known to follow a nonlinear path, in the sense that children go through phases of rapid development interspersed with phases of slow or no apparent development (e.g., Ilari et al., 2016). These "ups and downs" of musical development have been interpreted in multiple ways (see, e.g., Delalande & Cornera, 2010; Bamberger, 2013). Another aspect of musical development that has been explored is the notion of music learning plateaus, or periods when the development of specific skills appears to stabilize (Bailey & Penhune, 2013; Ilari et al., 2016).

Studies on the development of skills are obviously linked to researchers' conceptualizations of the musical child. The sociology of childhood invites scholars to reflect on taken-for-granted categories such as infancy, childhood, and development (see Young, 2018). Musical development researchers usually conceptualize childhood in three distinct ways: being ("in the moment"), becoming ("as future adults"), or a combination of both. As an example, in our study of musical improvisation, we uncovered age contrasts in the conceptualization of children's improvised products and processes. We learned that young children's improvisation skills are typically interpreted in a holistic fashion and in light of their overall development, whereas improvisation skills in middle childhood are often compared with those of adults (see Ilari et al., 2018). A challenge for future scholarship is to integrate these contrasting views of childhood with an understanding of development as a biopsychosocial process, with context playing a much larger role than previously thought.

Self-Identification

Despite evidence that music perception, cognition, and engagement begin very early in life (Trehub, 2003), it is not until middle childhood (around age eight) that children start to self-identify as musicians (Rickard & Chin, 2017). Around this time, children's musical skills become differentiated from those of their peers, raising self-awareness (Rickard & Chin, 2017). Although years of musical training are often used as a measure of musicianship, it is interesting that music lessons alone are not enough for children to self-identify as musicians (Rickard & Chin, 2017). Identifying as a musician happens through sustained and engaged participation in music in different settings, such as home and school (Hargreaves & Lamont, 2019). Parents and caregivers not only shape the home musical environment but also play central roles in supporting children's musical potential (Trehub, 2019). Schools and musical programs, in turn, offer opportunities for children to develop musically. Still, the idea of being "unmusical" may haunt many children and adolescents, impeding their development. As Hargreaves and Lamont suggest:

> Although provision is made in schools and other institutions for children who show promise to develop their musical skills, many others start to see themselves as being "unmusical," fail to develop their early potential and follow a downward spiral in which lack of musical self-esteem and motivation leads to lower levels of performance, which leads to still lower self-esteem, and so on. (2017, p. xvii)

Studies focusing on children's representations of music and the musician offer additional insights into their self-identification as musicians and nonmusicians. In one study, Australian children (aged nine to ten years) were asked to write a sentence beginning with "Music is . . ." and then produce a drawing for the prompt "Music and me" (Southcott & Coisatis, 2015). The findings suggested some stark gender differences, with boys drawing more images of themselves involved in solitary music listening using different technologies, and girls representing themselves engaged in collective forms of musicking. Gender differences also emerged in another study of children (aged seven to ten years) from the UK (Colley et al., 2008). In the British study, while younger children drew figures of their own sex, older girls drew more male musicians than did younger girls and boys. The authors interpreted this finding as older girls' awareness of male domination in musical performance. What these findings suggest is that children are already susceptible to social forces as they construct their identities in and through music.

The construction of identities is clearly a complex process. Rickard and Chin (2017) argue that research on musical identity has focused heavily on active music making, often overlooking how musical identities may be shaped by human participation in receptive activities such as listening to music. They have called for a broadening of perspectives on musicianship and for the inclusion of different musical activities such as listening, musical engagement in everyday life, and emotional engagement with music. The examination of activities beyond active music making is particularly relevant where so-called nonmusicians are concerned (Rickard & Chin, 2017).

A final issue is the ubiquity of music in childhood and adolescence. Children are typically more exposed to music than adults are, and adolescents listen to more music than any other age group (Hargreaves & Lamont, 2019). Adolescents are also known to be passionate about music, with musical preferences providing a way to experiment with varied identities (Hargreaves & Lamont, 2019). Interestingly, repertoires that are appreciated in adolescence may leave "imprints" in autobiographical memory and constitute a reminiscence bump later in life (Krumhansl & Zupnick, 2013). Identifying as a musician, therefore, has implications for the development of nonmusical aspects of identity in adolescence and beyond.

Ways Forward: Interdisciplinary Thinking and Research

We began this chapter by outlining Christopher Small's (1998) contribution to music scholarship, particularly his notion of musicking, or music as a form of action. We also highlighted the systems view of development, in which individual attributes (or person), proximal processes, context, and time interact in multiple ways (Bronfenbrenner &

Morris, 2006). Musical development is clearly influenced by the interactions among these factors, as children engage with music in multiple ways. As we argued throughout the chapter, it is impossible to speak of musical potential, musicality, and musicianship in the singular (see also Kemp & Mills, 2002). Musical potentials and opportunities for skill development—from unintentional, less intentional, and informal means of learning to more deliberate and formal means—play significant roles in self-identification as a musician. It is also vital to recognize the roles of individual differences, cultural context, and time ("in the moment" and the zeitgeist) in the construction of musical identities. Acknowledging the many articulations between musical potential, skill development, and self-identification is imperative for future work, particularly developmental studies in which research findings may have direct applications in education and child care. Thus, we believe that the musician-nonmusician binary does little to clarify the musical experiences of infants and children. Apart from being reductionist, the musician-nonmusician binary leaves out central aspects of the relationship between musical experiences and child development, including agency, identity work, and the enjoyment of music. Identity work and musical development are known to be intrinsically linked (Hargreaves & Lamont, 2019). Furthermore, it is important to consider the other end of the developmental spectrum, or late adulthood. Although this chapter focused primarily on childhood and adolescence, we recognize the need for more work on musicking, musicianship, and musicality in older adults to advance our understanding of musical development across the life span.

We conclude this chapter by reinforcing the need for interdisciplinary thinking and collaboration in studies on human musicking, musicality, and musicianship (see also chapter 18). Such an approach is imperative as our world continues to experience the forces of globalization, technological advances, climate change, and human migration. These forces have been influencing human musicking for some time now, as many have been keen to point out. Thus, a reconceptualization of musicians, including the very young, is long overdue and beyond urgent.

References

Bailey, J. A., & Penhune, V. (2013). The relationship between the age of onset of musical training and rhythm synchronization performance: Validation of sensitive period effects. *Frontiers in Neuroscience, 29*(7).

Bamberger, J. (2013). *Discovering the musical mind: A view of creativity as learning.* Oxford University Press.

Bengtsson, S. L., Nagy, Z., Skare, S., Forsman, L., Forssberg, H., & Ullén, F. (2005). Extensive piano practicing has regionally specific effects on white matter development. *Nature Neuroscience, 8*(9), 1148–1150.

Bennett, H. (2017). *On becoming a rock musician.* Columbia University Press.

Blacking, J. (1973). *How musical is man?* University of Washington Press.

Brasil, A. (2020). *Música e periferia: O sonho e o real em um mundo negro chamado Bahia.* Appris.

Bronfenbrenner, U., & Morris, P. A. (2006). The bioecological model of human development. In W. Damon & R. N. Lerner (Eds.), *Handbook of child psychology: Theoretical models of human development* (5th ed., pp. 993–1027). Wiley & Sons.

Brown, R. M., Zatorre, R. J., & Penhune, V. B. (2015). Expert music performance: Cognitive, neural, and developmental bases. *Progress in Brain Research, 217,* 57–86.

Cho, E. (2019). Sensitive periods for music training from a cognitive neuroscience perspective: A review of the literature with implications for teaching practice. *International Journal of Music in Early Childhood, 14*(1), 17–33.

Cogo-Moreira, H., & Lamont, A. (2018). Multidimensional measurement of exposure to music in childhood: Beyond the musician/non-musician dichotomy. *Psychology of Music, 46*(4), 459–472.

Colley, A., Mulhern, G., Relton, S., & Shafi, S. (2008). Exploring children's stereotypes through drawings: The case of musical performance. *Social Development, 18*(2), 464–477.

Daly, H. R., & Hall, M. C. (2018). Not all musicians are created equal: Statistical concerns regarding the categorization of participants. *Psychomusicology: Music, Mind, and Brain, 28*(2), 117–126.

DeCasper, A., & Fifer, W. P. (1980). Of human bonding: Newborns prefer their mothers' voices. *Science, 208*(4448), 1174–1176.

Delalande, F., & Cornera, S. (2010). Sound explorations from the ages of 10 to 37 months: The ontogenesis of musical conducts. *Music Education Research, 3,* 257–268.

Drake, C., Jones, M., & Baruch, C. (2000). The development of rhythmic attending in auditory sequences: Attunement, referent period, focal attending. *Cognition, 77*(3), 251–288.

Durán, L. (2017). Growing into music in Mali: Perspectives on informal learning from West Africa. In N. Economidou & M. Stakelum (Eds.), *European Perspectives on Music Education* (pp. 49–64). Helbling Verlag.

Elpus, K., & Abril, C. (2016). High school music ensemble students in the United States: A demographic profile. *Journal of Research in Music Education, 59*(2), 138–145.

Ericsson, K. A., Krampe, R., & Tesch-Romer, C. (1993). The role of deliberate practice in the acquisition of expert performance. *Psychological Review, 100*(3), 363–406.

Folkestad, G. (2005). Formal and informal learning situations or practices vs. formal and informal ways of learning. *British Journal of Music Education, 23*(2), 135–145.

Gaser, C., & Schlaug, G. (2003). Brain structures differ between musicians and non-musicians. *Neuroimage, 13*(6), 1168.

Geertz, C. (1973). *The interpretation of cultures: Selected essays.* Basic Books.

Gjerdingen, R. (2020). *Child composers in the old conservatoires: How orphans became elite musicians.* Oxford University Press.

Gordon, E. (1971). The source of musical aptitude. *Music Educators Journal, 57*(8), 35–37.

Green, L. (2017). *How popular musicians learn: A way ahead for music education.* Routledge.

Gudmundsdottir, H. (2020). Revisiting singing proficiency in three-year-olds. *Psychology of Music, 48*(2), 283–296.

Habibi, A., Damasio, A., Ilari, B., Sachs, M., & Damasio, H. (2018). Music training and child development: A review of recent findings from a longitudinal study. *Annals of the New York Academy of Sciences, 1423*(1), 73–81.

Hargreaves, D. J. (1986). *The developmental psychology of music.* Cambridge University Press.

Hargreaves, D. J., & Lamont, A. (2019). *The psychology of musical development.* Cambridge University Press.

Heinrich, J. (2020). *The WEIRDest people in the world: How the West became psychologically peculiar and particularly prosperous.* Farrar, Straus & Giroux.

Henley, J. (2017). The musical lives of self-confessed nonmusicians. In R. Mantie & G. D. Smith (Eds.), *Oxford handbook of music making and leisure.* Oxford University Press.

Hepper, P. G. (1991). An examination of fetal learning before and after birth. *Irish Journal of Psychology, 12*(2), 95–107.

Hesmondalgh, D. (2013). *Why music matters.* Wiley Blackwell.

Hill, J. (2018). *Becoming creative: Insights from musicians in a diverse world.* Oxford University Press.

Hofstede, G. (2011). Dimensionalizing cultures: The Hofstede model in context. *Online Readings in Psychology and Culture,* Unit 2.

Honing, H. (2018). *The origins of musicality.* MIT Press.

Ilari, B. (2006, August). Mapping musical development in Brazil: Children's musical practices in Maranhão and Pará. In *Proceedings of ICMPC9.* Bologna.

Ilari, B. (2015). Rhythmic engagement with music in early childhood: A replication and extension. *Journal of Research in Music Education, 62*(4), 332–343.

Ilari, B., Fesjian, C., Ficek, B., & Habibi, A. (2018). Improvised song endings in a developmental perspective: A mixed methods study. *Psychology of Music, 46*(4), 500–520.

Ilari, B., Keller, P., Damasio, H., & Habibi, A. (2016). The development of musical skills of underprivileged children over the course of one year: A study in the context of an El Sistema–inspired program. *Frontiers in Psychology, 7*(62).

Kemp, A., & Mills, J. (2002). Musical potential. In R. Parncutt & G. McPherson (Eds.), *The science and psychology of music performance: Creative strategies for teaching and learning.* Oxford University Press.

Kirschner, S., & Ilari, B. (2014). Joint drumming in Brazilian and German preschool children: Cultural differences in rhythmic entrainment, but no prosocial effects. *Journal of Cross-Cultural Psychology*, *45*(1), 137–166.

Krumhansl, C. L., & Zupnick, J. A. (2013). Cascading reminiscence bumps in popular music. *Psychological Science*, *24*(10), 2057–2068.

Leadbeater, C., & Miller, P. (2004). *The pro-am revolution: How enthusiasts are changing our economy and society*. Demos.

Lehmann, A. C., Gruber, H., & Kopiez, R. (2018). Expertise in music. In K. A. Ericsson, R. R. Hoffman, A. Kozbelt, & A. M. Williams (Eds.), *Cambridge handbooks in psychology. The Cambridge handbook of expertise and expert performance* (pp. 535–549). Cambridge University Press.

Lerner, R. M., & Benson, J. B. (2013). Introduction: Embodiment and epigenesist: A view of the issues. *Advances in Child Development and Behavior*, *44*, 1–19.

Lerner, R. M., Harris, E., & Stacey, D. C. (2018). Developmental science. In M. Bornstein (Ed.), *The Sage encyclopedia of lifespan human development*. Sage.

Malloch, S., & Trevarthen, C. (Eds.). (2009). *Communicative musicality: Exploring the basis of human companionship*. Oxford University Press.

McPherson, G., & Hallam, S. (2008). Musical potential. In S. Hallam, I. Cross, & M. Thaut (Eds.), *Oxford handbook of music psychology*. Oxford University Press.

Moore, J. K., & Linthicum F. H., Jr. (2007). The human auditory system: A timeline of development. *International Journal of Audiology*, *46*(9), 460–478.

Müllensiefen, D., Gingras, B., Musil, J., Stewart, L., & Snyder, J. (2014). The musicality of non-musicians: An index for assessing musical sophistication in the general population. *PLOS ONE*, *9*(2), e8964–e89642.

Penhune, V. B. (2011). Sensitive periods in human development: Evidence from musical training. *Cortex*, *47*(9), 1126–1137.

Peretz, I., Gosselin, N., Nan, Y., Caron-Caplette, E., Trehub, S. E., & Béland, R. (2013). A novel tool for evaluating children's musical abilities across age and culture. *Frontiers in Systems Neuroscience*, *7*(30).

Rickard, N. S., & Chin, T. (2017). Defining the musical identity of "non-musicians." In R. MacDonald, D. J. Hargreaves, & D. Miell (Eds.), *Handbook of musical identities* (pp. 288–303). Oxford University Press.

Rogoff, B., Moore, L., Najafi, B., Dexter, A., Correa-Chávez, M., & Solís, J. (2007). Children's development of cultural repertoires through participation in everyday routines and practices. In J. E. Grusec & P. D. Hastings (Eds.), *Handbook of socialization* (pp. 490–515). Guilford Press.

Sloboda, J. A. (2005). *Exploring the musical mind: Cognition, emotion, ability, function*. Oxford University Press.

Sloboda, J. A., & Howe, M. (1991). Biographical precursors of musical excellence: An interview study. *Psychology of Music, 19,* 3–21.

Small, C. (1998). *Musicking: The meanings of performing and listening.* Wesleyan University Press.

Southcott, J., & Coisatis, W. (2015). Drawing "music and me": Children's images of musical engagement. *Australian Journal of Music Education, 2,* 78–90.

Tardif, C. L., Gauthier, C. J., Steele, C. J., Bazin, P. L., Schäfer, A., Schaefer, A., . . . Villringer, A. (2016). Advanced MRI techniques to improve our understanding of experience-induced neuroplasticity. *Neuroimage, 131,* 55–72.

Trehub, S. E. (2003). The developmental origins of musicality. *Nature Neuroscience, 6*(7), 669–673.

Trehub, S. E. (2019). Nurturing infants with music. *International Journal of Music in Early Childhood, 14*(1), 9–15.

Turino, T. (2008). *Music as social life: The politics of participation.* University of Illinois Press.

Watson, C. (2012). *How musicians develop the ability to improvise: A cross-cultural comparison of skill development in Egyptian, Hindustani classical and jazz traditions* [Unpublished doctoral dissertation]. University of New South Wales.

Winkler, I., Haden, G. P., Ladinig, O., Sziller, I., & Honing, H. (2009). Newborn infants detect the beat in music. *PNAS, 106*(7), 2468–2471. https://doi.org/10.1073/pnas.0809035106.

Young, S. (2018). *Critical new perspectives in early childhood music: Young children engaging and learning through music.* Routledge.

Zatorre, R. J., Chen, J. L., & Penhune, V. B. (2007). When the brain plays music: Auditory-motor interactions in music perception and production. *Nature Reviews Neuroscience, 8*(7), 547–558.

Zentner, M., & Gingras, B. (2019). The assessment of musical ability and its determinants. In J. Rentfrow & D. Levitin (Eds.), *Foundations in music psychology: Theory and research* (pp. 641–683). MIT Press.

Zhang, J. D., Susino, M., McPherson, G., & Schubert, E. (2020). The definition of a musician in music psychology: A literature review and the 6-year rule. *Psychology of Music, 48*(3), 403–409.

18 Building Sustainable Global Collaborative Networks: Recommendations from Music Studies and the Social Sciences

*Patrick E. Savage, *Nori Jacoby, *Elizabeth H. Margulis, Hideo Daikoku, Manuel Anglada-Tort, Salwa El-Sawan Castelo-Branco, Florence Ewomazino Nweke, Shinya Fujii, Shantala Hegde, Hu Chuan-Peng, Jason Jabbour, Casey Lew-Williams, Diana Mangalagiu, Rita McNamara, Daniel Müllensiefen, Patricia Opondo, Aniruddh D. Patel, and Huib Schippers

*These authors contributed equally.

Introduction

Diversity is one of the key challenges facing many societies in the twenty-first century. In scholarly research, this challenge has crystallized around the overrepresentation of and overreliance on societies that are WEIRD: Western, educated, industrialized, rich, and democratic (Henrich et al., 2010). This acronym has been critiqued (Clancy & Davis, 2019; Barrett, 2020), and even its creators emphasize that WEIRD is a rhetorical device not intended to suggest a binary opposition with non-WEIRD (Apicella et al., 2020; Muthukrishna et al., 2020). However, the acronym has become popular for framing issues of inclusion and representation in academia. In music studies, this involves the historical overrepresentation of music by European classical composers and the overrepresentation of undergraduate students at Western universities in participant samples (Thompson et al., 2019; Jacoby et al., 2020; Savage, in press).

These issues have gained visibility within the mainstream, particularly following calls for decolonial research approaches (Mignolo, 2011) and the rise of the Black Lives Matter movement. Efforts to decolonize music studies and make them more inclusive and equitable (e.g., Ewell, 2020; Brown, 2020; Iyer & Born, 2020; Diamond & Castelo-Branco, 2021; Sauvé et al., 2021) have been covered by the *New York Times* (Powell, 2021) and *Fox News* (Betz, 2020). In the United States, they have triggered important changes at the highest levels of the organizational structure of the Society for Ethnomusicology (SEM) and prompted the board of the Society for Music Perception and Cognition to publish an antiracism statement (Baker et al., 2020). These changes are part of a broader, long-term international trend, as evidenced by the International Council for Traditional Music (2021) issuing a statement on the topic and instituting a year-long series of dialogues about the decolonization of music and dance studies.

A number of music science publications have highlighted both the momentum for change and the challenges that remain. For example: Thirty-five authors published the results of rhythm perception experiments involving 923 participants from thirty-nine participant groups in fifteen countries (Jacoby et al., 2021); nineteen authors published analyses of 4,709 ethnographic documents and 118 audio recordings of music from around the world (Mehr et al., 2019); eighteen authors published a global database of performing arts, including analyses of 5,779 songs from 992 societies (Wood et al., 2021); and twenty authors published a critical discussion of the challenges of and potential for cross-cultural work in music cognition (Jacoby et al., 2020). Some have praised the ambition of these global multidisciplinary collaborations, but others have voiced concern that they may actually reinforce preexisting power structures and hierarchies through the overrepresentation of authors from well-funded science programs in elite Euro-American universities and through the use of scientific methods to identify potential cultural "universals" (see Russonello, 2017; Rasmussen & Cowdery, 2018; Savage, 2018; Yong, 2018; Woo, 2019; Jacoby et al., 2020; Loughridge, 2021; Sauvé et al., 2021).

Similar challenges are shared by fields outside of music studies, which have also grappled with the WEIRD concept, its relationship to race and racism (Clancy & Davis, 2019), and the related issue that WASP (Western, academic, scientific, psychology) researchers tend to be overrepresented in cross-cultural research (Sinha, 2002; Berry, 2015). Social science fields such as anthropology, economics, and psychology are already making progress on practical solutions to enable sustainable global collaborative research (e.g., Henrich et al., 2005; Banerjee et al., 2015; Jabbour & Flachsland, 2017; Purzycki et al., 2022; Moshontz et al., 2018; Broesch et al., 2020; Byers-Heinlein et al., 2020; Urassa et al. 2021; Parker & Kingori, 2016; Barrett, 2020; Haelewaters et al., 2021).

The aim of this chapter is to provide concrete recommendations for moving beyond the traditional overreliance on Western music and musicians and toward sustained collaborations that include members of diverse societies throughout the world as equal partners in shared research practices and as part of an ecology of knowledge (de Sousa Santos, 2007; Sardo, 2017; Schippers & Grant, 2016). These recommendations are not intended to be onerous, prescriptive rules but rather suggestions to encourage progress and create excitement about future opportunities. Our goal is not to discourage cross-cultural research that doesn't follow these recommendations but rather to encourage more research and provide practical guidance to help realize this goal.

Based on the lessons of an earlier symposium focused on bridging ethnomusicology and music cognition (Jacoby et al., 2020), the first three authors of this chapter (PES, NJ, and EHM) organized a symposium entitled "Building Sustainable Global Collaborative Research Networks," with a goal of attracting global participants and a desire

Figure 18.1
The twenty-three participants at the February 7, 2021, virtual symposium "Building Sustainable Global Collaborative Research Networks" (https://www.ae.mpg.de/glo-co).

to learn best practices from fields outside of music studies. This symposium featured a group of twenty-three researchers and practitioners whose expertise was roughly equally distributed among (ethno)musicology, music cognition, and other social sciences (figure 18.1).[1]

Prior to the symposium, invitees were asked to submit ideas and resources related to best practices. After a careful review by the symposium organizers, four overarching themes emerged: (1) diversity, inclusion, and equity; (2) logistics; (3) reproducibility and standardization versus cultural specificity; and (4) incentives, attribution, and leadership. Participants discussed these ideas in groups of five to six people. The following sections synthesize and summarize our shared conclusions about best practices for each of these four key themes.

Diversity, Inclusion, and Equity

How do we enhance representation in global collaborations? The importance of diversity is widely recognized, but achieving inclusive and equitable representation in global collaborations is easier said than done. Many of the documents cited by ourselves and others that emphasize diversity in cross-cultural research were coauthored mostly or entirely by researchers from elite Euro-American universities (e.g., Broesch et al., 2020; Jacoby et al., 2020). Such imbalance reflects a variety of power structures, including extractive legacies of colonialism as well as practical barriers of language, politics, economics, and disciplinary conventions. Overcoming such legacies and barriers requires rethinking research methods that many of us have come to take for granted. It requires affirmative action to compensate for historical power imbalances and underrepresentation, and it requires us to ensure that research goals serve the interests of both the researchers and the communities. Many companies, governments, academic societies, indigenous communities, and other organizations have begun to develop best-practices guidelines for diversity and inclusion, and these can vary substantially, depending on the specific goals and needs of the organization (e.g., Kirkness & Barnhardt, 1991; Awesti et al., 2016; Boatright et al., 2018; Chambers et al., 2017; Laland et al., 2022; Nature Editors, 2020; Muru-Lanning, 2020). Below we outline some best practices specifically tailored to build sustainable global collaborative networks.

Goals and leadership: The goals of the collaboration should be aligned with the needs of the communities involved. The best way to accomplish this alignment is to ensure that the relevant communities' voices are heard and reflected at the highest levels and earliest stages of a planned collaboration, ideally by involving representatives of these communities in the initial decision making. The benefits of early collaborative planning need to be balanced against the realities that (1) involving too many people can reduce the ability to make high-level decisions efficiently, and (2) members of underrepresented communities are often disproportionately burdened with requests to represent that community. Such constraints can result in unintended negative consequences, such as incentivizing the inclusion of so-called diverse members in ways that "tick the boxes" for diversity on paper only. At a minimum, we recommend identifying and recruiting stakeholders representing diverse communities at all levels, beginning at the outset of a project and proceeding through shared research practices. To facilitate such recruitment, some organizations have created informal lists (e.g., a list of evolutionary human sciences researchers belonging to underrepresented minority groups: https://diversifyehs.mystrikingly.com/) or formal networks (e.g., the International Council for Traditional Music [ICTM] network of music and dance researchers from more than 100 countries: http://ictmusic.org/world-network). These lists and networks should be combined with discussions with local stakeholders, and care should be taken to ensure that power imbalances are not perpetuated locally (e.g., research on lower-caste musicians performed exclusively with higher-caste local collaborators).

Interdisciplinary collaboration: Global research collaborations are often driven by the interests and funding of scientists, sometimes at the expense of researchers from the humanities or members of the public outside of academia. This is particularly pertinent when the focus is a research area, such as music, that involves practices unfamiliar to the scientific community. Thus, it is essential that researchers sincerely engage with other methods and values. For example, qualitative and quantitative methodologies should be considered, as well as research outputs such as artistic performances, community workshops, and blog posts, in addition to academic outputs such as peer-reviewed journal articles. For such outputs, it is important to negotiate issues such as authorship and compensation early on. In some societies or disciplines, having one's name listed as a coauthor on a scientific article has little value compared to being paid as a consultant or research assistant, while in others, the reverse may be true. Even within scientific communities, norms regarding authorship and attribution are heterogeneous and rapidly evolving. Describing the process taken to involve local researchers and advisers can be helpful, regardless of whether they are acknowledged as coauthors (Thompson et al., 2019).

We recommend recognizing contributions to research networks through both financial (e.g., consultation fees, grants, experiment costs) and intellectual (e.g., coauthorship, author contribution statements, named acknowledgments) mechanisms, as well as ensuring access to, and credit for, research-related outputs (e.g., making archival or field recordings available to community members or providing high-quality video recordings for musicians to use in their own promotional materials). For example, a project measuring global diversity in rhythm perception led by one of us (NJ; Jacoby et al., 2021) includes thirty-four authors—scientists, (ethno)musicologists, and musicians—from fifteen countries (Germany, Austria, Sweden, USA, UK, Canada, Japan, South Korea, China, Chile, Bolivia, Uruguay, Mali, India, Turkey) and names fifty-one individuals and organizations in the acknowledgments.

It is imperative to ensure that the recognition received by researchers and participants is specifically of value to them. This does not necessarily mean formal coauthorship (e.g., Araújo & Cambria, 2013; Miguel, 2018). For example, in a project exploring musical diversity in India (Daikoku et al., 2020), the graduate student leading the project (our coauthor HD) is from India and receives both financial support (a stipend and tuition funded by Yamaha) and intellectual credit (first authorship). He is working with musicians in India to take music lessons and conduct interviews and experiments, and these musicians receive financial compensation but not coauthorship. It is also important to recognize that in some communities it may be considered inappropriate to explicitly discuss such rewards; as always, these suggestions should be adapted to the norms of the local context.

Language, geography, and accessibility: The current concentration of academic power in English-speaking countries incentivizes us to organize events and collaborations in

English, such as the symposium that led to this chapter (which is also written in English). This marginalizes members of non-Anglophone communities and creates barriers to their inclusion in global research networks. Providing travel funding can minimize some economic barriers but does not solve other problems, such as language barriers, visa restrictions, and other factors that can limit participation. While it may seem inefficient to hold meetings in other countries using languages other than English, given the additional costs for travel, translation, and the like, these short-term costs are necessary to build long-term sustainability. ICTM is an example of an academic society that has successfully organized in-person world conferences in diverse countries with multilingual translation (e.g., South Africa in 2009, Kazakhstan in 2015, China in 2018, Thailand in 2019 featuring papers in English and the local language), as well as virtual events in English and other languages (e.g., "ICTM Dialogues 2021: Towards Decolonizing Music and Dance Research"; ICTM, 2021). The rapid normalization of virtual participation due to the COVID-19 pandemic may help reduce barriers and costs associated with travel, but it will not solve language issues and may create other imbalances.

To actively reduce such barriers to participation, we recommend organizing events in diverse geographic locations using diverse languages, providing opportunities for translation, and making virtual participation as accessible as possible (e.g., for participants with disabilities, those with caregiver obligations). This may go beyond the literal translation of language to include the conceptual translation of ideas, which may need to be entirely rethought and reformulated in terms that are more relevant to the participant communities.

Logistics

How can we minimize logistical challenges in global collaborations? Even research within a single society involves substantial logistical challenges, and these are amplified drastically when conducting global collaborative research. Different societies have different rules, norms, and institutional structures. Collaborating in a meaningful way therefore requires careful planning, including considerations such as organizational structure, funding, and ethical review.

Project management: Building and sustaining a global collaborative network requires a clear leadership structure that balances flexibility and agency for individual researchers and labs in different societies with a clear, unifying vision and strategy. In service of this aim, we recommend balancing top-down (e.g., standardized protocols) and bottom-up (e.g., local adaptations) approaches to project management. For example, the Evolution of Morality Project developed standardized and validated methods of measuring cooperation and morality cross-culturally but adapted these methods according to the relevant belief systems of the fifteen societies investigated (Purzycki et al., 2022). Likewise, Jacoby et al. (2021) provided researchers in different societies with standardized, thoroughly piloted, and well-documented equipment and protocols for rhythm experiments, but they partially delegated decisions about translation and appropriate participant sampling to local researchers (while maintaining a consultation role to ensure that the sampling rationale remained consistent across societies). The equipment was designed to be portable and flexible, allowing researchers to conduct experiments in remote areas with unreliable infrastructure.

Funding: Funding logistics can be particularly complicated for global collaborations. Economic and geopolitical power imbalances mean that some countries offer more funding than others, and they may limit the countries to which funds can be transferred. The complexity of global collaborations often requires retracing steps and pivoting to different approaches at key junctures in the research, making it challenging to specify and follow long-term funding timelines. In addition, extra funding needs may arise that are difficult to fully anticipate at the time of funding applications. For example, the ManyBabies Consortium (Byers-Heinlein et al., 2020), in an ongoing collaboration with scientists from various nations in Africa, did not originally budget for institutional review board (IRB) fees, which are not usually charged in Western universities but are common in some academic communities. After obtaining initial funding, they learned that many collaborators would need to pay the equivalent of US$500 to each of their institutions. This unanticipated expense impacted other components of the project's budget. We recommend identifying collaborators prior to writing grants and then jointly crafting detailed budgets that accommodate the range of expenses involved in global collaborations.

Ethical review: Many of the logistical issues involved in global collaborations intersect with ethical issues related to disparities across different sites. These range from specific practical issues (e.g., compensation, data management, and anonymity of participants) to more general ones, such as how to ensure the research is helping and not hurting the local community. IRBs are a formal mechanism for evaluating such issues, although they have been criticized for being "more concerned with protecting the institution than research participants" (Grady, 2010). However, if they are well stewarded, IRBs can clarify the rights and obligations of everyone involved in the project early on, avoiding unfortunate situations later. For example, of the approximately 6,000 music recordings at the Global Jukebox (Wood et al., 2021), about 1,000 from indigenous groups in North America and Australia will not be available for listening until time-consuming negotiations with representatives of each individual group have been completed. These problems might have been avoided if such issues had been clarified in IRB protocols at the beginning of the project (although the Global Jukebox project began more than half a century ago, before IRB input had become standard research practice). For societies without local IRBs, we recommend applying the highest standards to protect the rights of participants and avoid ethics dumping (Schroder et al., 2018; Nature Editors, 2022). It is also crucial to realize that the principles of IRBs may not be appropriate in all cultures, such as a hierarchy or consensus culture that makes individual consent meaningless. As in the "project management" section above, we recommend adopting a combined top-down–bottom-up approach in which general IRB protocols are prepared in consultation with diverse team members, and these standardized protocols are then adapted to local institutions as necessary.

Accessibility: Setting up easily accessible online data collection can reduce the logistical costs of traveling to remote areas, especially when travel is not physically possible (e.g., during the COVID-19 pandemic). Some communities benefit from the ability to use mobile phones, where experiments, questionnaires, and the like can be implemented using responsive web-based applications. These methods are sometimes insufficiently embedded in the cultural context and don't allow much control over the conditions in which the experiment takes place. However, such concerns can be mitigated by taking appropriate precautions (e.g., prescreening tasks, data quality checks, bonus payments). This is true even for highly controlled experiments, such as infant research (Tran et al., 2017), language production tasks (Vogt et al., 2021), or iterated tapping experiments (Jacoby et al., 2021). Online data collection has had considerable success (Kohavi & Thomke, 2017) and is likely to become a mainstay of research methodology. We recommend including online options when feasible to enhance accessibility and diversity.

Reproducibility and Standardization

How can we ensure meaningful, repoducible, and standard comparisons on a global scale?

Reproducibility: Increasingly popular open science practices enhance transparency and reproducibility through the free sharing of data, analysis code, stimuli or protocols, preregistered hypotheses, and research reports via repositories such as Open Science Framework, Github, Zenodo, arXiv, and related preprint servers. However, the need to preserve and promote diversity sometimes works against this tendency toward openness and standardization. Many historically underrepresented minorities are wary of having their personal data documented and shared in forms they cannot control, given the atrocities and humiliation they have experienced in the name of science

(Brandt, 1978; van Noorden, 2020). Here again, we recommend using the IRB process as an opportunity to specify data-sharing procedures and grapple with the associated ethical considerations early on. In addition, we recommend prioritizing the sharing of stimuli, protocols, and analysis code, even when sharing participants' data is more complex.

Standardization and translation: Even when diverse participants provide informed consent, cross-cultural differences mean that standardized research metrics, such as IQ, can at best be considered meaningful only when interpreted cautiously and at worst can be meaningless or actively harmful (Pawlowski et al., 2020). The same caveat likely applies to attempts to measure other things that lack cross-culturally universal definitions, such as "music" (Savage, 2019), "musical sophistication" (Müllensiefen et al., 2014), or "musical IQ" (Neely, 2020). Building sustainable global collaborations requires us to constructively address such challenges. We recommend collaborating with local researchers to develop, translate, and adapt questionnaires, experimental stimuli, and protocols to ensure that the resulting data can be used to make meaningful comparisons. The limitations of existing inventories should be acknowledged, and the possibility of completely reframing ideas from alternative perspectives rather than simply translating them should be explored (cf. Harris, 1976). When possible, subjective self-report measures (e.g., daily practice time) should be combined with objective measures (e.g., beat synchronization; Müllensiefen et al., 2014). Although no index for terms such as *musicality* and *musician* will ever be perfect, we believe that creating indices that are more thoroughly cross-culturally validated than existing ones is a constructive goal. Promising steps have already been made through cross-cultural collaborations (e.g., a Chinese translation of the Gold-MSI musicality index; Lin et al., 2019). (For discussion and critical analyses of these concepts, see the chapters in this volume by Patel, Mundy, and Ilari and Habibi.)

One possibility is to aim for comparability at the conceptual level of the latent construct to be measured in different cultures. For example, for the construct "musical expertise," researchers from different musical cultures might agree that measurement on a unidimensional scale ranging from low to high would be meaningful. Once this is agreed on, it might be possible to create different inventories with questions specific to each culture, thus measuring the same construct by asking different questions. Similarly, researchers might agree that a specific musical skill (e.g., intonation accuracy) is important in their cultures. Several different versions of a perceptual test could then be developed, each version using stimuli that are culturally meaningful to each culture, and each version being validated with a sample of participants from the corresponding musical culture. Scores of the task could be made comparable by using a scoring metric that does not depend on the individual test items (e.g., item response scoring or Rasch modeling). Such efforts should take advantage of existing methods for establishing comparability of scale or questionnaire-based measures across groups, using techniques such as measurement invariance (e.g., Fischer, 2004; Chen, 2008; Fischer & Poortinga, 2018; Boer et al., 2018; Jeong & Lee, 2019).

Sampling: A core scientific principle is that the sample population in a given study should be representative of the population to which the conclusions will be generalized. But given the extreme cultural diversity within and between populations, what does it mean to say that one group of humans is "representative" of another? The WEIRD problem described earlier is increasingly recognized as a major sampling limitation, but simply sampling from non-WEIRD societies does not solve the problem and may in fact exacerbate it (e.g., if the groups are essentialized in misleading ways). Such overly simplistic approaches also risk failing to acknowledge the massive diversity within societies and failing to capture the full range of human cultural diversity (Barrett, 2020). There are also major, theoretically relevant differences within a given country or society, such as age, gender, race, or musicianship (Taras et al., 2009). Controlling for all these variables in cross-cultural research is often impossible, but we recommend documenting and justifying sampling and inclusion criteria to increase the generalizability and reproducibility of a given study and to prevent overinterpretation. For example, because Jacoby et al. (2021) were interested in cross-cultural diversity in rhythm perception, they recruited participants with extensive training in local non-Western musical traditions, as well as two types of control participants with similar demographics who had either training in Western musical traditions or no formal musical training. Ultimately, it is impossible to control for all demographic factors, but acknowledging such limitations is an important part of sustainability. Concepts such as "cultural distance" (Muthukrishna et al., 2020) may be useful to control for cross-cultural similarities and differences (such approaches can, in principle, simultaneously address diversity within and between societies; cf. Rzeszutek et al., 2012, for a musical example).

Incentives, Attribution, and Leadership

How can we design systems that will promote sustainable global collaborations? Many of the barriers to sustainability stem from the systemic nature of research incentive systems such as publication, funding, and hiring practices. While such global systems cannot be easily changed, a number of strategies may help researchers work effectively within them while gradually increasing their equity and sustainability.

Leadership and credit: Historical research assessment systems emphasizing first-author or sole-authored academic publications disincentivize truly interdisciplinary and global collaborations, which require a sustained investment from several individuals from multiple disciplines. It can be hard to interest researchers in collaborations if their names will end up in the middle of the author list, where evaluation committees see little value. Within the current system, effective strategies for incentivizing collaboration include negotiating financial, intellectual, and data-sharing mechanisms that allow coauthors to receive credit for aspects of the project. For example, local researchers can be given priority data access or first or shared first authorship on related papers (e.g., journal special issues, edited volumes; cf. Henrich et al., 2004; Apicella et al., 2020) based on the same data. Financial incentives can also help, such as paying consulting fees upon the completion of data collection. Ultimately, however, solving these problems will require a fundamental reevaluation of the nature of research credit attribution (cf. Kiser, 2018; Holcombe, 2019).

Multidisciplinarity: Effective global collaborations require researchers to work across disciplines within academia and to work with local communities, government funders, nongovernmental organizations, private industry, and other nonacademic stakeholders. Communicating across disciplines and across diverse stakeholders is challenging, and it can take extra time to ensure that everyone feels included and valued and believes the collaboration is in their own interest. Nevertheless, we recommend developing shared research practices that include and synthesize the diverse value systems of community stakeholders to maximize long-term sustainability (Sardo, 2017). For example, the 2018 workshop that preceded our 2021 symposium involved multiple days of long and sometimes heated discussions between ethnomusicologists and music cognition researchers (Jacoby et al., 2020). Ultimately, though, it led to greater interdisciplinary goodwill and collaborative spirit, as well as the realization that important voices were missing from the discussion, an omission the organizers of our follow-up symposium attempted to address. At that symposium, we discussed the lack of voices of musicians and performers and the need to accommodate the different goals and incentives of performers and academics. Such iterated dialogues will be necessary to facilitate sustainable long-term collaborations.

Intergenerational sustainability: Building sustainable global collaborations is a long-term goal that requires long-term strategies. By adopting the recommendations listed here, we can build infrastructures and systems to make global collaborations easier over time, as existing networks grow and stimulate additional funding and opportunities for members of underrepresented communities to become involved. To ensure such long-term intergenerational sustainability, we recommend that senior members actively recruit and incentivize junior members from diverse backgrounds. This can include recruiting and securing funding for graduate students and postdocs from developing countries, coauthoring grant applications led by researchers at institutions in these countries, and creating incentives to encourage and reward the next generation of researchers for investing in global collaborations.

Conclusion

Box 18.1 condenses and summarizes the fourteen key recommendations provided in this chapter. These recommendations are ambitious, and we have all failed to achieve them in the past. For example, having three white researchers organize the symposium that led to this chapter and using only English violated recommendations 1.1 and 1.3. But the perfect should not be the enemy of the good. We also believe it's important to learn from past failures and to set goals that may not be attainable but should nevertheless be strived for. As stated earlier, these recommendations are not intended to be onerous, prescriptive rules; rather, they are meant to encourage progress and create excitement about future opportunities.

The 2021 symposium participants did not represent any formal consortium and did not plan any joint projects with this group as a whole. Rather, they were invited to attend and accepted that invitation based on their shared interests and experiences in cross-cultural research and their unique perspectives. In choosing invitees, the organizers attempted to balance representation across multiple dimensions, including gender, ethnicity, geography, seniority, and discipline. We were not attempting to establish an exclusive power clique but rather to invite and encourage anyone who shared similar ideas and interests to do cross-cultural research.

We recognize that our list of recommendations reflects our own priorities and experiences, which have been shaped by our backgrounds as researchers in music studies and the social sciences. These may not necessarily reflect the full range of recommendations and priorities we might have come up with had we included an even more diverse range of stakeholders (e.g., representatives from community interest groups, professional artists, corporations, government). We hope to include and learn from such perspectives and voices in the future.

We hope that by the time the next generation is organizing similar symposia, these recommendations will seem so obvious as to be hardly worth stating. We look forward to seeing future developments toward equitable and sustainable global research collaborations.

Box 18.1
Fourteen key take-home recommendations

1. **Diversity: How can we enhance representation in global collaborations?**
   - 1.1. Identify and recruit stakeholders representing diverse communities at all levels of organization and all stages of a project.
   - 1.2. Recognize contributions to research networks by means of financial (e.g., consultation fees) and intellectual (e.g., coauthorship) mechanisms. Ensure access to, and credit for, research-related outputs (e.g., audiovisual recordings).
   - 1.3. Organize events in diverse geographic locations using diverse languages, providing accessible options for translation and virtual participation.
2. **Logistics: How can we minimize logistical challenges in global collaborations?**
   - 2.1. Balance top-down (e.g., standardized protocols) and bottom-up (e.g., local adaptations) approaches to project management.
   - 2.2. Work with a diverse team to draft detailed but flexible budgets that can accommodate the expenses involved in global collaborations.
   - 2.3. Prepare standardized IRB protocols in consultation with diverse team members, and adapt these standardized protocols to local institutions as necessary.
   - 2.4. Include online options for data collection when feasible to enhance accessibility.
3. **Comparison: How can we ensure meaningful, reproducible comparisons on a global scale?**
   - 3.1. Use the IRB review process to specify data-sharing procedures and associated ethical considerations early on.
   - 3.2. Collaborate with local researchers to develop, translate, adapt, and reframe questionnaires, experimental stimuli, and protocols.
   - 3.3. Document and justify sampling and inclusion criteria.
4. **Incentives: How can we design systems that will promote sustainable global collaborations?**
   - 4.1. Negotiate financial, intellectual, and data-sharing mechanisms that allow coauthors to receive appropriate credit.
   - 4.2. Develop shared research practices that include and synthesize the value systems of diverse stakeholders.
   - 4.3. Encourage senior members to actively recruit and incentivize junior members from diverse backgrounds.
   - 4.4. Fundamentally reevaluate the nature of research credit attribution.

Author Contributions

PES, JN, and EHM conceived, planned, and organized the symposium, and PES, JN, EHM, and SF obtained funding to hold it. All the authors contributed recommendations and ideas prior to and during the symposium and took part in discussing and editing the resulting manuscript. EHM drafted the initial outline, PES wrote the first full draft, and NJ and EHM edited this draft before circulating it to the other authors. HD and MA-T assisted with symposium preparation and logistical support. Other authors are listed in alphabetical order.

Acknowledgments

We thank Dana Bevilacqua for the extensive initial planning for our symposium entitled "Building Sustainable Global Collaborative Research Networks" (https://www.ae.mpg.de/glo-co). We thank Joseph Henrich, David Huron, Aprille Knox, Josh McDermott, and Xavier Serra for their participation at the symposium and Dorsa Amir, who contributed initial ideas but was unable to attend the symposium after it was postponed. We thank Bill Thompson, Alyssa Crittenden, Gabe Zuckerberg, Jim Sykes, Psyche Loui, Deirdre Loughridge, and participants at the "Science-Music Borderlands" Zoom workshop on May 21, 2021, for their feedback on a previous draft of this chapter and discussion of other chapters in this volume.

The "Building Sustainable Global Collaborative Research Networks" symposium was funded by a Japan Society for the Promotion of Science Grant-in-Aid (Grant No. 19KK0064) and a Keio Global Research Institute Startup Grant. The "Science-Music Borderlands" workshop received funding from Princeton University and Northeastern University. PES and HD are supported by funding from the Yamaha Corporation.

Note

1. The symposium was originally intended to be a two-day in-person gathering at the Max Planck–NYU Center for Language, Music, and Emotion in New York City on March 15–16, 2020, but had to be modified due to the COVID-19 pandemic. It was eventually held virtually on February 7, 2021.

References

Apicella, C., Norenzayan, A., & Henrich, J. (2020). Beyond WEIRD: A review of the last decade and a look ahead to the global laboratory of the future. *Evolution and Human Behavior, 41*(5), 319–329.

Araújo, S., & Cambria, V. (2013). Sound praxis, poverty, and social participation: Perspectives from a collaborative study in Rio de Janeiro. *Yearbook for Traditional Music, 45*, 28–42.

Awesti, A., Flinders, M., & Savigny, H. (2016) Pursuing the diversity and inclusion agenda: The PSA in the UK. *European Political Science, 15*(4), 508–518.

Baker, D. J., Belfi, A., Creel, S., Grahn, J., Hannon, E., Loui, P., Margulis, E. H., Schachner, A., Schutz, M., Shanahan, D., & Vuvan, D. T. (2020). Embracing anti-racist practices in the music perception and cognition community. *Music Perception, 38*(2), 103–105.

Banerjee, A., Duflo, E., Goldberg, N., Karlan, D., Osei, R., Parienté, W., Shapiro, J., Thuysbaert, B., & Udry, C. (2015). A multifaceted program causes lasting progress for the very poor: Evidence from six countries. *Science, 348*(6236). https://doi.org/10.1126/science.1260799.

Barrett, H. C. (2020). Towards a cognitive science of the human: Cross-cultural approaches and their urgency. *Trends in Cognitive Sciences, 24*(8), 620–638.

Berry, J. W. (2015). Global psychology: Implications for cross-cultural research and management. *Cross Cultural Management: An International Journal, 22*(3), 342–355.

Betz, B. (2020, August 4). Music faculty at Texas college stokes ire for suggesting music theory isn't white supremacist. *Fox News.*

Boatright, D., Branzetti, J., Duong, D., Hicks, M., Moll, J., Perry, M., Pierce, A., Samuels, E., Smith, T., Angerhofer, C., & Heron, S. (2018). Racial and ethnic diversity in academic emergency medicine: How far have we come? Next steps for the future. *AEM Education and Training, 2,* S31–S39.

Boer, D., Hanke, K., & He, J. (2018). On detecting systematic measurement error in cross-cultural research: A review and critical reflection on equivalence and invariance tests. *Journal of Cross-Cultural Psychology, 49*(5), 713–734.

Brandt, A. M. (1978). Racism and research: The case of the Tuskegee syphilis study. *Hastings Center Report, 8*(6), 21–29.

Broesch, T., Crittenden, A., Beheim, B., Blackwell, A., Bunce, J., Colleran, H., Hagel, K., Kline, M. A., McElreath, R., Nelson, R., Pisor, A., Prall, S., Pretelli, I., Purzycki, B. G., Quinn, E., Ross, C., Scelza, B., Starkweather, K., Stieglitz, J., & Borgerhoff Mulder, M. (2020). Navigating cross-cultural research: Methodological and ethical considerations. *Proceedings of the Royal Society B Biological Sciences, 287.* https://doi.org/10.1098/rspb.2020.1245.

Brown, D. (2020, June 12). An open letter on racism in music studies: Especially ethnomusicology and music education. *My People Tell Stories.* https://www.mypeopletellstories.com/blog/open-letter.

Byers-Heinlein, K., Bergmann, C., Davies, C., Frank, M. C., Hamlin, J. K., Kline, M., Kominsky, J. F., Kosie, J. E., Lew-Williams, C., Liu, L., Mastroberardino, M., Singh, L., Waddell, C. P. G., Zettersten, M., & Soderstrom, M. (2020). Building a collaborative psychological science: Lessons learned from ManyBabies 1. *Canadian Psychology/Psychologie canadienne, 61*(4), 349–363.

Chambers, D., Preston, L., Topakas, A., de Saille, S., Salway, S., Booth, A., Dawson, J., & Wilsdon, J. (2017). Review of diversity and inclusion literature and an evaluation of methodologies and metrics relating to health research. https://wellcome.org/sites/default/files/review-of-diversity-and-inclusion-literature.pdf.

Chen, F. F. (2008). What happens if we compare chopsticks with forks? The impact of making inappropriate comparisons in cross-cultural research. *Journal of Personality and Social Psychology, 95*(5), 1005–1018.

Clancy, K. B. H., & Davis, J. L. (2019). Soylent is people, and WEIRD is white: Biological anthropology, whiteness, and the limits of the WEIRD. *Annual Review of Anthropology, 48*, 169–186.

Daikoku, H., Wood, A. L. C., & Savage, P. E. (2020). Musical diversity in India: A preliminary computational study using Cantometrics. *Keio SFC Journal, 20*(2), 34–61.

de Sousa Santos, B. (2007) Beyond abyssal thinking: From global lines to ecologies of knowledge. *Review, 30*(1), 45–89.

Diamond, B., & Castelo-Branco, S. E.-S. (Eds.). (2021). *Transforming ethnomusicology*. Vol. 2. *Political, social & ecological issues*. Oxford University Press.

Ewell, P. A. (2020). Music theory and the white racial frame. *Music Theory Online, 26*(2).

Fischer, R. (2004). Standardization to account for cross-cultural response bias: A classification of score adjustment procedures and review of research in JCCP. *Journal of Cross-Cultural Psychology, 35*(3), 263–282.

Fischer, R., & Poortinga, Y. H. (2018). Addressing methodological challenges in culture-comparative research. *Journal of Cross-Cultural Psychology, 49*(5), 691–712.

Grady, C. (2010). Do IRBs protect human research participants? *JAMA—Journal of the American Medical Association, 304*(10), 1122–1123.

Haelewaters, D., Hofmann, T. A., & Romero-Olivares, A. L. (2021). Ten simple rules for Global North researchers to stop perpetuating helicopter research in the Global South. *PLOS Computational Biology, 17*(8), e1009277.

Harris, M. (1976). History and significance of the emic/etic distinction. *Annual Review of Anthropology, 5*(1), 329–350.

Henrich, J., Boyd, R., Bowles, S., Camerer, C., Fehr, E., & Gintis, H. (Eds.). (2004). *Foundations of human sociality: Economic experiments and ethnographic evidence from fifteen small-scale societies*. Oxford University Press.

Henrich, J., Boyd, R., Bowles, S., Camerer, C., Fehr, E., Gintis, H., McElreath, R., Alvard, M., Barr, A., Ensminger, J., Henrich, N. S., Hill, K., Gil-White, F., Gurven, M., Marlowe, F. W., Patton, J. Q., & Tracer, D. (2005). "Economic man" in cross-cultural perspective: Behavioral experiments in 15 small-scale societies. *Behavioral and Brain Sciences, 28*, 795–855.

Henrich, J., Heine, S. J., & Norenzayan, A. (2010). The WEIRDest people in the world? *Behavioral and Brain Sciences, 33*, 61–135.

Holcombe, A. (2019). Farewell authors, hello contributors. *Nature, 571*, 147.

International Council for Traditional Music. (2021). *ICTM statement and activities in view of decolonization of music and dance studies*. http://ictmusic.org/documents/decolonization-music-dance-studies.

Iyer, V., & Born, G. (2020, February 18). Of musicalities and musical experience: Vijay Iyer and Georgina Born in conversation. *Wigmore Hall Podcasts*. https://wigmore-hall.org.uk/podcasts/of-musicalities-and-musical-experience-vijay-iyer-and-georgina-born-in-conversation.

Jabbour, J., & Flachsland, C. (2017). 40 years of global environmental assessments: A retrospective analysis. *Environmental Science and Policy*, *77*, 193–202.

Jacoby, N., Margulis, E., Clayton, M., Hannon, E., Honing, H., Iversen, J., Klein, T. R., London, J., Mehr, S., Pearson, L., Peretz, I., Perlman, M., Polak, R., Ravignani, A., Savage, P. E., Steingo, G., Stevens, C., Trainor, L., Trehub, S., Veal, M., & Wald-Fuhrmann, M. (2020). Cross-cultural work in music cognition: Methodologies, pitfalls, and practices. *Music Perception*, *37*(3), 185–195.

Jacoby, N., Polak, R., Grahn, J. A., Cameron, D. J., Lee, K. M., Godoy, R., Undurraga, E. A., Huanca, T., Thalwitzer, T., Doumbia, N., Goldberg, D., Margulis, E., Wong, P. C. M., Jure, L., Rocamora, M., Fujii, S., Savage, P. E., Ajimi, J., Konno, R., . . . McDermott, J. H. (2021). Universality and cross-cultural variation in mental representations of music revealed by global comparison of rhythm priors. *PsyArXiv*. https://doi.org/10.31234/osf.io/b879v.

Jeong, S., & Lee, Y. (2019). Consequences of not conducting measurement invariance tests in cross-cultural studies: A review of current research practices and recommendations. *Advances in Developing Human Resources*, *21*(4), 466–483.

Kirkness, V. J., & Barnhardt, R. (1991). First nations and higher education: The four R's—respect, relevance, reciprocity, responsibility. *Journal of American Indian Education*, *30*(3), 1–15.

Kiser, G. L. (2018). No more first authors, no more last authors. *Nature*, *561*(7724), 435.

Kohavi, R., & Thomke, S. (2017). The surprising power of online experiments. *Harvard Business Review*, *95*(5), 74–82.

Laland, K., Kendal, R., Jordan, F., & Creanza, N. (2020). *Cultural evolution society: Guidelines for organising a diverse conference or workshop*. https://culturalevolutionsociety.org/files/ces_conferenceworkshop_guidelines_3_nov2020-20201112075628.pdf.

Lin, H.-R., Kopiez, R., Müllensiefen, D., & Wolf, A. (2019). The Chinese version of the Gold-MSI: Adaptation and validation of an inventory for the measurement of musical sophistication in a Taiwanese sample. *Musicae Scientiae*, *25*(2), 226–251.

Loughridge, D. (2021). "Always already technological": New views of music and the human in musicology and the cognitive sciences. *Music Research Annual*, *2*, 1–22.

Mehr, S. A., Singh, M., Knox, D., Ketter, D. M., Pickens-Jones, D., Atwood, S., Lucas, C., Jacoby, N., Egner, A. A., Hopkins, E. J., Howard, R. M., Hartshorne, J. K., Jennings, M. V., Simson, J., Bainbridge, C. M., Pinker, S., O'Donnell, T. J., Krasnow, M. M., & Glowacki, L. (2019). Universality and diversity in human song. *Science*, *366*, eaax0868.

Mignolo, W. D. (2011). *The darker side of Western modernity: Global futures, decolonial options*. Duke University Press.

Miguel, A. F. (2018). O que é o Musicultura ? Um estudo de caso sobre um grupo de pesquisa participativa na Maré, Rio de Janeiro. *Rio de Janeiro*, *31*(2), 143–167.

Moshontz, H., Campbell, L., Ebersole, C. R., IJzerman, H., Urry, H. L., Forscher, P. S., . . . Chartier, C. R. (2018). The Psychological Science Accelerator: Advancing psychology through a distributed collaborative network. *Advances in Methods and Practices in Psychological Science, 1*(4), 501–515.

Müllensiefen, D., Gingras, B., Musil, J., & Stewart, L. (2014). The musicality of non-musicians: An index for assessing musical sophistication in the general population. *PLOS ONE, 9*(2), e89642.

Muru-Lanning, M. (2020). Vision Mātauranga, eclectic anthropology and the fading empire. In L. George, J. Taori, & L. T. A. MacDonald (Eds.), *Indigenous research ethics: Claiming research sovereignty beyond deficit and the colonial legacy* (pp. 53–65). Emerald.

Muthukrishna, M., Bell, A. V., Henrich, J., Curtin, C. M., Gedranovich, A., McInerney, J., & Thue, B. (2020). Beyond Western, educated, industrial, rich, and democratic (WEIRD) psychology: Measuring and mapping scales of cultural and psychological distance. *Psychological Science, 31*(6), 678–701.

Nature Editors. (2022). Nature addresses helicopter research and ethics dumping. *Nature, 606*, 7.

Neely, A. (2020). *What is your musical IQ?* https://www.youtube.com/watch?v=b15BrgrctPs.

Parker, M., & Kingori, P. (2016). Good and bad research collaborations: Researchers' views on science and ethics in global health research. *PLOS ONE, 11*(10), e0163579.

Pawlowski, B., Boothroyd, L., Sheppard, P., van Duijn, M., Street, S., Galbarczyk, A., Zhang, M., & Sear, R. (2020). *European Human Behaviour and Evolution Association statement on national IQ datasets*. https://www.cambridge.org/core/membership/ehbea/principals-and-ethics.

Powell, M. (2021, February 14). Obscure musicology journal sparks battles over race and free speech. *New York Times*. https://www.nytimes.com/2021/02/14/arts/musicology-journal-race-free-speech.html.

Purzycki, B. G., Lang, M., Henrich, J., & Norenzayan, A. (2022). The Evolution of Religion and Morality project: Reflections and looking ahead. *Religion, Brain & Behavior, 12*(1–2), 190–211.

Rasmussen, A., & Cowdery, J. (2018). Forum: The lure of universals. *SEM Newsletter, 52*(2), 7–8.

Russonello, G. (2017, July 11). The unfinished work of Alan Lomax's Global Jukebox. *New York Times*. https://www.nytimes.com/2017/07/11/arts/music/alan-lomax-global-jukebox-digital-archive.html.

Rzeszutek, T., Savage, P. E., & Brown, S. (2012). The structure of cross-cultural musical diversity. *Proceedings of the Royal Society B: Biological Sciences, 279*(1733), 1606–1612.

Sardo, S. (2017) Shared research practices *on* and *about* music: Toward decolonising colonial ethnomusicology. In J. Martí & S. R. Gútiez (Eds.), *Making music, making society* (pp. 217–238). Cambridge Scholars.

Sauvé, S. A., Phillips, E., Schiefelbein, W., Daikoku, H., Hegde, S., & Moore, S. (2021). Cross-cultural and anti-colonial/imperial research: ICMPC-ESCOM 2021 workshop. *PsyArXiv*. https://doi.org/10.31234/osf.io/bt6zn.

Savage, P. E. (2018). Alan Lomax's Cantometrics project: A comprehensive review. *Music & Science, 1*, 1–19.

Savage, P. E. (2019). Universals. In J. L. Sturman (Ed.), *Sage international encyclopedia of music and culture* (pp. 2282–2285). Sage Publications.

Savage, P. E. (in press). An overview of cross-cultural music corpus studies. In D. Shanahan, A. Burgoyne, & I. Quinn (Eds.), *Oxford handbook of music and corpus studies*. Oxford University Press.

Schippers, H., & Grant, C. (Eds). (2016). *Sustainable futures for music cultures: An ecological approach*. Oxford University Press.

Schroeder, D., Cook, J., Hirsch, F., Fenet, S., & Muthuswamy, V. (Eds.). (2018). *Ethics dumping: Case studies from North-South research collaborations*. Springer Open.

Sinha, D. (2002). Culture and psychology: Perspective of cross-cultural psychology. *Psychology & Developing Societies, 14*(1), 11–25.

Taras, V., Rowney, J., & Steel, P. (2009). Half a century of measuring culture: Review of approaches, challenges, and limitations based on the analysis of 121 instruments for quantifying culture. *Journal of International Management, 15*(4), 357–373.

Thompson, W. F., Sun, Y., & Fritz, T. (2019). Music across cultures. In P. J. Rentfrow & D. J. Levitin (Eds.), *Foundations in music psychology: Theory and research* (pp. 503–541). MIT Press.

Tran, M., Cabral, L., Patel, R., & Cusack, R. (2017). Online recruitment and testing of infants with Mechanical Turk. *Journal of Experimental Child Psychology, 156*, 168–178.

Urassa, M., Lawson, D. W., Wamoyi, J., Gurmu, E., Gibson, M. A., Madhivanan, P., & Placek, C. (2021). Cross-cultural research must prioritize equitable collaboration. *Nature Human Behaviour, 5*, 668–671.

van Noorden, R. (2020). The ethical questions that haunt facial-recognition research. *Nature, 587*(7834), 354–358.

Vogt, A., Hauber, R., Kuhlen, A. K., & Abdel Rahman, R. (2021). Internet-based language production research with overt articulation: Proof of concept, challenges, and practical advice. *Behavior Research Methods*. https://doi.org/10.3758/s13428-021-01686-3.

Woo, M. (2019, November 21). Music may really be a universal language. *Inside Science*. https://www.insidescience.org/news/music-may-really-be-universal-language.

Wood, A. L. C., Kirby, K. R., Ember, C. R., Silbert, S., Daikoku, H., McBride, J., Passmore, S., Paulay, F., Flory, M., Szinger, J., D'Arcangelo, G., Guarino, M., Atayeva, M., Rifkin, J., Baron, V., El Hajli, M., Szinger, M., & Savage, P. E. (2021). The Global Jukebox: A public database of performing arts and culture. *PsyArXiv* preprint.

Yong, E. (2018, January 25). A study suggests that people can hear universal traits in music: But some music scholars have doubts. *Atlantic*. https://www.theatlantic.com/science/archive/2018/01/the-search-for-universal-qualities-in-music-heats-up/551447/.

19 Conversations with Steven Feld

Steven Feld, Nori Jacoby, Deirdre Loughridge, Psyche Loui, and Elizabeth H. Margulis

[*Editors' note*: Anthropologist Steven Feld has been thinking about sound as a site of nature and culture since at least his work on the Kaluli and birdsong, published in 1982. In a piece coauthored with Aaron Fox for the *Annual Review of Anthropology* in 1994, Feld laid out an optimistic vision of a future in which the sciences and the humanities worked together to understand music and language. Two coeditors of this volume, Deirdre Loughridge and Elizabeth Margulis, interviewed Feld on January 13, 2021. Psyche Loui and Nori Jacoby joined for a follow-up interview on June 22, 2021. A condensed version of those conversations appears here.]

Reintegrating Cognitive Approaches with Grounded Investigation

EM: I'm going to quote you to yourself now, if that's a kosher interview strategy. In your 1994 article with Aaron Fox, you say that "an important agenda for contemporary musical anthropology is the reintegration of sophisticated cognitivist approaches with grounded investigation" [Feld & Fox, 1994]. Is there some ideal version of the relationship between the cognitive sciences and anthropology of music that failed to come to fruition?

SF: I think I was reacting to certain things around me that I didn't like, and certain things that I did. What I didn't like was the increasing use of anthropological or ethnomusicological or musicological work just to construct a kind of contrary, what in anthropology we call the "but not among the Bongo-Bongo" syndrome. Work would be dismissed as simplistic or not grounded because somebody can pull one counterexample out of the air, as if ethnographers are in combat with psychologists in some ring called "Universalism vs. Something. . . ." And you know, I hated that. I just thought that was a kind of descent into a kind of anti-intellectualism, and a form of professional stupidity. I felt it was a kind of closing up of an intellectual perspective as people rallied around the idea that you had to be a card-carrying ethnographer to make certain kinds of statements about music and anybody who wanted to generalize beyond the specifics of ethnography was somehow suspect.

In the 1990s there was a lot of that, because there was a really important and interesting resurgence of certain kinds of dialogue that was probably very threatening to musical ethnographers. The science was getting considerably more complex, depending on levels of specialization that most people didn't have access to. It started to become harder for ethnomusicologists to really understand what the experimental agendas were about, and why they were exciting to the people that were doing them, because of the specialization. At the same time, there was an increased level of funding for cognitive and neurosciences and an enlarging gap between science-rich and humanities-poor sectors at many universities, especially at the state level, that left people doing ethnographic work feeling like there wasn't really a conversation. So in that article, I was making a kind of contrary, anticontrarian statement: why not be open to the possibility that there could be a sophisticated conversation, and a different kind of integration of these approaches, where ethnographers could be valued because of the kinds of depth they might have, and people who don't do ethnography could be valued because they come up with good questions.

DL: My sense is that this "naysayer" attitude that you are describing in the 1990s may have really set back and delayed the ability of cognitive science and more experimental approaches to deal with culture as part of something that is not just the arbitrary variable happening above the natural innate, but rather to think about how the cultural is part of the biological.

If I Were a Cognitive Psychologist

DL: I feel like that line about "for you it's a bird, for me they're voices in the rain forest" [Feld, 2012] encapsulates so well the knowledge or insight I would hope to get at. But you have this whole other experience of the world [from your fieldwork]; how does a cognitive scientist or somebody designing an experiment bring that kind of experience or insight into that domain of knowledge production?

SF: Well, if I was a cognitive psychologist at the same point I was a graduate student in Melanesian ethnography, the kinds of questions that I would have been wanting to ask would have had to do with Why certain birds? Why were people obsessed with fruit doves and not other birds? What was so salient about birds that have very melodious calls, with specific descending intervals, that live close to people, that have interesting displays of color, and whose sounds are ubiquitous? I would have tried to ask a question about: What is it that predisposes people to certain kinds of interests in particular families of sound makers or particular kinds of sound making? Does it have anything to do with the nature of hearing in a place where you can't see more than three feet in front of you, in an environment which is very dense? I think I would have tried to go in the direction of what these days is something called ecological musical perception or the relationship between ecological factors and perceptual foci. If I had been there with somebody who was really trained as an acoustician or scientist in perception, we would have had a hell of a great time because I now see much more how all those questions connect to each other.

From Meta-Language to Epistemological Conversation

EM: Both of the examples you've raised about questions that seemed interesting for you to pursue are very situated within a particular context and place. Right now, within and around music cognition, there is a resurgence of a desire to look comparatively across cultures in a way that has been problematic in the past, and to develop questions and methods that are usable in multiple places. When we start to have these conversations about comparative work at a large scale, fundamental definitional approaches to topics like song or music or speech itself rapidly become vexing. I'm curious what you think the prospects are for the comparative enterprise, broadly speaking, and for the prospect of a shared vocabulary, just to begin with?

SF: It's a really important question because as interdisciplinarity becomes more important and, at the same time, the highly specific nature of training in individual disciplines becomes more differentiated, the question becomes what meta-language do we share? Where are we talking about commensurate things, or compatible things? That's really quite an issue.

To me, I think the conversation that's really missing might be the epistemological one. I mean, the conversation that says, "Okay, I'm a music psychologist, the things that turn me on are these questions, now what would it take for you to realize why I follow this path if I want to know about this thing, or devise this experiment, or imagine that this result or this statement is authoritative versus qualifiable or arguable? What kinds of knowledge do I like to produce that I would call authoritative knowledge? What kinds of knowledge do I like to produce that I would call knowledge that is out there because it needs further refinement, knowledge because the end result of the experiment I did, or the thinking that I did, only made things more complex and were really useful to me because of the way they allowed me to understand how something was more complex?"

Comparative Approaches

EM: I wonder if maybe one specific kind of example might help here, which is a problem that is facing a lot of people who are doing this kind of comparative work within music cognition, where you want to have a demography, a demographic inventory that will give you some kind of means for comparing these categories across places. So you want to have something about, say, musical training. Well, how are you possibly going to have some kind of inventory about musical training that is valid and not nonsensical when applied to all the different kinds of contexts where musical training might be comprised of many different kinds of things? Is this even doable? It's this kind of question that seems to get at something deep and hard to reconcile between psychology and anthropology.

SF: Well, I think that the classical anthropological approach to this question is to talk about how comparables cluster and how relationships are rendered. What are we comparing anyway, and in whose interest is it to compare something? How do we define a meaningful relationship as opposed to one that is less so? So again, I go back to this epistemological question: Comparison? Yeah sure, let's talk about it, but on what epistemological grounds can we talk about comparables and relationships? I think there are a number of necessary conversations yet to be fleshed out to ask why compare? Compare what? Related how? Related why? It seems to me that those questions are foundational to the place where music cognition can speak to cultural difference.

What's the Payoff of Saying Something Is Universal?

DL: I wonder about your thoughts on the move that some psychologists want to make with the cross-cultural comparative work, which is to get at universals of music and underlying psychological mechanisms that might account for those, what is found musically across cultures.

SF: When I think about anthropology and I think about other fields and universals, it seems to me that it's a perfectly worthwhile exercise. I mean people have thought about it for a long time, but in the organization of sound, this is a variety of the question, What is pan-human and species-specific, and what has to be discussed in some other kind of evolutionary framework? What's in the background as I speak that question is recently rereading the review of musical protolanguage in W. Tecumseh Fitch's *The Evolution of Language* [2010] and recognizing an interdisciplinary synthesis that is linguistically, musically, and biologically sophisticated *and* is not afraid to seriously listen to birds. So the universal question is always there, and it's always going to be important, but the potential now is to really raise the interdisciplinary bar for discussing the relational evolution of all forms of sonic interaction, and in doing so cognitively and comparatively, not simply privileging the musicological or linguistic framework.

I think is also important to pose the universal questions on more solid epistemological ground. What's the payoff of saying something is universal? I mean, what does it mean to say that the perceptual unit we call the octave is universal? Okay, that's really interesting, but tell me why I should be turned on by that conclusion, with or without deeper comparative experimentation. Or tell me what kind of counterexample would really turn you on. Or not. I'm simply returning to Gregory Bateson's memorable question—"how does difference makes a difference?"—and wondering whether we can create a more interesting cognitive-ethnographic conversation grounded in his postulates [Bateson, 2000, 2002].

I think about my dad, who was a professional pianist, who could do math and music in his head so fast it would astound me. I think about standing next to him, at his left side, as he played standard tunes with standard chord progressions, teaching me how to hear the relationship between II–V–I and III–VI–II–V. As he played he made me ride my left hand on top of his and hear and feel each one of those bass notes. This was when I was five or six years old. Or he'd sing the bass notes to me and I would have to pick them out and play them so I would be able to play with any Tin Pan Alley song. And I think about how he was computing a sonic macrosystem, but he also had an idea that if I knew that, that if I knew the relationship between II–V–I and III–VI–II–V in every key and in all kinds of popular song forms, then that would empower me to hear a certain way and to be able to play a certain way no matter what instrument I ended up playing. He knew that. He had this idea that there was a payoff for this little bit of computing that he was teaching me to do.

I keep coming back to simple experiential things like that when I think of learning about musical thought or musical systems. I can understand why, if I were a cognitive scientist, it would be really important to study this kind of thing. What does it mean to think in jazz, that you utilize the knowledge of these things in a certain kind of way, and they offer you profound and limitless possibilities for improvising? Negotiating that relationship between familiarity and the unknown is what the real joy and what the core of the jazz aesthetic are about [Berliner, 1994, 2020].

Suppose I was a cognitive scientist and my experimental population was jazz students trying to learn chord substitutions, and my questions were: What do they have to know about flat Vs? What do they have to know about 9s and 11s? What do they have to be able to hear about 9 and 11? What do they have to hear about that in order to understand the kinds of possibilities of how you get back to, like standard types of questions, like how do you get back to I? Like how do you get to I if you don't have V? What else can you do? You are trying to get to I, and you understand that there's a series of possibilities that can substitute for V. Can it really be blues if it's just I–IV–I and a bunch of substitutions? This is all about question construction, but question construction in relation to a payoff.

EM: I think that's a really telling example of how thinking about the stakes and the payoff can help ground the conversation and move it forward.

DL: Yes, and finding common ground where projects share something and where it's okay for them to be different.

SF: If I encounter a particular kind of music for the first time, I think about a question like, What do people do to get back to I? I mean, if we want to hypothesize that there is something called a tonal center and that it might be universal, we want to have some kind of constructive dialogue about what one does within any system of musical knowledge to get back to it, no matter where else you've gone. What does it mean that we could possibly describe in cognitive terms a set of procedures and pathways, a kind of production of knowledge, which is the knowledge we would call "getting to I"? Why is it important? We want to relate the question of "Why is it important?" to the question "What are all the different ways you can do it with whatever musical materials you have available to you?" So when you ask about universals and that kind of thing, this is the sort of research I would cherry-pick, because if I see something about getting to I, it's like, "Yeah, that's interesting; I want to know about that." As a musician, and as somebody who does a different kind of research from the kind you all do, I would want to know about that. It's an example of where very pragmatic, very epistemologically rich, cognitively significant kinds of factors would come together, and where a metalanguage or vocabulary for posing that across disciplinary bounds could be useful, so that we could have a serious conversation about this and think about how we would all work across experimental and natural historical divides to really talk about this. I think stuff like that is very cool.

[*Editors' note*: Nori Jacoby and Psyche Loui joined Deirdre Loughridge and Elizabeth Margulis for a second conversation on June 22, 2021, with a focus on two specific projects. One was a collaboration between Steven Feld and Ghanaian multi-instrumentalist Nii Otoo Annan, in which Nii Otoo improvised a set of variations as overdubs to a six-minute soundscape recording to yield an album—*Bufo Variations*

(VoxLox, 2008)—that "sound[s] the generative mathematics of many rhythm families" and provides a way of listening to Nii Otoo's "histories of listening" (Feld, 2015, pp. 95, 102). The other project was a collaboration between Nori Jacoby and thirty-three other researchers who studied mental representations of rhythm by measuring how people tapped along to audio that began as a random "seed" rhythm created from a repeating cycle of three clicks but was iteratively adjusted to match the exact timing with which the participant tapped to it—in essence, a game of musical telephone with oneself (Jacoby et al., 2021).]

Iterative Processes

EM: I was struck by the degree to which these kinds of iterative processes played a role in Nori's paper and also in some of the projects described in your chapter on the Bufo Variations.

SF: Multitrack recordings started around the time I started to be a musician getting a bit of work in recording studios. I had an intuitive but not analytic understanding of the meaning of playback. And so I developed a fascination first as a mixture of a musician, player, and then recordist and analyst. And then, in 1974 in Paris, I met Simha Arom and became really fascinated with his early experimental techniques for using playback for analysis [Feld, 1976]. After many years and developments with his method, I think that many people have focused on the way he's analyzed the music and thought about it, yet very few people picked up on how far out that experimental methodology was and how much it offered. So, in my own funny way, I've done lots of experiments with playback for a long time, bringing my studio and field knowledge together. When I met Nii Otoo and started studying with him, it was amazing, because it was like, wow playback, this is like someone giving you a window into their musical brain. What a gift.

PL: I think it actually opens so many more questions that an empirical psychologist might be curious about. Like how much are these recordings a result of your sensitivities, and your interactions with each other, and your trust in each other, and how much is it something that you might have when you're an infant? I think it raises more questions than it answers. It gets at lots of intuitions, but I think we can go on to ask lots more questions about those intuitions, and where they came from, and how come they are there.

SF: I really agree with that. I really like to be amazed. I really love to be surprised. I love it when something happens that I can't anticipate in any kind of way. My experience of

Accra over fifteen years continues to be a wonderful experience because there is a music called jazz that I think I know very well, because I've been a deep listener to it since I was young and I grew up with it, in many kinds of ways. It was, in a certain sense, my first music—in other words, my first musical language was improvisation. So I thought I knew quite a bit about improvisation, or embodied it as a result of many many years of improvising in contexts ranging from avant-garde classical new music to straight-ahead jazz to many other kinds of popular music. Then Accra and musicans like Nii Otoo made me rethink everything about jazz improvisation.

When I went to Accra I knew how people had previously studied and written about West African music. And I knew the ideological background: if one can articulate how the level of local complexity matches Western European art music, then one is implicitly not just analyzing a music but countering racism, and that is somehow politically good.

The downside of this is that it gives us a rigid idea of African musical traditions in order to create strict parallels to Western European art music practices. And where this falls apart politically is that it doesn't grant Africans the agency of being experimentalists. It would be like saying the only people I want to study a language with are the people who are the grammarians of the language, not its speakers. In West Africa you can meet the grammarians, of course. But you also meet someone like Nii Otoo, who is not just a percussionist but a prodigy bass player, guitar player, and singer who plays in everything from church bands to gospel bands, who is the drummer of choice for avant-garde groups, jazz groups that visit, all kinds of things. So the agency of music and musicians in West Africa is much more complex when you explore it with someone like Nii Otoo as your guide. So why would you want to focus on an experiment with a musician like Nii Otoo rather than somebody who is a master drummer at the University of Ghana? Sure, Nii Otoo taught me the basic variations on the standard bell patterns and timelines the way master drummers teach those things to their students. But he did not teach me as a grammarian; he taught me as a master who knows that the most creative thing about "tradition" is that it is unruly.

Nii Otoo taught me to be interested in what it is about his knowledge that would surprise me. And that encouraged me to interact musically with him to reveal the more surprising dimensions of his musical agency, rather than just focus on his deep mastery of principles of multipart interactions.

There is a tendency to want to make difference into this kind of master trope for everything. I really can't stand this ethnomusicological discourse about how it's all about difference. If you're a musician and you've played musics across different parts of the world and worked with many kinds of musicians, you just deeply know that this is bullshit. Musical difference, when I play with Nii Otoo, surely connects with the way you, Nori, experimentally question core sensibilities and core perceptual things that are just ontologically species specific and human. Overdoing the question of difference is just like overdoing this issue of greatness, trying to make Western European art music commensurate to and equal with everything else. These two things have been ideologically pounded into the history of ethnomusicology, and they are actually two things that completely diminish the intellectual agenda and intellectual character of ethnomusicology.

NJ: I've actually had some passionate arguments with contemporary ethnomusicologists on those exact points. And I always try to convince them that a lot of big ideas are not about universals; they are about central issues in cognition, like what is the meaning of a concept or what kinds of words do people use to describe song or birdsong.

Measurement and Experimentation

NJ: One thing that I have found very challenging is trying to understand what aspects of our lives contribute to different musical experiences. Experimentally, we have demographics, but these are low dimensional and reflect only the kinds of features that can be answered by simple questions. Another option is to analyze music, similar to what you describe in your paper, but this is still a very limited way to describe what actually happens in a musical performance. Do you have some ideas about how we could develop experimental methods to enrich the vocabulary with which we describe people's musical experience?

SF: Yeah, I mean that's fantastic! Look, if you want to measure the electrical potential of a crystal, you get a voltmeter. But everybody knows when you touch a voltmeter to a crystal, you change the electrical potential of the crystal. If you ask a physicist about this, or someone who's a rigorous experimentalist, they say it doesn't matter; all that matters is if you have a good voltmeter. Okay, well, what's the difference between the good voltmeter and the $1 Radioshack voltmeter? The good voltmeter takes you deeper into the plus or minus of a measurement; it is going to be much more refined. It's going to be much more subtle. So in a sense, what we are talking about here is the possibility that we could get into the zone of having much better voltmeters.

I don't mean to oversimplify how much we can do in a rigorously controlled way that really addresses the subtlety of difference. So a new form of conversation would be between people who focus on subtlety, that is, people who focus on changing what kind of voltmeter we could have, and people who refine the voltmeter because of the idea that there are still necessary refinements that have to be made even on the side of what is the core set of understandings we want to have about experience and listening and memory and cognition and so forth.

PL: So can I share this paper, Nori do you know this one? "Musical Friends and Foes" [Aucouturier & Cannone, 2017], I think it is actually closer to the kind of experience of two people improvising together, except the experimenter adds a dimension to it. The experimenter says, can you play conciliatory, or can you play trust? And so one person starts, and the other person improvises on top of them, and then the researcher's role is to record everything, see whether independent third parties can afterwards decode from the interaction what they were trying to get at and then try to understand what acoustic properties or what musical properties went into that emergent interaction that gave it that meaning.

I really like that paper. There are very few like it. There are very few people in a position where they can recruit all these advanced avant-garde improvising musicians to do these studies and also have that kind of interest and care to try to understand these very subtle interactions. I think most of the field is still trying to debate why we have emotions or what are the bread-and-butter ways in which we can tease apart cognition, which is fine. But this is a more nuanced way.

SF: Well, I really agree with that. I mean it goes back to what I said at the beginning. The stuff that's amazing to me is not writing articles about how the people who study Western European art music are wrong or people who say this or that about universals are wrong. The zone of amazing that I'm talking about is like getting the opportunity to really get close year by year, day by day, over a long period of time with a Nii Otoo . . . like now I take bass lessons with him. I know how to play the bass, but I am taking bass lessons with him because I want to learn bass from somebody who taught himself how to play the bass in an African context. I want to understand why that's different. I mean, that's amazing to me in terms of understanding what this knowledge of the hands is about. So this idea, for me, this has nothing to do with ethnomusicology versus musicology or music psychology versus anthropology. It just has to do with musicianship and understanding what musical knowledge consists of and how can I contribute something to whatever that conversation is about. What does it mean to know a music? What does it mean to know a music closely? What does it mean to feel and embody it?

NJ: There is something very interactive about a tradition like jazz, where you are generating something creative with a style that gives you unbelievable richness. Culture can be seen as some kind of process rather than as a conservative, static tradition.

SF: You know, you've nailed a really core problem. The term *culture* has been made brain dead. When I write about Nii Otoo, I am not writing about culture. I am writing about the relationship between how knowledges are always being negotiated. He has a knowledge of time. If you want 6/4, you'll get it, but there's a lot more to West Africa than doing that and hearing underneath it where the beats are placed so that it pushes up against the next bar line. Okay, no big deal, we get it. You know, he's got that, but what else is there? . . . We have to talk about skill. . . .

PL: I feel like a lot of people do get into these local minima, right? I think that neuroscience gets into these like popularity contests, cognitive science gets into these popularity contests too.

SF: You're sounding like an ethnographer, Psyche!

PL: Oh, I'm learning from the best. But I do think that there are different kinds of structural pressures that shape the field because music cognition is a relatively young field. I got into the field as an undergrad because I wanted to know why V had to go to I, and music theory classes told me: it just does. Right, and then you start to read some papers that were coming out at the time, and there are these brain potentials that happen when V goes to I and when you put a Neapolitan six instead of a I or a Neapolitan six instead of a IV that came before the V; they look different, and that tells you something about expectancies and predictions, and that tells us something about meanings and expectations. That got me super interested and was kind of why I went to grad school, and then in postdoc times I was told, well, people want to know why music makes you smarter. If you keep studying why V goes to I, you will never get funded and never get a job, so instead you should study why music makes you smarter. I mean, I'm totally mischaracterizing my mentors, but there are these structural pressures for certain questions to be prioritized within a field. I'm not sure they're great, but they are what they are.

SF: You know, what you're saying reminds me of the 1970s when I was a linguistics graduate student and transformational grammar was the thing and MIT was Mount Olympus and everything had to emanate from the Chomskyan paradigm. I was kind of cheeky, and as students we were learning how to write rules for everything and turn everything in linguistics into mathematics. And so I get to the end of my rope one day and I said, "You know, you guys are acting like what's really at stake here is that the human mind needs to economize on storage space. Is there any empirical evidence for that?" Transformational linguistics formalized through mathematics all the different properties of forms of grammatical description, and that was an extraordinary intellectual advance, but it never really did ask or answer the question of what this idea of grammar says about the mind.

From a musical perspective, what do we have to contribute to the question "Does the mind economize on storage space?" What could we talk about, or how could we talk about the idea of an economy of listening, an economy of mind. So when I'm fooling around with Nii Otoo, I'm thinking about, how much does this guy simultaneously think about 6 on 4 or 4 on 3 or 3 on 2, and what's the play, what's going on, how many metronomes can this guy keep happening in his mind and be tracking all these kinds of interactions taking place that I could possibly numerize or understand?

NJ: That goes back to the overdubbing, because the overdubbing allows us to analytically separate processes. If we can think about the question of the economy of the mind and see if we can somehow reduce it to the overdubbing process, then we have a degree of control over the process and can see it gradually building up in the mind of a person who is creating it.

SF: Well, I mean for me, it just raises the question of how useful Pro Tools is as a piece of laboratory equipment rather than a mixing console.

EM: I keep thinking about your foundational question about what constitutes knowing something about music. You're excited about knowledge that emerges out of repeated encounters making this music collaboratively with one person. How might that connect to the kind of abstracted knowledge of the sort that might come out of a scientific paradigm? It seems like you just drew an explicit link there in your questions around an economy of the mind. What are the space limitations? That's a certain kind of abstract thing to know about. So I guess my question here is around the possibility for a linkage between the kind of knowledge that you're pursuing and that kind of question.

SF: It goes back to the questions that you've asked and really addressed in what I consider to be a pragmatic and down-to-earth way in the paper that you participated in with Nori. Can one talk about the influences of culture, history, environment, listening preferences, listening patterns, listening backgrounds, and things like that, and not completely nuke core capacity with difference? Is it possible to deal seriously with difference *and* bracket its all-too-frequesnt use as an ideological submarine to bash into the idea of core capacities and core knowledges?

Culture and Relationality

SF: The question I really want to ask is: how big is culture and how small is culture? I like your paper because it gives us some tools to think about that. That culture can be very very big, and we need to think about it in very very big ways in terms of the dynamic capacity for expansive agency—Culture with a capital *C*. But at the same time, we have to recognize that culture is in many cases more about the level of deep nuance in the way of doing things, in the way of listening, in the way of understanding things—culture with a lowercase *c*.

It just happens that there are some people in the world who are only interested in that particular form of nuance and other people who are interested in the story of what that nuance and the persistence of that nuance and the power of that nuance have to do with the importance of the core that we're talking about, and understanding how to refine experiments and refine our discourse so we are not overwhelmed by culture and at the same time we are not diminishing it into a kind of default basket category of everything that's outside what we really want to understand about the mind or about knowing or about listening or about sound. I think that Lisa's little book [Margulis, 2018] gives a sense that the reason why there is much more to do is because the good questions are yet to be asked and to be explored. I think that's the generosity that's intellectually necessary in order for any kind of conversation to take place across disciplines.

NJ: Yes, it seems like one of the recurring ideas here is the interaction between details, nuance, and structure. I think that is very difficult to articulate: How exactly can we be very sensitive to details while keeping the big questions in view? How can we be more sensitive to and aware of the nuances? How can we go from the small nuances to the big ideas without losing both?

SF: I don't know. I am a worker, I'm not a shaman. I just consider myself a worker who learns things incrementally and tries to remain humble about it but at the same time tries to keep insight into those bigger questions I'm posing to you: Does the human mind need to economize on storage space? What level of subtlety or nuance are we talking about when we talk about culture? How do we need to define *culture* as really big and really small?

PL: I think that there's something really key here. I am not trying to answer Nori's question on behalf of Steven, that would be very silly, but I mean that trust is such a key component. Right? And Nori, I'm sure you're super familiar with this, but going into a different location and establishing the connections that you establish with the research participants. If they don't trust you, they are not going to give you the nuance that you're looking for, right? It reminds me of this gorgeous neuroscience study about the trust game. You can play these economic games where I give you some money and you invest that money and then you decide how much to give back to me, and if I play with someone who's trustworthy, over time I learn to trust them. They then showed the statistical contrast in brain activity between when you're looking at someone that you trust versus when you're looking at someone who consistently screws you over in the game. You get these reward system differences, and as you learn to trust someone, there is a shift in the reward system activity [King-Casas et al., 2005], so that you become more attuned to their actions, you actually know what their actions are going to be, or you anticipate what their actions are going to be before they even do it, and I think that's what makes a successful musical interaction possible. I think that actually implementing that study would be really really hard, but probably Nori is at the forefront of folks who can do this.

SF: I love that. In Melanesia, there's a lot that's been written about exchange and gender and circulation. One of my first and most formative experiences working in Papua New Guinea was with the person who taught me the most about birds, who also taught me the most about song composition. This led me to constantly question why the people who are the most expert ornithologists are also the most expert composers. I mean, how does culture produce that? What kind of history of coevolution do I have to begin to grok in order to be able to even pose that question in some kind of empirically interesting way? So I go hunting with this guy, and we're looking at his traps and he's showing me this, because it's in the context of just hanging out with people and finding out about routine knowledges that all kinds of things happen, you learn about birds, you learn about the forest, ecology, and so forth. So I'm just hanging out with him while he is checking his traps. And he turns to me and says, "In your place, what do you do if another guy gives you two smoked marsupials?" So I look at the guy, and I've been there six or eight months and can barely speak the language well, and like a wiseass I say, "I give him two back." This is a version of Psyche's game. And the guy looks at me like, "What planet are you from, white man!" and he says, "I would give seven." Okay, welcome to New Guinea Sociology 101: if it doesn't grow, it's not a social relationship. Duhhh, I read all these books about this, so why am I not getting this? Okay, so one of the hardest things to adapt to when doing fieldwork in Melanesia is the idea of understanding the importance of inequality. If you want to be in good relationships with people, if you want to understand your place in the world, you have to understand that being owed a lot and owing a lot at the same time is a good way to be, that is *the* way to be. Okay, this is what I would say is interesting about culture. What does it mean to inhabit a body, what does it mean to inhabit an orientation to the world where being owed a lot and owing a lot and never thinking about getting the score to zero is the way you are in everything you do? That this is trust and this is security. This is like the social security system. This is why everybody knows: no food, no problem. Pigs trash your garden, no problem; you can be fed. So, in terms of thinking about the expressive capacity of people and the way in which this kind of fundamental orientation to the world really puts its mark on expressive capacity like, for example, in the production of poetics, or ceremonialism, or art making, or anything like that.

It's essential to know that getting to zero is not the game. The game is remaining productively unequal. Being productively unequal is a way of reimagining the relationship between generosity and security. This is exactly what Psyche is describing. It's like an extraordinary game in calibration, like how humans calibrate each other, how they tune each other, and we can use the word *trust*. I think that that is an important word in this, but it's also . . . I use the word *culture* because, for me, it's about the system of calibrations. It's about the core idea of relationality. Culture is like endless variations on the idea of relationality. What is this particular variation on relationality saying to us about what this place is, who these people are? What does it mean that the way you are about one thing is the way you are about everything, and that this is that way? So I think that an experiment like that, a game like that, I can imagine playing that game in New Guinea. I can imagine trust games as a really great way to do that kind of experimentation in a Melanesian context. People would love to play that game.

NJ: There is something also about this relationality that opens its way to understanding this ever-evolving process. That's something we really don't grasp very well in music cognition. We understand snapshots, but we still don't understand how things transform, and that is essential for tackling the issue we are talking about right now.

SF: Yeah. The thing that made me first start thinking about this question and this idea of relationality—or, to put it in big words, acoustemology as relational ontology—was this experience in New Guinea. In the world of the Melanesian people and the rain forest they live in, where I worked for twenty-five years, one never has the experience of unison. One never has the listening experience of unison. It just does not exist. My sister studied the socialization, the language socialization, of Bosavi children and learned that, for example, mothers never scold children about interrupting, and in fact, there isn't a word in the Bosavi language for interrupting [Schieffelin, 1990]. And when kids are preverbal, the way mothers and elder siblings deal with those kids in social contexts is to face the child, the preverbal child, outward into a group and, as people are speaking, go "eeehhhh" and move the child's body so that the child learns to overlap and interlock and alternate with other voices sonically. I've got a video of this, it's incredible. They become so tuned to overlapping, interlocking, and alternating, right? So overlapping, alternating, and interlocking is the world that you hear. I mean, that's the world of the forest, it's the world of everything around you. It not only becomes stylized and aestheticized in speaking, but it's also really deeply socialized. And so, of course, this idea that everybody can be talking at the same time, and in a society that's pretty much egalitarian in many ways, this becomes another kind of token of this type, a kind of relationality which is about owing and being owed, about everybody constantly negotiating the tense relationality of not being exactly equal but not being above or really below because you're always just . . . it's just a game of moving sweet potatoes instead of moving money. So, what I started to listen to was a history of listening, of overlapping and alternating and interlocking, for which there is actually one term in this language. And when I started working with singers and started to learn how to sing, every time I would get too close and sing in unison, somebody would knock me off by jumping ahead or whatever. So I had to resocialize my whole sense of timing and tuning to be a split second ahead or a split second behind, and then I observed the missionaries trying to teach people to sing hymns in unison, and immediately it was all over the place, and the missionary's asking me, "Why are you here? These people are so unmusical, they can't even sing together." And I realize, this is culture, right? These people are constantly in this world of seven smoked marsupials in return for two, and it really does have something to do with the way they talk, and the way they sing, and the way they listen, and the way they are attuned to one another. Understood this way, culture is a kind of relationality which predisposes you to the stylization of interaction. So there it is, acoustemology is a relational ontology, but, yes, a lot of it is very nuanced, it's very micro.

The rain forest is the perfect place for doing experimental psych with that sort of thing. Here you see this kind of way in which it is; this kind of core way of being with the world becomes elaborated in every cultural modality—speaking, interacting, exchanging, hunting, relations between women and women, men and women, men and men. I mean, there are all these different ways we see culture as the way in which new threads get woven into the fabric. It seems to me that this is my local way of summarizing in an ethnographic snapshot Nori's critical and foundationally rich question: Is the culture big or is the culture small? Does small *c* culture trump big *C* Culture ever or always? Or vice versa? It's the weaving of the threads on top of the threads that's the interesting story here. But really, the exciting part for me is to read experimentalists ready to ask, and find new ways to grapple with, this critical question in a cognitive framework: where and how is *c*ulture/Culture the difference that makes a difference?

Acknowledgments

Thanks to Johanna Linna for transcribing the original interviews.

References

Aucouturier, J.-J., & Canonne, C. (2017). Musical friends and foes: The social cognition of affiliation and control in improvised interactions. *Cognition*, *161*, 94–108.

Bateson, G. (2000). *Steps to an ecology of mind*. University of Chicago Press.

Bateson, G. (2002). *Mind and nature: A necessary unity*. Hampton Press.

Berliner, P. (1994). *Thinking in jazz: The infinite art of improvisation*. University of Chicago Press.

Berliner, P. (2020). *The art of Mbira*. University of Chicago Press.

Feld, S. (1976). The use of play-back techniques in the study of oral polyphonies. *Ethnomusicology, 20*, 483–519.

Feld, S. (2012). *Sound and sentiment: Birds, weeping, poetics, and song in Kaluli expression* (3rd ed.). Duke University Press.

Feld, S. (2015). Listening to histories of listening: Collaborative experiments in acoustemology with Nii Otoo Annan. In G. Borio (Ed.), *Musical listening in the age of technological reproduction* (pp. 91–104). Routledge.

Feld, S., & Fox, A. A. (1994). Music and language. *Annual Review of Anthropology, 23*, 22–53.

Fitch, W. T. (2010). *The evolution of language*. Cambridge University Press.

Jacoby, N., Polak, R., Grahn, J., et al. (2021). Universality and cross-cultural variation in mental representations of music revealed by global comparison of rhythm priors. https://psyarxiv.com/b879v.

King-Casas, B., Tomlin, D., Anen, C., Camerer, C. F., Quartz, S. R., & Montague, P. R. (2005). Getting to know you: Reputation and trust in a two-person economic exchange. *Science, 308*(5718), 78–83.

Margulis, E. H. (2018). *The psychology of music: A very short introduction*. Oxford University Press.

Schieffelin, B. B. (1990). *The give and take of everyday life: Language socialization of Kaluli children*. Cambridge University Press.

Contributors

Manuel Anglada-Tort, Computational Auditory Perception Group, Max Planck Institute for Empirical Aesthetics, Frankfurt am Main, Germany

Salwa El-Sawan Castelo-Branco, Ethnomusicology Institute, Center for Studies in Music and Dance, NOVA University of Lisbon, Portugal

Hu Chuan-Peng, School of Psychology, Nanjing Normal University, Nanjing, China

Laura K. Cirelli, Department of Psychology, University of Toronto, Scarborough, Ontario, Canada

Alexander W. Cowan, Department of Music, Harvard University, Cambridge, MA, USA

Hideo Daikoku, Faculty of Environment and Information Studies, Keio University, Fujisawa, Japan

Jonathan De Souza, Don Wright Faculty of Music, Western University, London, Ontario, Canada

Diana Deutsch, Department of Psychology, University of California, San Diego, CA, USA

Diandra Duengen, Comparative Bioacoustics Group, Max Planck Institute for Psycholinguistics, Nijmegen, The Netherlands

Sarah Faber, Department of Psychology, University of Toronto, Toronto, ON, Canada

Steven Feld, Department of Anthropology, University of New Mexico, Albuquerque, NM, USA

Shinya Fujii, Faculty of Environment and Information Studies, Keio University, Fujisawa, Japan

Assal Habibi, Brain and Creativity Institute, University of Southern California, Los Angeles, CA, USA

Erin E. Hannon, Department of Psychology, University of Nevada, Las Vegas, NV, USA

Shantala Hegde, Department of Clinical Psychology, National Institute of Mental Health and Neurosciences, Bengaluru, India

Beatriz Ilari, Department of Music Teaching and Learning, University of Southern California, Los Angeles, CA, USA

Jason Jabbour, UN Environment Program, Washington, DC, USA

Nori Jacoby, Computational Auditory Perception Group, Max Planck Institute for Empirical Aesthetics, Frankfurt am Main, Germany

Haley E. Kragness, Department of Psychology, University of Toronto, Scarborough, Ontario, Canada

Grace Leslie, College of Music, ATLAS Institute, University of Colorado, Boulder, CO, USA

Casey Lew-Williams, Department of Psychology, Princeton University, Princeton, NJ, USA

Deirdre Loughridge, Department of Music, College of Arts, Media and Design, Northeastern University, Boston, MA, USA

Psyche Loui, Department of Music, College of Arts, Media and Design, Northeastern University, Boston, MA, USA

Diana Mangalagiu, Neoma Business School, Paris, France and Environmental Change Institute, University of Oxford, UK

Elizabeth Hellmuth Margulis, Department of Music, Princeton University, Princeton, NJ, USA

Randy McIntosh, Institute for Neuroscience and Neurotechnology, Department of Biomedical Physiology and Kinesiology, Simon Fraser University, Burnaby, BC, Canada

Rita McNamara, School of Psychology, Victoria University of Wellington, New Zealand

Eduardo Reck Miranda, Interdisciplinary Centre for Computer Music Research (ICCMR), School of Society and Culture, University of Plymouth, Plymouth, UK

Daniel Müllensiefen, Department of Psychology, Goldsmiths, University of London, UK

Rachel Mundy, Department of Arts, Culture and Media, Rutgers University, Newark, NJ, USA

Florence Ewomazino Nweke, Department of Creative Arts, University of Lagos, Nigeria

Patricia Opondo, Department of Music, University of KwaZulu-Natal, Durban, South Africa

Aniruddh D. Patel, Department of Psychology, Tufts University, Boston, MA, USA, and Canadian Institute for Advanced Research, Toronto, Ontario, Canada

Andrea Ravignani, Comparative Bioacoustics Group, Max Planck Institute for Psycholinguistics, Nijmegen, The Netherlands, and Center for Music in the Brain, Department of Clinical Medicine, Aarhus University & The Royal Academy of Music Aarhus/Aalborg, Aarhus, Denmark

Carmel Raz, Histories of Music, Mind, and Body, Max Planck Institute for Empirical Aesthetics, Frankfurt, Germany

Matthew Sachs, Center for Science and Society, Columbia University, New York, NY, USA

Marianne Sarfati, Comparative Bioacoustics Group, Max Planck Institute for Psycholinguistics, Nijmegen, The Netherlands

Patrick E. Savage, Faculty of Environment and Information Studies, Keio University, Tokyo, Japan

Huib Schippers, Creative Arts Research Institute, Griffith University, Queensland, Australia

Jim Sykes, Department of Music, University of Pennsylvania, Philadelphia, PA, USA

Gary Tomlinson, Department of Music, Yale University, New Haven, CT, USA

Jamal Williams, Princeton Neuroscience Institute, Princeton, NJ, USA

Maria A. G. Witek, Department of Music, University of Birmingham, UK

Pamela Z, Composer, San Francisco, CA, USA

Index

Note: Page numbers followed by *n* indicate a note, followed by the note number. Page numbers followed by *f* and *t* indicate figures and tables, respectively.

absolute pitch, 291, 312
  cross-species research in, 75–77, 76t, 77t
  instrument-specific, 148
accessibility, in global research collaborations, 351–352, 354
acoustemology, 216n5
acoustic multinaturalism, 7
acoustics, historical studies of, 295
action potentials, 163–164
Activating Memory project, 228–230, 231
active engagement, musical development and, 191–193
active inference, 168–169, 186
adaptations, emergence of, 188
adaptive function, 25–26
adaptive theories of human musicality, 15–18, 21–22
  alternatives to, 16–18
  Darwin-James debate, 15–17
Aethereal Medium, Newton's theory of, 117
Affective Computing Group, 281
affective responses to music, 249–251, 253–254
affordances, 41
  and found objects, 306
  musical development and, 185, 188, 330, 336, 337
  radical niche-constructive plasticity and, 46
  social, 150
  tool use and, 147
agency
  animal, 102
  art-as-agency, 204, 213, 215
  in global research collaborations, 352
  musical development and, 341
  of sound, 212
  of West African music and musicians, 374
agile gibbons (*Hylobates agilis agilis*), duets in, 73
aging, music and
  age effects, identification of, 272–274
  complex system framework, characteristics of, 237, 253, 264–266
  complex system framework, implications for music neuroscience, 269–271, 270f
  hidden Markov modeling for, 271–274
  information processing in space and time, 266–269, 267f, 268f
  multiscale entropy curve, 269–271, 270f
  overview of, 263–264
  segregation/integration processes, 265–266
Alzheimer's disease, music and, 263, 282
  age effects, identification of, 272–274
  complex system framework, characteristics of, 237, 253, 264–266

Alzheimer's disease, music and (cont.)
  complex system framework, implications for music neuroscience, 269–271, 270f
  hidden Markov modeling for, 271–274
  information processing in space and time, 266–269, 267f, 268f
  multiscale entropy curve, 269–271, 270f
  overview of, 263–264
  segregation/integration processes, 265–266
amateur musicians, 331
amplification of sound, cross-species research in, 71
amusia, 312, 318
amygdala, 246, 250
analogues, animal, 58
Anang Ibibo of Nigeria, musical participation among, 333
Anatomy of the Human Body (Cheselden), 118
Anatta (not-self), 207
ancient cognitive traits, 25
angular gyrus, 244
anhedonia, musical, 31n4, 312
animal laboratory ethics
  animal musicality studies, history of, 100–102
  Animal Welfare Act, 97, 102–103
  human *versus* animal needs in, 105–107
  implications for studies of musical capacity, 105–107
  kinship principle in, 108n7
  liberation school of, 98
  origins of, 97–99
  translation and, 106–107
  welfare school of, 98
Animal Musicalities (Mundy), 100
animal musicality, 7, 12–13
  animal homologues/analogues, 58
  anthropocentric approach to, 58–60
  biocentric approach to, 58–60
  biomusicology components in, 58
  definition of, 57
  directions for future inquiry in, 80–81
  goals of research on, 57–59
  history of, 100–102
  instrumental music and tool use, 67–71
  perspectives from, 104
  pioneer studies in, 60–63
  potential pitfalls in, 80–81
  questions in, 58
  spectral sound parameters, 74–80
  synchronization, 71–73
  temporal sound parameters, 73–74
  See also animal laboratory ethics; animal song; cross-species research; transspecies perspective, musical meaning in
animal song, 58
  bats, 65
  birds (*see* birdsong)
  calls *versus*, 63–64
  definition of, 63–64
  seals, 66–67
  whales, 65–66
animal spirits, ancient Galenic theory of, 117
Animal Welfare Act (AWA), 97, 102–103
anisochrony, cross-species research in, 74
Annan, Nii Otoo, 5, 372–375
antennal lobes, in honeybee brain, 45, 46f
anterior cingulate cortex (ACC), 246
anthropocentric approach, 58–60
anthropological perspectives, 203–216
  capitalism and Christianity, 209–210
  comparative approaches, 370
  culture and relationality, 378–382
  distributed person, 204, 213–215
  integration of cognitive and cultural understandings, 313, 367–368
  interdisciplinarity and epistemological conversation, 369
  iterative processes, 373–375
  measurement and experimentation, 375–378
  musical transcendence, 210
  music-self relation, 208
  porous nature of humans and, 212–214
  relationship between description and explanation, 210–215
  Sinhala Buddhist ritual drumming, 205–208

sound-as-gift, 215
sound-as-medium, 204, 212–214
sound-as-relations, 203–205, 215
universalism, 12, 367, 370–373
antirepresentationalism
directions for future inquiry, 176
embodied cognition and, 173–176
aphasia, nonfluent, 16
Aristoxenus, 292–293
Arom, Simha, 70, 373
art-as-agency, 204, 213, 215
art-as-relationality, 213
artificial intelligence, 165, 223, 254
artificial neural networks, 254
art objects, agency of, 213
arXiv, 354
aspectual nature of signs, 43
associative learning, 42
asymmetry, hemispheric, 291
attention
bee behavior and, 46–47
birdsong production and, 45
definition of, 42–43
perceptual organization and, 291
attribution, in global collaborations, 356–358
Audio and Music Aptitude Profile, 337
audition, birdsong production and, 45
auditory cortex, 147, 230, 243–244, 247, 336
auditory pathways, in avian brain, 45f
auditory pattern perception tests, 26–27
auditory perceptual illusions, 135, 183, 278, 291, 299–301
auditory working memory, 291
"Aus tiefer Not" (Bach), 148–149, 149f
Australian magpie-larks, pseudoduetting by, 47–49
autopoiesis, 170, 176n1
avant-garde music, 19–20, 30, 374, 376
avian brain structure
auditory and vocal pathways, 44–45, 45f
semiosis and interpretant formation and, 47–49
See also birdsong
awe, musically induced, 251
Ayurvedic medicine, 208, 215

Babbitt, Milton, 296
Babinski, Joseph, 115
Bach, Johann Sebastian, 136
"Aus tiefer Not," 148–149, 149f
"Christus, der ist mein Leben," 137–143, 138f, 145
Ich habe genug, 150
"Nun komm, der Heiden Heiland," 141–142, 142f
baleen whales (Mysticetes), song in, 65–66
Barcelona Music Reward Questionnaire, 271
Bartkowski, Frances, 108n7
basal ganglia, 124, 144, 164
Basonge people, word for music among, 19
Bateson, Gregory, 371
bats
bat song, 65
instrumental music by, 69, 71
pitch height discrimination by, 75
BaYaka of Central Africa, musical participation among, 333
Bayesian methods, 140, 168
beat
beat perception and synchronization, 25, 189–191
cross-species research in, 71–72
Mersenne's studies on, 294
movement in synchrony with, 19–20
See also entrainment
Beattie, James, 120
Becker, J., 19
Beethoven, Ludwig van, 222–225, 226, 230–231
beliefs, supernatural, 25
Bell Labs, 299
bellows, theory of lungs as, 117
Bengalese finches, birdsong communication in, 49, 50
Bennett, J., 213

Bentley Measures of Musical Abilities (BMMA), 337
Beravā, Sinhala Buddhist ritual drumming, 205–208
Berlioz, Hector, 121
Beston, Henry, 57
beta oscillations, 164
bias, implicit, 21–22
big data studies, 27–28
biocentric approach, 58–60
biometric studies of inherited traits, 317–319
biomusicality, cross-species research in
  animal homologues/analogues, 58
  animal song (*see* animal song)
  anthropocentric approach to, 58–60
  biocentric approach to, 58–60
  biomusicology components in, 58
  directions for future inquiry in, 80–81
  instrumental music, 67–71
  need for, 57–58
  pioneer studies in, 60–63
  potential pitfalls in, 80–81
  questions in, 58
  spectral sound parameters, 74–80
  synchronization, 71–73
  temporal sound parameters, 73–74
biomusicology
  components of, 58
  definition of, 57
birdsong
  avian brain structure and, 44–45, 45f
  daily and seasonal patterns in, 64
  duetting, 64–65
  functions of, 64–65
  gender differences in, 64
  history of studies of, 100–102
  hyperindexicality of, 49–52
  meanings of, 47–49
  overview of, 64–65
  Pamela Z on, 308–309
  radical niche construction and, 44–47
  semiosis and interpretant formation for, 47–49
  situational learning in, 42
  vocalization types in, 64
Black, J. E., 22
blackbird, early documentation of, 100
black-capped chickadees (*Poecile atricapillus*)
  octave equivalence in, 77t
  pitch height discrimination by, 76t
  processing of consonant *versus* dissonant intervals in, 79
  relative pitch discrimination and generalization by, 78
  timbre perception in, 80
Black Lives Matter movement, 347
Black Panthers, 99
Black studies, 104
Blackwell, H. R., 76
Blake, William, 298
blobology, 240
Blue Book (US Department of Agriculture), 97
blue-capped cordon-bleus (*Uraeginthus cyanocephalus*), dance in, 58
Boas, Franz, 210–211
body/mind relationship
  impact of musical instruments on, 147–151
  in music cognition, 143–147
Boethius, 143, 291–292
bonobos (*Pan paniscus*)
  coordinated group rhythmic vocalizations in, 23
  drumming by, 70
borderline tools, 68
boreal chickadees (*Poecile hudsonicus*), pitch height discrimination by, 76t
Bornean orangutans (*Pongo pygmaeus*), instrumental music by, 69–70
Bosavi culture, language socialization in, 381–382
Boswall, J., 67–68
bottlenose dolphins (*Tursiops truncatus*)
  animal musicality studies with, 62
  octave equivalence in, 76–77, 77t
  pitch contour discrimination by, 77
  song in, 65–66

Bourneville, Désiré-Magloire, 115
bowhead whales (*Balaena mysticetus*), song in, 66
Bowling, D. L., 78
Boyd, R., 25
Braid, James, 117
brain-body performance practice, 20, 277–284
  brain music, 281–283
  breathing music, 280–281
  conceptual framework for, 278–280
  goals in, 278–280
  heart music, 283–284
  practices informed by, 133
  relationship between artistic and scientific practice, 236, 277–280
brain-computer interfaces (BCIs)
  Activating Memory project, 228–230, 231
  brain-computer music interfacing, 225–227, 231, 232n4
  experimental music traditions and, 221
  potential of, 222
  purist *versus* utilitarian approaches to, 223
  questions addressed by, 221–222
  steady-state visual evoked potential, 227–230, 228f
  Symphony of Minds Listening project, 222–225, 230–231
  The Paramusical Ensemble, 228–230, 229f
brain-computer music interfacing (BCMI)
  science-inspired *versus* science-informed music, 231
  Symphony of Minds Listening project, 225–227
brain injury, 226
brain music, 281–283
brain structure
  action potentials in, 163–164
  amygdala, 246–247, 250
  angular gyrus, 244
  auditory cortex, 147, 230, 243–244, 247, 336
  basal ganglia, 124, 144, 164
  caudate, 247–248, 250
  cerebellum, 164
  cingulate, 246–251, 255n5
  as complex adaptive system, 264–266
  cortical brain regions, 42, 164, 246
  deep brain stimulation, 124
  default mode network, 248–249
  dorsal medial prefrontal cortex, 245
  dorsal posterior cingulate, 255n5
  dorsal striatum, 250
  frontal gyri, 244
  higher-order cortical areas, 244, 248–249, 255n6
  hippocampus, 42, 245–251
  inferior colliculus, 250
  insula, 246, 250
  medial prefrontal cortex, 244
  motor networks in, 163–166
  nucleus accumbens, 246–248
  orbitofrontal cortex, 248
  pallidum, 246
  parietal cortex, 164
  parietal gyri, 244
  posterior medial cortex, 244, 255n5
  precuneus, 255n5
  prefrontal cortex, 164, 245–247
  premotor cortex, 144, 164
  retrosplenial cortex, 255n5
  striatum, 247
  subcortical brain regions, 164
  superior temporal gyrus, 244
  supplementary motor area, 164
  thalamus, 45, 246–251
  ventral pallidum, 246
  ventral posterior cingulate, 255n5
  ventral putamen, 246
  ventral striatum, 24, 246, 250
  ventromedial prefrontal cortex, 247
  voxels, 240, 242
Brauer, J., 58
Braunlein, P., 213
breathing music, 280–281
Bregman, M. R., 80
Brelet, Gisèle, 143
Brittan, Francesca, 121

Brown, Steven, 104
brown rats (*Rattus norvegicus*)
  octave equivalence in, 77t
  pitch height discrimination by, 76t
budgerigars (*Melopsittacus undulates*)
  octave equivalence in, 77t
  pitch height discrimination by, 76t
  timbre perception in, 80
Bueno-Guerra, N., 59
"Building Sustainable Global Collaborative Research Networks" symposium, 348–349, 349f, 360n1
Byrne, David, 304

Cabral, J., 272f
calls, animal, 64
Campbell's monkeys (*Cercopithecus campbelli*), interval processing in, 79
Campos, J. J., 185
capacity, musical, 23–24
capitalism, and Christianity, 209–210
capuchin monkeys (*Cebus apella*), pitch contour discrimination and, 77
caregiver song, 191–193
Carolina chickadees (*Poecile carolinensis*), frequency ratios of vocalizations, 78
Carpentier, S. M., 253
caste, music as, 209
Castro, Viveiros de, 212–213
catalepsy, 115–117, 116f, 121, 123
categorization
  animal musicality studies with, 60–63
  definition of, 60
Cattell, James McKeen, 318
caudate, music-evoked emotions and, 247–248, 250
causal covariance, 39
causal information, 39–41
Center for Interdisciplinary Research in Music Media and Technology, 284
cerebellum, 164
cerebrum, 44–45, 45f
cetaceans, song in, 65–66
cetacean vocalization, 108n5
Chalmers, David, 167–168
Chambaud, Ménuret de, 119
change, experience-based, 194
Charcot, Jean-Martin, 115–117, 123
Chase, A. R., 61
Chavez, Alex, 203
Cheselden, William, 118
chickadees
  interval processing in, 79
  octave equivalence by, 77t
  pitch height discrimination by, 76t
  relative pitch discrimination by, 78
  timbre perception by, 80
child musicians, 336–340
  childhood precociousness, 312
  musical potential in, 337–338
  self-identification of, 339–340
  skill development in, 338–339
chills, musically induced, 248, 251
chimpanzees (*Pan troglodytes*)
  coordinated group rhythmic vocalizations in, 23
  dance in, 58
  drumming by, 70
  musical tool use by, 67
  pioneer studies with, 62
  processing of consonant *versus* dissonant intervals in, 79
Chin, T., 340
Chladni, Ernst, 295
Chomet, Hector, 125n2
Chomskyan linguistics, computational approaches based on, 131, 137–143, 165
choruses, cross-species research in, 72–73
Chowning, John, 299
"Christus, der ist mein Leben" (Bach), 137–139, 145
chroma, pitch, 76–77
chromaticism, development of, 296
cingulate, 246–251, 255n5
Civil Rights Act (1964), 108n4
Clark, Andy, 167–168

Clarke, E. F., 132
Clifford, James, 211
climate crisis, 7, 12, 98, 341
cognitive aging, music and
  age effects, identification of, 272–274
  complex system framework, characteristics of, 237, 253, 264–266
  complex system framework, implications for music neuroscience, 269–271, 270f
  hidden Markov modeling for, 271–274
  information processing in space and time, 266–269, 268f
  multiscale entropy curve, 269–271, 270f
  overview of, 263–264
  segregation/integration processes, 265–266
cognitive science, 136
  antirepresentational definition of, 173–176
  artificial intelligence, 165, 223, 254
  behavioral genetics of, 27
  body-instrument interaction in, 147–151
  cognitive abilities, 185–186
  cognitivism, 165
  computational approaches to, 137–143
  developmental science perspectives on, 183–194
  directions for future inquiry in, 133
  enactivism and, 170–173
  evolved neural specializations and, 25
  historical context of, 131–132
  information processing, 165
  integration with cultural understandings, 313, 367–368
  iterative nature of discovery in, 133
  mind/body relationship in, 143–147
  music/mind association in, 135–137
  social context of, 150–151
  statistical learning, 140
  See also embodied cognition
cognitivism, 165
Coleridge, Hartley, 120
Coleridge, Samuel Taylor, 120
collaborative research. *See* global collaborative networks
colonial world order, 3
common chaffinch (*Fringilla coelebs*), 107
comparative approaches, 3
  Feld on, 370
  reproducibility and standardization of, 354–356, 359t
complex systems framework, 237, 253
  age effects, identification of, 272–274, 272f
  characteristics of, 264–266
  implications for music neuroscience, 269–274, 270f
  space-time structure of, 266–269, 267f
composition, music cognition research as, 20, 277–284
  brain music, 281–283
  breathing music, 280–281
  conceptual framework for, 278–280
  goals in, 278–280
  heart music, 283–284
  practices informed by, 133
  relationship between artistic and scientific practice, 236, 277–280
  Symphony of Minds Listening project, 223–225, 225f, 230–231
compositional syntax, 49
computational approaches to music cognition, 137–143, 138f, 165
computer synthesizers, 299
computer technology, 8, 237, 299–301
conscious effort, manipulating EEGs with, 226–227
conscious selection, 322–323
consonance, cross-species research in, 78–79
construction grammar–inspired approaches, 141–142
contagion, 284
contingent negative variation (CNV), 281
contour, pitch, 77
Cook, P., 72
Cook, R. G., 74
coqui frogs (*Eleutherodactylus coqui*), pitch height discrimination by, 75
correspondence, 212

cortical brain regions, 42, 164, 246
cortisol, participatory music making and, 150
cotton-top tamarins (*Saguinus oedipus*), processing of consonant *versus* dissonant intervals in, 79
coupling, 166
  embodied cognition and, 170–173
  synchronization of, 175–176
  See also entrainment
covariance, causal, 39
COVID-19
  pandemic, 352, 354, 360n1
  vaccines, 108n7
cowbirds (*Molothrus ater*), pitch contour discrimination and, 77
Cox, Arnie, 150, 166
Cox, Catharine Morris, 136
creativity, role of experimentation in, 304–305, 312
Creativity Research Journal, 312
credit and attribution, in global collaborations, 356–358
crickets, musical tool use by, 67–69, 71
cross-cultural comparisons, 3, 8, 12–13. *See also* anthropological perspectives; global collaborative networks; relationality
cross-species research
  animal homologues/analogues, 58
  anthropocentric approach to, 58–60
  biocentric approach to, 58–60
  biomusicology components in, 58
  calls *versus* song in, 63–64
  directions for future inquiry, 80–81
  instrumental music, 67–71
  need for, 57–58
  pioneer studies in, 60–63
  potential pitfalls in, 80–81
  questions in, 58
  spectral sound parameters, 74–80
  synchronization, 71–73
  temporal sound parameters, 73–74
  See also animal song
Crowther-Heyck, Hunter, 131
cult of sensibility, 121
culture and relationality
  cultural knowledge, importance of, 140, 235–236
  cultural variation in music, 27–28
  developmental science perspectives on, 184–185
  in embodied cognition, 146–147
  musicality and, 17, 184
  musician-nonmusician dichotomy and, 330–331
  relationality and, 378–382
  self-identification as musician and, 334–335
  See also gene-culture coevolution
Cusick, Suzanne, 143, 148
Cynx, J., 74, 76–77, 80

dance
  in children, 184
  cross-species research in, 58
  embodied cognition and, 143, 175
  musical development and, 186, 193
  relationship with music, 19, 21
  shared experiences and, 186
  Sinhala Buddhist ritual drumming and, 205, 206
  See also entrainment
Darwin, Charles
  evolutionary biology, 216n7
  theories of aesthetics, 100
  theory of human musicality, 15–17, 23–24, 51, 104
Darwinism, social, 325
Davenport, Charles B., 318, 323
Davis, Janet, 99
Dayasheela, Herbert, 206
DC stimulation, 282
deaf listening/musicality, 146
Debussy, Claude, 296
decolonial research approaches, 347
deep brain stimulation, 124
default mode network (DMN), 248–249, 271
"defective," as archetype, 317–319

degeneracy, 132, 242
de Jonge, F. H., 62
Deleuze, G., 213
demon drum, 205
Dennett, D. C., 42
Descartes, René, 117, 143, 294
Descent of Man, The (Darwin), 15, 23
Deutsch, Diana, 2, 291, 292f, 299–301
development, musical
  active engagement and feedback loops in, 191–193
  adaptations, emergence of, 188
  conceptualization of music in, 183–184
  culture in, 184–185
  developmental processes throughout life, 188–191
  development in other domains compared to, 185–188
  directions for future inquiry, 193–194
  entrainment behaviors and, 189–191
  experience-based perception and change, 194
  infancy, 191–193
  influence of movement on, 187–188
  language ability and, 186–187
  neural plasticity and, 188
  notion of musicality in, 184
developmental prosopagnosia, 24–25
developmental research, musician-nonmusician dichotomy in
  ambiguity in, 329–330
  child musicians, 336–340
  developmental perspectives on, 336–340
  education perspectives on, 331–332
  ethnomusicology perspectives on, 332–333
  interdisciplinarity points of convergence in, 336
  interdisciplinary thinking and research, 341–342
  neuroscience perspectives on, 335–336
  psychology perspectives on, 333–335
Diderot, Denis, 120
Dimrock, Nathaniel, 98–99
direct-fit models, 237
disability, embodied cognition and, 146
discrimination
  animal musicality studies with, 60–63
  definition of, 60
Disklavier piano, 226
dissonance, cross-species research in, 78–79
distributed person, 204, 213–215
diversity
  challenges of, 347–349
  in global collaborations, 349–352, 359t
doctrine of vibrations, 118–119
dolphins (Delphinidae), song in, 62, 65–66
domestics chicks (*Gallus gallus*), processing of consonant *versus* dissonant intervals in, 79
dopaminergic neurons, 246
dorsal medial prefrontal cortex (DMPFC), 245
dorsal posterior cingulate, 255n5
dorsal striatum, 250
drumming
  cross-species research in, 58, 70–71
  Sinhala Buddhist ritual, 205–208
dual-coil scanning, 252
Du Bois-Reymond, Emil, 123
duets
  in birdsong, 47–48, 64–65
  cross-species research in, 72–73
  pseudoduets, 47–48
dynamic attending theory of entrainment, 163
dynamic systems theory, 166, 264. *See also* complex systems framework
dyslexia, 16

eared seals (Otariidae), song in, 66–67
earless seals (Phocidae), song in, 66–67
ecologically valid paradigms, combatting reductionism with
  artificial neural networks, 254
  characteristics of, 240–241
  conclusions of, 253–254
  directions for future inquiry in, 253–254
  emotions and feeling, 249–251, 253–254

ecologically valid paradigms, combatting reductionism with (cont.)
hidden Markov modeling, 244
long-timescale processing of musical structure, 243–244, 255n2
memory, music's influence on, 245–246
methodology and statistical approaches, 241–243
non-Western populations and stimuli, 254
overview of, 6, 8, 132, 237, 239–241, 278
reward processing, 246–249, 253
social context in, 251–253
ecological psychology, 147–151
ecomusicology, 216n8
education, musician-nonmusician dichotomy in, 331–332
effector world (*Wirkwelt*), 59
Einstein on the Beach (Glass), 231
electrocardiogram (ECG), 281
electrocorticography (ECoG), 283
electrodermal activity (EDA), 281
electroencephalogram (EEG), 124–125, 164, 269–271, 270f
brain-body performance practice and, 281–283
brain-computer music interfacing, 226–227
capabilities of, 239–241
hyperscanning studies with, 252–253, 255n9
limitations of, 241
reward processing studies with, 253
electronic-based instruments, 304–305
embedded thinking, 165–166, 249–251
embodied, definition of, 165–166
embodied cognition, 5, 7
4E framework of, 165–166
antirepresentationalism and, 173–176
autopoiesis and, 170, 176n1
bodily imitation, 150
body-instrument interaction in, 147–151
contemporary theories of, 165–166
cultural differences in, 146–147
definition of, 132
degree of embodiment in, 162
directions for future inquiry in, 176
embedded thinking in, 165
embodied predictive processing, 167–169
enacted mind in, 165–166
enactivism, 169–173, 176n1
extended mind theory, 165–166, 167–169
groove, 146, 186
interdisciplinary nature of, 165
listening as embodied cultural practice, 132
mind/body relationship in, 143–147
music theory and, 166
Pamela Z on, 305–306
predictive processing, 167–169
systems dynamics, 169–173
See also Rhythmic entrainment
emergentist views of musicality, 4, 367–382
acoustemology, 216n5
comparative approaches, 370
culture and relationality, 378–382
integration of cognitive and cultural understandings, 313
interdisciplinarity and epistemological conversation, 369
iterative processes in, 373–375
measurement and experimentation, 375–378
universalism, 12, 367, 370–373
emotions, music-evoked, 246–251, 253–254, 255n7
enacted mind, 165–166
enactivism, 166, 169–173, 176, 176n1
encultured system of knowledge, musicality as, 5, 367–382
acoustemology, 216n5
comparative approaches, 370
culture and relationality, 378–382
integration of cognitive and cultural understandings, 313
interdisciplinarity and epistemological conversation, 369
iterative processes in, 373–375
measurement and experimentation, 375–378
universalism, 12, 367, 370–373

Eno, Brian, 304
entrainment, 5
  cross-species research in, 71–72
  definition of, 161–163
  embodied cognition and, 166, 175
  musical development and, 189–191
  musically induced, 251
  neural entrainment and motor networks in brain, 163–164
  oscillatory nature of, 163–164
  types of, 163
entropy
  definition of, 266
  multiscale, 266–269, 267f
environment, role in shaping musical behavior, 3
"Eolian Harp, The" (Coleridge), 120
epilepsy, stimulation protocols for, 282, 284
Epilepsy and Cognition Lab, 282
episodic memory, 42, 245–246
epistemological conversations, Feld on, 369
equal temperament, cross-species research and, 79
equity, in global collaborations, 349–352, 359t
equivalence relations between pitch structures, 296, 297f
Essays: On Poetry and Music, as They Affect the Mind (Beattie), 120
ethereal vibrations, Newton's theory of, 117–118
ethical review, for global research collaborations, 353–354
ethics of animal research
  animal musicality studies, history of, 100–102
  Animal Welfare Act, 97, 102–103
  definition of, 108n4
  human *versus* animal needs in, 105–107
  implications for studies of musical capacity, 105–107
  kinship principle in, 108n7
  origins of, 97–99
  translation and, 106–107
  welfare school of, 98
ethnic fetish, 214
ethnography, ethnomusicology as, 216n4
ethnomusicology, 18–19, 203–216
  capitalism and Christianity, 209–210
  ethnomusicological self-critique, 3
  ethnomusicology-as-ethnography, 216n4
  musical transcendence, 210
  musician-nonmusician dichotomy in, 332–333
  music-self relation, 208
  porous nature of humans and, 212–214
  relationship between description and explanation, 210–215
  Sinhala Buddhist ritual drumming, 205–208
  sound-as-medium, 212–214
  sound-as-relations, 203–205, 215
eugenics, notions of musical ability and, 6, 8, 101
  biometric studies of inherited traits and, 317–319
  concept of musicality and, 315–317
  conscious selection, 322–323
  impact and legacy of, 323–326
  Mendelism in, 318, 320–321, 323, 325
  musical biography, critique of, 317–319
  Seashore's Measures of Musical Talent, 319–322, 337
  social psychology, 323–326
European Society for the Cognitive Sciences of Music, 301
European starlings (*Sturnus vulgaris*)
  octave equivalence in, 77t
  perception of isochronous patterns by, 74
  pitch contour discrimination by, 77
  processing of consonant *versus* dissonant intervals in, 79
  relative pitch discrimination and generalization by, 78
  timbre perception in, 80
euthanasia, 97

evolutionary psychology
  evolutionary specialization, 25–26
  musician-nonmusician dichotomy in, 333–335
  notion of musicality in, 184
  See also gene-culture coevolution; transspecies perspective, musical meaning in
Evolution of Language, The (Fitch), 370
Evolution of Morality Project, 352
Excursions in World Music, 210
experience-based change, 194
experience-based perception, 194
experience-dependent neuroplasticity, 21–22, 335–336
experience-expectant plasticity, 21–22
experimental music
  brain-computer interfaces informed by, 221
  brain music, 281–283
  breathing music, 280–281
  conceptual framework for, 278–280
  goals of, 277–280
  heart music, 283–284
  relationship between artistic and scientific practice in, 20, 133, 236, 277–284
experimental science, history of, 237, 291–301
  Aristoxenus, 292–293
  auditory perceptual illusions, 299–301
  Babbitt, Milton, 296
  Chladni, Ernst, 295
  Chowning, John, 299
  Descartes, René, 294
  Deutsch, Diana, 291, 292f, 299–301
  experimental techniques, 8
  Faraday, Michael, 295
  Galilei, Galileo, 294
  Galilei, Vicenzo, 293–294
  Helmholtz, Hermann von, 298
  Kepler, Johannes, 294–295, 295f
  Mathews, Max, 299
  Mersenne, Marin, 294
  Poussin, Nicolas, 298
  Pythagoras, 291–292, 293f
  Rameau, Jean-Philippe, 295–296
  Risset, Jean-Claude, 299
  Romantic movement, 298
  Schoenberg, Arnold, 296, 297f
  Shepard, Roger, 299
  Wheatstone, Charles, 295
experimentation
  cultural knowledge in, 235–236
  definition of, 5–6
  experiment design, 7
  Feld on, 375–378
  measurement and, 375–378
  Pamela Z on, 304–305
Exposure to Music in Childhood Inventory (EMCI), 334
extended mind theory, 165–166, 167–169

facial recognition, 24–25
fais de moi un instrument (Leslie), 283
Faraday, Michael, 295
Farbood, M. M., 243
Fay, R. R., 73
feedback loops, 7–8, 191–193
feelings, influence of music on, 249–251, 253–254
Feld, Steven, 216n5, 367–382
  acoustemology concept of, 216n5
  cognitive science-related questions of, 368–369
  on culture and relationality, 378–382
  on interdisciplinarity and epistemological conversation, 369
  iterative processes of, 373–375
  measurement and experimentation, 375–378
  reintegration of cognitive approaches with grounded investigation, 313, 367–368
  on universalism, 12, 367, 370–373
feminism, perspectives from, 104
ferrets (*Mustela furo*), pitch contour discrimination by, 77
finches
  birdsong communication in, 49–50
  gene-environment interactions in, 23

pitch height discrimination by, 76t
timbre perception in, 80
fin whale (*Balaenoptera physalus*), song in, 65–66
Fiorelli, P., 66
Fisher, S. E., 23
Fitch, W. Tecumseh, 57–58, 63, 64, 81n1, 367
Five Precepts, 206
flows, structured flows on manifolds (SFM), 273–274
FM synthesis, 299
folk ideologies of the person, 212
found objects, 306–307
Fox, Aaron, 367
free energy principle, 176n1
frequency ratio, pitch, 51, 77–78
Freud, Sigmund, 115
frisson, 251
Fritsch, Gustav, 123
frogs, 75
instrumental music by, 68–69
processing of consonant *versus* dissonant intervals in, 79
frontal gyri, 244
Fuchs, Thomas, 170
function, localization of, 240, 242
functional magnetic resonance imaging (fMRI), 164, 272–274
age effects, identification of, 272f
analysis of, 240–243
capabilities of, 240
episodic memory retrieval studies, 245–246
experimental symphonic piece based on, 230–231
hyperscanning studies with, 252–253, 255n9
limitations of, 241
long-timescale processing of musical structure with, 243–244
reward processing studies with, 246–249, 253
functional near-infrared spectroscopy (fNIRS), 252
funding of global research collaborations, 353
Gafurius, 293f
Galilei, Galileo, 294
Galilei, Vicenzo, 293–294
Galton, Francis, 136, 317
gamma oscillations, 164, 282
Gell, Alfred, 213
gender differences
in birdsong, 64
in self-identification as musician, 340
gene-culture coevolution, 7
abilities based on evolved neural specializations, 24–25
adaptive function, 25–26
for beat perception and synchronization, 25, 27
capacity *versus* proclivity in, 23–24
cultural variation, study of, 27–28
evolutionary specialization, 25–26
gene-environment interactions in, 23
genetically influenced traits, 26–27
history and origins of, 18
overview of, 11
purely cultural inventions in, 23
testing of, 18
universal cognitive traits, 25
variance in musicality and, 28–29
gene-environment interactions, 23
General Scholium (Newton), 118
Generative Theory of Tonal Music (Lerdahl and Jackendoff), 137, 165
genetically influenced traits, 26–27
Geneva Convention (1949), 108n3
genius archetype, 136, 312, 317–319
genomic studies of musicality, 17. *See also* gene-culture coevolution
Gentner, T. Q., 80
Geographical barriers, in global research collaborations, 351–352
Ghandeharioun, Asma, 281
Ghosh, Amitav, 98
gibbons, song in, 64, 72–73
Gibson, James J., 147
gift, sound as, 215

giftedness, experimental designs for, 312
Gill, K. Z., 78
Gingras, B., 333
Gish, S., 66
Github, 354
Glass, Phillip, 231
glass harmonica, 121
Glennie, Evelyn, 146
glissando illusion, 300
global collaborative networks, 8, 347
  attribution and leadership in, 356–358
  "Building Sustainable Global Collaborative Research Networks" symposium, 348–349, 349f, 360n1
  criticisms of, 348
  directions for future inquiry, 358
  diversity, equity, and inclusion in, 349–352, 359t
  incentives in, 356–358, 359t
  logistics of, 352–354, 359t
  need for, 347–349
  overrepresentation of WEIRD societies and, 347–348, 356
  reproducibility and standardization of comparisons, 354–356, 359t
  summary of recommendations for, 359t
Global Jukebox, 353–354
glossogeny, 81n1
Goldsmiths Musical Sophistication Index (MSI), 312, 334
Goldstein, A., 3, 237
Goodall, Jane, 67
Goodman, Steve, 174–176
Gordon, Edwin, 337
Granville, Joseph Mortimer, 122–123
gray seals (*Halichoerus grypus*), song in, 67
Greenough, W. T., 22
Griots in Mali, musicianship among, 333
groove, 146, 186

Hagmann, C. E., 74
harbor seals (*Phoca vitulina*)
  rhythm in, 74
  song in, 66–67
Hargreaves, D. J., 339
harmonica, glass, 121
Harmonic Elements (Aristoxenus), 292
harmonic series, Mersenne's studies on, 294
harmony, conceptualization of, 295–296
Hartley, David, 118–119, 125
Hasson, U., 3, 237
Hatara varam deviyo (four warrant gods), 206
Haynes, Richard, 98
heartbeat evoked potential (HEP), 283
heart music, 283–284
height, pitch
  cross-species research in, 76t
  image schemas for, 145–146
Helmholtz, Hermann von, 78, 135, 298, 319
hemispheric asymmetry, 291
Henrich, J., 25
Herder, Johann Gottfried, 120
hereditary genius. *See* genius archetype
hermit thrush, 100
hidden Markov modeling (HMM), 244, 255n4, 271–274
Higgins, C. I., 185
higher-order cortical areas, 244, 248–249, 255n6
high-frequency periodic stimulation, 282
hippocampus, 42, 245–251
historiometric approaches, 312
Hitzig, Eduard, 123
Hoeschele, M., 64, 77t
hollow pipes, theory of nerves as, 117
Holmes, Jessica, 146
homologues, animal, 58
honeybee (*Apis melifera*)
  associative learning in, 42
  attention in, 42–43
  brain structure of, 45–47, 46f
  causal information used by, 47
  radical niche construction and, 44–47
  sign and meaning making among, 42–43
Honing, Henkjan, 104
house mice (*Mus musculus*), pitch height discrimination by, 75
Howell, Philip, 102

Hulse, S. H., 74, 80
humane treatment
  animal musicality studies, history of, 100–102
  Animal Welfare Act, 97, 102–103
  definition of, 108n4
  human *versus* animal needs in, 105–107
  implications for studies of musical capacity, 105–107
  kinship principle in, 108n7
  origins of, 97–99
  translation and, 106–107
  welfare school of, 98
humanities-sciences divide, 7, 97–98
  concept of musicality in, 4–5
  consequences of, 1–4
  embodiment and antirepresentationalism, 170–173
  experimentation and, 5–6
  possibilities for dialogue and collaboration in, 6–9
humanizing education, 97–98, 107n2
human musicality, 12–13, 203–216
  adaptationist theories of, 21–22
  adaptive theories of, 15–18
  cognitive sense of, 19
  concept of, 19
  cultural view of, 17, 23, 184
  deaf musicality, 146
  definition of, 4–5, 330
  developmental science perspectives on, 184
  emergentist view of, 4
  as encultured system of knowledge, 367–382
  evolutionary history of, 25–26
  evolutionary psychology perspectives on, 184
  gene-culture coevolution of (*see* gene-culture coevolution)
  genomic studies of, 17
  importance of research on, 29–30
  music *versus*, 18–19
  nativist view of, 4
  Pamela Z on, 308–309
  variance in, 24, 27–29
  See also animal musicality
human song
  caregiver song, 191–193
  definition of, 63
  speech-to-song illusion, 135
Humpal, J., 74
humpback whales (*Megaptera novaeangliae*), song in, 66, 108n5
hyperalignment, 242
hyperindexicality, 49–52
hyperscanning studies, 252–253, 255n9
hypnosis, 117
hysteria, 115, 122

Ich habe genug (Bach), 150
Iconographie photographique de la Salpêtrière, 115
idiot savant trope, 319
Igbo people, word for music among, 19
illusions, auditory perceptual, 135, 183, 278, 291, 299–301
image schemas, 145–146
imitation, 150
implicit bias, 21–22
improvisation, 269, 373–375
incentives, for global collaborations, 356–358, 359t
inclusion, in global collaborations, 349–352, 359t
indexes of agency, 213
indexical commons, 13, 51–52
industrial psychology, 324
inequality, culture and, 380–381
infant auditory perception, 183–184, 187–188
infant-directed song, 191–193
inference
  active, 168–169, 186
  unconscious, 300
inferior colliculus, 250
information
  causal, 39–41
  informational space, 242, 255n3
  information processing, 165
  relational, 39–40
  semiotic, 40–41

Ingold, Tim, 211–212, 216n3, 216n7
"Inheritance of Musical Talent, The" (Seashore), 316, 321
inherited traits, biometric studies of, 317–319
Institute de Recherche et Coordination Acoustique Musique, 284
Institutional Animal Care and Use Committees (IACUCs), 103
institutional review boards (IRBs), 353, 355
instrumental music
  cross-species research in, 58, 67–71
  definition of, 67
  electronic-based *versus* acoustic, 304–305
  resonance effect in, 68
insula, 246, 250
integration, 265–266
Intelligence, 312
intelligence testing, 136, 312
interdisciplinarity, 1–4
  developmental science perspectives, 183–194
  in global research collaborations, 350–351
  musician-nonmusician dichotomy and, 341–342
intergenerational sustainability, of global research collaborations, 357–358
International Conference on Music Perception and Cognition, 301
International Council for Traditional Music (ICTM), 350, 352
interpersonal entrainment, 163
interpretants
  creation of, 41–42
  radical niche construction and, 43–47
intervals, 77–79
Introduction to Psychology (Seashore), 323
inversion, 296
isochrony, cross-species research in, 74
Iterative processes, 133, 373–375
Iyer, V., 235

Jackendoff, Ray, 137–143, 165
Jackson, Zakkiyah Iman, 105
James, William, 15–17, 25
Janata, P., 245
Janik, V. M., 67
Japanese monkeys, processing of consonant *versus* dissonant intervals in, 79
Javanese musicians, perception of pitch relationships by, 145–146, 148
Java sparrows (*Padda oryzivvora*), 60, 62
jazz improvisation, 373–375
Jobst, Barbara, 282
Johnson, Mark, 145
Juslin, P. N., 283
just intonation, cross-species research and, 79

kama muta, 251
Kane, Brian, 174
Kassabian, A., 132
Kataragama, 206
Keller, Evelyn Fox, 11, 12
Keller, Peter, 149
Kelso, J. A. Scott, 171
Kepler, Johannes, 294–295, 295f
Kermoian, R., 185
keyboard instrumentalists, experience of musical space, 148–151
King, Martin Luther, Jr., 99, 108n4
kinship principle, 108n7
kiss-squeak vocalization, orangutans, 381–382
Köhler, Johann Friedrich, 150
koi carps (*Cyprinus carpio*), 60, 61
Kozak, Mariusz, 166
Krumhansl, C. L., 145–146
Kwaito, 214

labor unions, 324
Laland, K. N., 29–30
Lamont, A., 339
language/music intersection
  in global research collaborations, 351–352
  history of studies of, 101
  musical development and, 186–187
  Pamela Z on, 307–308
  relationality and, 381–382

semiosis and symbolism in, 51
Sinhala Buddhist ritual drumming as, 206–207
Latour, B., 213
Laurel-Yanny phenomenon, 1
leadership, of global collaborations, 350, 356–358
leading eigenvector decomposition analysis (LEIDA), 272–274, 272f
learning, 246
associative, 42
bee behavior and, 46–47
birdsong production and, 44–45
MusEng software, 223–224
situational, 42
left brain/right brain concept, 298
Legislation, Animal Welfare Act (AWA), 97, 102–103
Leman, Marc, 166
Lerdahl, Fred, 137–143, 145–146, 165
liberation school of animal laboratory ethics, 98
Lichtenthal, Peter, 125n2
liking, wanting *versus*, 246
Lind, Jenny, 319
linguistics, transformational, 377
listening, as embodied cultural practice, 132
Liszt, Franz, 121
literacy
as cultural invention, 15–16, 21
as example of experience-dependent plasticity, 22
spread of, 31n2
LiveLAB, McMaster University, 252, 284
localization of function, 240, 242
Locke, John, 118
locked-in syndrome, music neurotechnology for. *See* music neurotechnology
locomotor ataxia, 123
logistics, of global collaborations, 352–354, 359t
long-timescale processing of musical structure, 243–244
low-country drum, 205–208
low-frequency periodic stimulation, 282
Lucier, Alvin, 221
lyrebirds (*Menura novaehollandiae*), dance in, 58

Madeiran chaffinch (*Fringilla coelebs maderensis*), 107
magnetic resonance imaging (MRI), hyperscanning studies with, 252–253, 255n9
magnetoencephalography (MEG), 124–125, 164
magpie-larks, pseudoduetting by, 47–49
Mahler, Gustav, 296
Maitreya, 206
manipulable boundary, 8
ManyBabies Consortium, 353
Marler, P., 59, 80
Mathews, Max, 278, 299
Mauss, M., 213
Max Planck Institute for Empirical Aesthetics, 284
McAllester, D. P., 235
McDermott, J., 78
McMaster University, LiveLAB, 252, 284
Mead, Margaret, 210
meaning, musical
causal *versus* semiotic information in, 39–41
challenges of defining, 39
hyperindexicality, 49–52
metarelations and, 40–41
radical niche construction, 43–47
sign-making process, 41–43
measurement, experimentation and, 375–378
Measures of Musical Abilities, 337
Measures of Musical Talent (Seashore), 319–322, 337
medial prefrontal cortex (MPFC), 244–245
medium, sound as, 204, 212–214
Mehr, S. A., 27–28
Melanesians, 212
Melodic cognition, neural processing for, 17
Melodic enculturation, 186

memory
 auditory working, 291
 bee behavior and, 46–47
 birdsong production and, 44–45
 episodic, 42
 exceptional, 312
 music's influence on, 245–246, 253
 Pamela Z on, 307–308
Mendel, Gregor, 318
Mendelism
 eugenics and, 318, 320–321, 323, 325
 social, 323, 325
Merkwelt, 59
Mersenne, Marin, 294
metarelations, 40–41, 43–47
metastability, 171–172
Meyer, Leonard, 143–144
Middle Ages, numerological approach to music theory in, 293
Mientien tree frogs (*Kurixalus idiootocus*), instrumental music by, 68–69
Million Years of Music, A (Tomlinson), 51
mimetic hypothesis, 150
mind, theory of, 25
mind/body relationship, 132
 impact of musical instruments on, 147–151
 in music cognition, 143–147
mind/music association, 135–137
Mingle, M. E., 62
Misaki, M., 255n9
missing fundamental effect, 135
mockingbirds (*Mimus polyglottos*), pitch contour discrimination and, 77
mole crickets, instrumental music by, 68, 71
Montague, E., 166
Montreal Battery for Evaluation of Musical Abilities (MBEMA), 334
Morley, I., 19
motor development
 motor entrainment, 163
 motor habit formation, 16
 musical development and, 185–188
mountain chickadees (*Poecile gambeli*), pitch height discrimination by, 76t
movement, influence on music development, 187–188
Mozart effect, 136
Muezzins, 333
Müller, Johannes Peter, 123
multidisciplinarity, in global research collaborations, 357
multinaturalism, 7, 213
Multiscale entropy (MSE), 266–269, 268f
Munn, Nancy, 212
MusEng software, 223–224
mushroom bodies, in honeybee brain, 45, 46f, 47
music, 18–19
 age and ubiquity of, 25
 definition of, 39
 hyperindexicality and, 49–52
 indexicality of, 50–52
 individual conceptualization of, 183–184
 as social construction, 19
 value of, 21–22
musical ability, eugenics and, 6, 8
 biometric studies of inherited traits and, 317–319
 conscious selection, 322–323
 impact and legacy of, 323–326
 Mendelism in, 318, 320–321, 323, 325
 musical biography, critique of, 317–319
 Seashore's concept of musicality, 315–317
 Seashore's Measures of Musical Talent, 319–322, 337
 social psychology, eugenics as, 323–326
musical affect, physiologically based accounts of, 125
musical anhedonia, 31n2, 312
musical biography, critique of, 317–319
musical development. *See* development, musical
musical expertise
 psychological models of, 333–334
 standardized comparisons of, 354–356
"Musical Friends and Foes" (Aucouturier and Cannone), 376
musical genius archetype, 312, 317–319

musical illusions, 135, 183, 278, 291, 299–301
musical instrument digital interface (MIDI), 226, 232n3
musical instruments, 147–151
musicality. *See* animal musicality; human musicality
musical savantism, 312
Musical Sophistication Index (MSI), 334
musical tension, listeners' ratings of, 139
musical transcendence, 210
music-as-caste, 209
Music at Hand (De Souza), 147
music cognition, 7–8
  body-instrument interaction in, 147–151
  cognitive abilities and, 185–186
  composition as, 277–284
  computational approaches to, 137–143
  developmental science perspectives on, 183–194
  directions for future inquiry, 133
  enactivism and, 170–173
  enculturation and, 140
  historical context of, 131–132
  iterative nature of discovery in, 133
  mind/body relationship in, 143–147
  music/mind association, 135–137
  overrepresentation of WEIRD societies and, 184
  social context of, 150–151
  statistical learning in, 140
  See also embodied cognition; rhythmic entrainment
music criticism, impact of vibrationalist theories on, 121–122
music education, musician-nonmusician dichotomy and, 331–332
MUSIC I, 299
musician-nonmusician dichotomy, 8, 311–313, 329–331
  ambiguity in, 329–330
  arbitrary nature of, 329–330
  child musicians, 336–340
  cultural bias in, 313, 330–331
  definitional conundrum of, 330–331
  developmental perspectives on, 336–340
  education perspectives on, 331–332
  ethnomusicology perspectives on, 332–333
  experimental designs for, 312
  interdisciplinary points of convergence in, 336
  interdisciplinary thinking and research, 341–342
  measures of talent, 312
  musicking and, 39, 50–52, 329, 333–335, 340–341
  neuroscience perspectives on, 335–336
  psychology perspectives on, 333–335
  self-identification in, 334–335
  as self-identified category, 315
  See also musical ability, eugenics and
musicianship, definitional conundrum of, 330–331
musician wren (*Cyphorhinus arada*), 63
music information retrieval (MIR), 271
musicking
  definition of, 39, 50, 329
  hyperindexicality in, 51–52
  interdisciplinary thinking and research, 341
  musician-nonmusician dichotomy and, 333–335, 340–341
music neurotechnology
  Activating Memory project, 228–230, 231
  brain-computer music interfacing, 225–227, 232n4
  experimental music traditions and, 221
  musician-nonmusician dichotomy in, 335–336
  origins of term, 232n1
  questions addressed by, 221–222
  science-inspired *versus* science-informed music, 231
  steady-state visual evoked potential, 227–230, 228f
  Symphony of Minds Listening project, 222–225, 230–231
  The Paramusical Ensemble, 228–230, 229f
music notes, 75

musicophilia, 312
music-related neuroplasticity, 335–336
music-selective brain regions, 17
music-self relation, 208
music structure, long-timescale processing of, 243–244
music studies, rifts in, 1–4
music theory
  embodied cognition in (*see* embodied cognition)
  experimental science approach to (*see* experimental science, history of)
music therapy
  history of, 121
  impact of vibrationalist theories on, 121, 125n2
mysterious melody illusion, 299–300

Näeketi, 205
Nastase, S., 3, 237, 243
Natha, 206
nativist view of musicality, 4
naturalistic music listening studies
  artificial neural networks, 254
  characteristics of, 240–241
  concept of "natural" in, 254n1
  conclusions of, 253–254
  definition of, 254n1
  directions for future inquiry in, 253–254
  emotions and feeling, 249–251, 253–254
  hidden Markov modeling (HMM), 244
  long-timescale processing of musical structure, 243–244, 255n2
  memory, music's influence on, 245–246
  methodology and statistical approaches, 241–243
  non-Western populations and stimuli, 254
  overview of, 6, 8, 80, 132, 237, 239–241, 278
  reward processing, 246–249, 253
  social context in, 251–253
natural selection, 23–24, 25, 29, 42, 43, 188
nature-culture distinction, 213, 216n8
nature-nurture dichotomy, 7
  areas of research shaped by, 12–13
  persistence of, 11–12
  reformulation of, 12
  See also cross-species research; gene-culture coevolution; sound experiments, humane treatment in; transspecies perspective, musical meaning in
Nerve-Vibration and Excitation (Granville), 122
nervous transmission, vibration and, 115–125
  brain rhythms, 122–124
  music criticism influenced by, 121–122
  music therapy influenced by, 121, 125n2
  psychiatric and physiological research based on, 122–124
  social and scientific context of, 122–124
  sympathetic resonance, 122–124
  tuning fork trials, 115–117, 116f, 123
  vibrationalist theories, 117–120
neural plasticity, 17, 188
  experience-dependent, 21–22
  experience-expectant, 21–22
neural sciences, role of music in development of, 8, 115–125
neural specializations
  abilities based on, 24–25
  ancient, universal cognitive traits and, 25
  for beat perception and synchronization, 25, 27
Neuringer, A., 60, 61
neurocentrism, 165
neurodegeneration, music and
  age effects, identification of, 272–274
  complex system framework, characteristics of, 237, 253, 264–266
  complex system framework, implications for music neuroscience, 269–271, 270f
  hidden Markov modeling for, 271–274
  information processing in space and time, 266–269, 267f
  multiscale entropy metrics for, 269–271, 270f
  overview of, 263–264
  segregation/integration processes, 265–266

neurotechnology, music. *See* music neurotechnology
New Materialism, 213
New Musicology, 214
Newton, Isaac, 117–118
Niche construction, radical, 43–47
Nietzsche, Friedrich, 121
nightingale, 100
noise-canceling headphones, 241
nonfluent aphasia, 16
nonhuman animals. *See* animal musicality
nonmusicians. *See* musician-nonmusician dichotomy
non-Western populations and stimuli, naturalistic music listening studies with, 254
"Normal Mind versus the Genius and the Defective, The" (Seashore), 317
North American Society for Music Perception and Cognition, 301
northern elephant seals (*Mirounga angustirostris*), 73–74, 80
Nostalgia, musically induced, 251
"Not among the Bongo-Bongo" syndrome, 205, 367
notes, music, 75
not-self, 207
Novembre, Giacomo, 149
nucleus accumbens (NAc)
  music-evoked emotions and, 246–251
  reward processing and, 247–248
numerology, 293, 295–296
"Nun komm, der Heiden Heiland" (Bach), 141–142, 142f

objectivity, 101, 211–212
objects, found, 306–307
Ochoa Gautier, Ana Maria, 7, 216n8
octave equivalence, 76–77
octave generalization, 76
octave illusion, 278, 291, 299
Odom, K. J., 64
Oliveros, Pauline, 304
Ollen Musical Sophistication Scale, 312
"On Humanizing Humanity" (Dimrock), 98–99
online data collection, accessibility of, 354
"On the Physiological Causes of Harmony in Music" (Helmholtz), 135
ontological consubstantiality, between humans and communities, 212
Open Science Framework, 354
operant conditioning, manipulating EEGs with, 226–227
Opticks (Newton), 117
optic lobes, in honeybee brain, 45, 46f
orangutans
  instrumental music by, 69–70
  musical tool use by, 67
orbitofrontal cortex (OFC), 246, 248
orcas (*Orcinus orca*), song in, 65–66
organ performance, embodied cognition and, 148–151
oscillatory nature of rhythmic entrainment, 163–164
oxytocin, participatory music making and, 150

padas (drum stanzas), 207
pahata rata beraya (low-country drum), 205
pallidum, 246
palm cockatoos (*Probosciger aterrimus*), drumming by, 70
Pamela Z, 303–309
  concept of musicality, 308–309
  found sound objects, use of, 306–307
  honors and awards, 303
  integration of technology into practice, 303–304
  on intersection of music and language, 307–308
  musical background of, 303–304
  performance goals of, 305
  perspective on embodied cognition, 305–306
  on role of experimentation in creativity, 304–305
Pandian, Anand, 210

parameters of sound, cross-species research in, 73–80
  spectral parameters, 74–80
  temporal parameters, 73–74
Paramusical Ensemble, The, 5, 228–230, 229f
parietal cortex, 164, 246
parietal gyri, 244
Parkinson's disease, 124
Partan, S., 59
partible person, 212–214
Pattini, 206
Pearson, Karl, 318
Peirce, Charles Sanders, 41
perception, definition of, 168
perceptual hysteresis, 1
perceptual world (*Merkwelt*), 59
percussion
  cross-species research in, 70–71
  Sinhala Buddhist ritual drumming, 205–208
percuteur, 122
personally significant music, behavioral responses to, 269
Petri, Olga, 102
phantom words illusion, 300
phase resetting, 282
Phillips-Silver, Jessica, 143
phonographs, history of, 101
Picard, Rosalind, 281
Piece for Solo Performer (Lucier), 221
pigeons (*Columba livia*)
  animal musicality studies with, 61
  perception of isochronous patterns in, 74
  pitch height discrimination by, 76t
  processing of consonant *versus* dissonant intervals by, 79
  relative pitch discrimination and generalization by, 78
  rhythm discrimination in, 73
piglets, animal musicality studies with, 62–63
pinnipeds, song in, 66–67
Pirahã, 141
pirith (chant), 206
pitch
  absolute, 75–77, 76t, 77t, 148, 291, 312
  behavioral responses to, 269
  chroma, 76–77, 77t
  contour, 77
  cross-cultural comparisons, 145–146
  cross-species research in, 74–78
  definition of, 74–75
  embodied cognition and, 145
  frequency ratio, 77–78
  height, 75–76, 76t, 145–146
  music/mind association and, 135–136
  pitch structures, theory of equivalence relations between, 296, 297f
  Pythagorean experiments on, 291
  register, 150–151
  relative, 77–78
  tonal perception, 140
  tonal pitch space, Lerdahl's theory of, 137
  transposition of, 19
planetary motion, third law of, 294–295, 295f
Planets Suite, The (Holst), 226–227
plasticity, 17, 21–22, 188
Platonic dualism, 165
pleasure, music-evoked, 246–249, 255n6
Pohl, Charles Ferdinand, 125
Poldrack, R. A., 243
population thinking, 216n7, 237
porous nature of humans, 212–214
Porter, D., 60, 61
posterior medial cortex (PMC), 244, 255n5
posthumanism, perspectives from, 104
potentiometers, 227
Poussin, Nicolas, 298
power spectrum analysis, EEG signal analysis with, 226–227
poya (full moon), 206
precapitalist musical labor, 209
precuneus, 255n5
prediction error, 168
predictive processing, 140–141, 166, 167–169
prefrontal cortex, 164, 245–247
premotor cortex, 144, 164

presymbolic syntaxes, 50
primary sources, importance of, 27–28
Principia (Newton), 118
Principles of Psychology, The (James), 15, 25
processes, in niche construction, 42
proclivity, musical, 23–24
professional-amateurs (pro-am), 331
project management, for global research collaborations, 352–353
prosopagnosia, developmental, 24–25
protomusic, 50
protosymbolic syntaxes, 50
pseudoduets, 47–48
psychiatric research, impact of vibrationalist theories on, 122–124
psychoacoustic research, history of, 101
Psychology of Music, The (Deutsch), 291
Psychology of Musical Talent, The (Seashore), 316
purist approach to computer-generated music, 22, 223
Purves, D., 78
Pythagoras, studies in experimental science, 291–292, 293f

Quon, Robert, 282

racism
  intelligence testing and, 136, 312
  racial stereotyping, 100–101
radical niche construction, 23, 43–47
rajakariya, 205
Ralls, K., 66
Rameau, Jean-Philippe, 295–296
rats
  octave equivalence in, 76
  pitch height discrimination by, 76t
  processing of consonant *versus* dissonant intervals in, 79
reading ability
  as cultural invention, 16, 21
  spread of, 31n2
recursive attention, 42–43
Redcay, E., 255n9
reductionism, 8, 235–237. *See also* aging, music and; naturalistic music listening studies; scientific practice, artistic creation and
redundant neuronal systems, 132
register, 150–151
Régnard, Paul, 115
reinforcement prediction errors (RPEs), 246–247
relating, embodied cognition and, 170–173
relationality
  culture and, 378–382
  metarelations, 40–41
relative pitch, cross-species research in, 77–78, 77t
reliable causal covariance, 39
Remarks upon Some of Mr. Norris' Books (Locke), 118
renaissance, empirical study of music perception in, 293–295
repositories, 354
representation, 40, 165. *See also* semiotic information
representational signs, 41
reproducibility of comparisons, 354–356, 359t
researchers, consultations with, 27–28
resonance, secular, 210
resonance, sympathetic, 118, 119–120, 122–124
resonance effect, 68, 175–176
response modeling, shared, 242
retrograde-inversion, 296
retrogression, 296
retrosplenial cortex, 255n5
reward processing, 246–249, 253
rhesus monkeys (*Macaca mulatta*), octave equivalence in, 76–77, 77t
rhythmic entrainment, 19–20, 284
  cross-species research in, 71–72
  definition of, 73, 161–163
  embodied cognition and, 145, 166, 175
  musical development and, 189–191
  neural entrainment and motor networks in brain, 163–164
  neural processing for, 17

rhythmic entrainment (cont.)
oscillatory nature of, 163–164
types of, 163
rhythmic vibration, 175–176
rhythm perception, language ability and, 187
Rickard, N. S., 340
Ridley, M., 23
Riedel, Friedlind, 166
right brain/left brain concept, 298
Risset, Jean-Claude, 279, 299
Robinson, Cedric, 325
Robinson, D., 132
rodents
animal musicality studies with, 60
octave equivalence in, 76
pitch height discrimination by, 76t
processing of consonant *versus* dissonant intervals in, 79
Roger, Louis, 125n2
Romantic movement, 298
Rosch, Eleanor, 170

Sabda pujava (sound offerings), 207
Salpêtrière, 115, 122–124
sampling, for global research collaborations, 356
sand gobies (*Pomatoschistus minutus*), instrumental music by, 68
Sansi, Roger, 212
Saunders, Aretas, 100
savantism, musical, 312
scale illusion, 299–300
Schenker, Heinrich, 139
Schilbach, L., 255n9
Schlosberg, H., 76
Schneider, Peter Joseph, 125n2
Schoenberg, Arnold, 30, 296, 297f
Schumann, Robert, 226
science-informed music, 231
science-inspired music, 231
scientific practice, artistic creation and, 20, 277–284
brain music, 281–283
breathing music, 280–281
conceptual framework for, 278–280
goals in, 278–280
heart music, 283–284
practices informed by, 133
relationship between artistic and scientific practice, 236, 277–280
Scientific Revolution, study of music perception in, 294
scientist-composers, 20, 277–284
brain music, 281–283
breathing music, 280–281
goals of, 278
heart music, 283–284
insights and practices, 133
Mathews, Max, 278
relationship between artistic and scientific practice, 236, 277–280
Risset, Jean-Claude, 279
Shepard, Roger, 278
Xenakis, Iannis, 279
Scudo, Paul, 121
sea lions (*Zalophus californianus*), entrainment and beat in, 72
Seashore, Carl E., 26–27
biometric studies of inherited traits, critique of, 317–319
concept of musicality, 315–317
conscious selection, 322–323
impact and legacy of, 323–326
Measures of Musical Talent, 319–322, 337
musical biography, critique of, 317–319
social psychology, eugenics as, 323–326
Second International Congress of Eugenics, 316, 322–323
secular resonance, 210
segregation, 265–266
self-entrainment, 163
self-identification as musician
child musicians, 339–340
cultural bias in, 334–335
semantic information, 40. *See also* semiotic information

semiotic information
in birdsong, 47–49
indexical semiotic commons, 51
reciprocal constraint in, 41
signs and, 40–41
sensibility, cult of, 121
shared response modeling, 242
Shepard, Roger, 278, 299
Shepard tones, 135
Shettleworth, S. J., 59
short-tailed cricket (*Anurogryllus muticus*), instrumental music by, 68
Siegel, E. H., 3, 237
signals, signs *versus*, 41
signs
aspectual nature of, 43
challenges of defining, 39
making of, 41–43
partial, aspectual nature of, 41
representational nature of, 41
semiotic information arising from, 40–41
signals *versus*, 41
Silver, Sean, 133
simple systems, 264–265
single pulses, 282
Sinhala Buddhist nationalism, 207
Sinhala Buddhist ritual drumming, 205–208
as sacred speech, 206–207
social context of, 207–208
as sound offerings, 207
sirens, mechanical, 135–136
situational learning, episodic memory and, 42
skill development, in child musicians, 338–339
Small, Christopher, 329, 340
Smotherman, M., 65
Snow, C. P., 298
social cognition, 150
social context
of music cognition, 150–151
in naturalistic music listening studies, 251–253
social music making, behavioral responses to, 269
social psychology, eugenics as, 323–326
social synchronization, cross-species research in, 58
Social Darwinism, 325
social entrainment, 163
Social Mendelism, 323
social sciences, embodiment and antirepresentationalism in, 170–173
Society for Ethnomusicology (SEM), 347
Society for Music Perception and Cognition (SMPC), 2, 231, 291, 347
software
MUSIC I, 299
TheVirtualBrain neuroinformatics platform, 274
solo musical behavior, 28
"sometimes behave so strangely" illusion, 183
songbirds
brain structure of, 44–45, 45f, 46f
daily and seasonal patterns in, 64
duets, 64–65
functions of song in, 64–65
gender differences in, 64
history of studies of, 100–102
hyperindexicality, 49–52
Pamela Z on, 308–309
radical niche construction and, 44–47
semiosis and interpretant formation in, 47–49
situational learning in, 42
song meanings, 47–49
vocalization types in, 64
Sonic Tonic (Leslie), 283
sound amplification, cross-species research in, 71
sound-as-gift, 215
sound-as-medium, 204, 212–214
sound-as-relations, 203–205, 215

sound experiments, humane treatment in
animal musicality studies, history of, 100–102
Animal Welfare Act, 97, 102–103
human *versus* animal needs in, 105–107
implications for studies of musical capacity, 105–107
kinship principle in, 108n7
liberation school of, 98
origins of, 97–99
translation and, 106–107
welfare school of, 98
sound offerings, Sinhala Buddhist ritual drumming as, 207–208
sound parameters, cross-species research in, 73–80
spectral parameters, 74–80
temporal parameters, 73–74
sound recording technology, 101
Sozialwelt, 59
space-time structure
age effects, identification of, 272–274, 272f
age-related changes in, 265–266
in complex systems, 264–266
information processing in, 266–269, 267f
long-timescale processing of musical structure, 243–244
sparse temporal sampling, 255n2
sparrows, 60, 62
frequency ratios of vocalizations, 78
history of studies of, 100
pitch height discrimination by, 76t
sparse temporal sampling, 241, 255n2
spatial normalizing, 242
spatiotemporal scales, 264–266
specialists, consultations with, 27–28
spectral analysis, 227
spectral parameters, cross-species research in, 74–80
consonance and dissonance, 78–79
pitch, 74–78
spectral shape, 79–80
timbre, 79–80
spectral shape, 79–80
speech-to-song illusion, 135, 291, 300
Spix's disc-winged bats (*Thyroptera tricolor*), instrumental music by, 69, 71
spoken language
acquisition of, 16–17
neural plasticity and, 22–23
speech accent patterns, 187
speech melodies, 20–21
Sri Lanka's Sinhala Buddhist drumming tradition. *See* Sinhala Buddhist ritual drumming
stable state, 171
Stadler, Gustavus, 319
Stafford, K. M., 66
standardization of comparisons, 354–356, 359t
standardized musical aptitude tests, 320, 334, 337
standard reductionist agenda, 2
Stansbury, A. L., 67
starlings
octave equivalence in, 77t
perception of isochronous patterns in, 74
pitch contour discrimination by, 77
processing of consonant *versus* dissonant intervals in, 79
relative pitch discrimination and generalization by, 78
timbre perception in, 80
state
metastability, 171–172
stable, 171
bias, in cognitive functions, 269
statistical approaches
hidden Markov modeling, 255n4, 271–274
for naturalistic music listening studies, 241–243
statistical learning, 140
steady-state visual evoked potential (SSVEP), 227–230, 228f
Activating Memory project, 228–230, 231
initial SSVEP-based BCMI system, 227–228, 228f
Steingo, G., 214

stimulus, salience of, 42–43
Stockhausen, Karlheinz, 277
stop-tension tasks, 139
Strathern, Marilyn, 212
Straus, Joseph, 146
striatum, reward processing and, 247
string analogy of nervous transmission, 118
structured flows on manifolds (SFM), 273–274
Stumpf, Carl, 318
Sturdy, C. B., 77t
subcortical brain regions, 164, 246–251
subcortical sound localization mechanisms, 21
subtraction method, 240
superior temporal gyrus (STG), 244
supernatural beliefs, 25
supplementary motor area, 164
surprise, reward processing and, 247
sustainable global research collaborations. *See* global collaborative networks
sympathetic resonance, 118, 119–120, 122–124
Symphony of Minds Listening project, 222–225, 225f, 230–231
synchronization
  of coupling, 175–176
  cross-species research in, 71–73
  synchrony to beat, 25, 27
  of vibration, 175–176
synesthesia, 312
synthesizers, computer, 299
systemic racism
  intelligence testing and, 136, 312
  racial stereotyping, 100–101
systems dynamics, embodied cognition and, 169–173
systems dynamics model, 170

Teicher, Amir, 323
temperament, Mersenne's studies on, 294
temporal parameters, cross-species research in, 73–74
tests, musical aptitude, 320, 334, 337
thalamus, 45
  in avian brain, 44–45, 45f
  music-evoked emotions and, 246–251
  thalamic stimulation, 282
Theorica Musicae (Gafurius), 293
theory of mind, 25
Theravada Buddhism, 206–207
TheVirtualBrain neuroinformatics platform, 274
third law of planetary motion, 294–295, 295f
Thompson, Evan, 170
Thompson, Marie, 174, 175
timbre
  cross-species research in, 79–80
  timbral processing, 51
Tinbergen, N., 25, 58, 81n1
Tipitaka, 207
tonal gravity, 145
tonal perception, 140
tonal pitch space, Lerdahl's theory of, 137
tool use, animal, 67–71
toothed whales (Odontocetes), song in, 65–66
Torti, V., 63
Tourette, Gilles de la, 115, 123
Trainor, Laurel, 143, 252
Tranchant, P., 28
transcendence, 210, 251
transformation
  MusEng software, 223
  narrative of, 210
translation, animal laboratory ethics and, 106–107
transposition, 78, 296
transspecies perspective, musical meaning in
  causal *versus* semiotic information, 39–41
  challenges of defining, 39
  hyperindexicality, 49–52
  metarelations, 40–41
  radical niche construction, 43–47
  sign-making process, 41–43
trapped agencies, 215
tree crickets, musical tool use by, 67
tree diagrams (Lerdahl), 137–139

tree-hole frogs (*Metaphrynella sundana*), instrumental music by, 69
Trehub, S. E., 19
tritone paradox, 291, 300
Truth, 212
Tsimane', 78, 140
"Tuesday Afternoon Sound Alternative, The" (radio program), 304
tungara frogs (*Physalaemus pustulosus*), processing of consonant *versus* dissonant intervals in, 79
tuning, Mersenne's studies on, 294
tuning fork trials, 115–117, 116f
twelve-tone system, 296
two-alternative forced choice (2AFC) paradigm, 61

Umwelt, 59
uncertainty, reward processing and, 247
unconscious inference, 300
universalism, 11–12, 25, 208, 325, 367, 370–373
University of Plymouth, Interdisciplinary Centre for Computer Music Research, 221–222
  Activating Memory project, 228–230, 231
  brain-computer music interfacing, 225–227, 232n4
  experimental music traditions and, 221
  science-inspired *versus* science-informed music, 231
  steady-state visual evoked potential, 227–230, 228f
  Symphony of Minds Listening project, 222–225, 230–231
  The Paramusical Ensemble, 228–230, 229f
US Department of Agriculture, *Blue Book* on animal welfare, 97
utilitarian approach to computer-generated music, 223

vagus nerve stimulation protocols, 282
Varela, Francisco, 170
variability, 29, 78, 237, 264, 280
variance in musicality, 24, 27–29
veeries (*Catharus fuscescens*), frequency ratios of vocalizations, 78
Venda of South Africa, musical participation among, 333
ventral occipitotemporal cortex, 17
ventral pallidum, 246
ventral posterior cingulate, 255n5
ventral putamen, 246
ventral striatum, 24, 246, 250
ventromedial prefrontal cortex (VMPFC), 247
verticality schema, 145
vibrating nerve theory. *See* vibrationalist theories
vibration, neural signals and, 8, 115–125
  brain rhythms, 122–124
  music criticism influenced by, 121–122
  music therapy influenced by, 121, 125n2
  psychiatric and physiological research based on, 122–124
  rhythmic vibration, 175–176
  social and scientific context of, 122–124
  sympathetic resonance, 122–124
  tuning fork trials, 115–117, 116f, 123
  vibrationalist theories, 117–120
vibrationalist theories, 117–120
  attributes of, 122
  conflation with notions of sympathetic resonance, 119–120
  Hartley's doctrine of vibrations, 118–119
  Newton's theory of ethereal vibrations, 117–118
Vishnu, 206
visual development, musical development and, 185–188
visual evidence, 101
visual word form area (VWFA), 17
vocalizations. *See* animal song; calls, animal; human song
vocal pathways, in avian brain, 45f
von Rueden, C., 27–28
voxels, 240, 242

Wagner, Richard, 121, 296
Waits, Tom, 304
Wallace, C. S., 22
walruses (Odobenidae), song in, 66–67
wanting, liking *versus*, 246
warrant gods, 206
Warrier, C. M., 75
WASP (Western, academic, scientific, psychology) researchers, overrepresentation of, 348
Weiner, Annette, 212
WEIRD (Western, educated, industrialized, rich, and democratic) societies, 209–210
  criticism of term, 216n6
  music in, 19
  overrepresentation of, 184, 347–348, 356
Weisman, R. G., 77t
welfare school of animal laboratory ethics, 98
Western music and musicians, overreliance on, 347–349. *See also* global collaborative networks
Western music theory, 296
whales, song in, 65–66, 108n5
Wheatstone, Charles, 295
white-throated sparrows (*Zonotrichia albicollis*)
  frequency ratios of vocalizations, 78
  pitch height discrimination by, 76t
Wiggins, Thomas "Blind Tom," 319
Willford, A., 214
Willis, Thomas, 117
Winters, Mike, 283
WIRED (white-dominated, imperialistic, rich, empires [that] don't always export democracy) societies, 209
Wirkwelt, 59
wonder, musically induced, 251
woodpeckers, drumming by, 70–71
Wright, Lindsay, 319
Wundt, Wilhelm, 319
Wynter, Sylvia, 105

Xenakis, Iannis, 279

yak beraya (demon drum), 205
yakku (beings of low karmic standing), 205, 206, 207
Yearsley, David, 148
yellow-bellied slider turtles (*Pseudemys scripta*), pitch height discrimination by, 75

Zatorre, R. J., 75
Zbikowski, Lawrence, 150
zebra finches (*Taeniopygia guttata*)
  gene-environment interactions in, 23
  pitch height discrimination by, 76t
  timbre perception in, 80
Zenodo, 354
Zentner, M., 333